# Machinery Lubrication Technician (MLT® I and II) Certification Exam Guide

# Machinery Lubrication Technician (MLT® I and II) Certification Exam Guide

Michael D. Holloway
and Sanya Mathura

Industrial Press

# Industrial Press, Inc.

1 Chestnut Street
South Norwalk, Connecticut 06854
Phone: 203-956-5593
Toll-Free in USA: 888-528-7852
Email: info@industrialpress.com

Authors: Michael D. Holloway and Sanya Mathura
Title: Machinery Lubrication Technician (MLT® I and II) Certification Exam Guide
Library of Congress Control Number: 2021944409

© by Industrial Press, Inc.
All rights reserved. Published in 2022.
Printed in the United States of America.

ISBN (print): 978-0-8311-3649-9
ISBN (ePUB): 978-0-8311-9544-1
ISBN (eMOBI): 978-0-8311-9545-8
ISBN (ePDF): 978-0-8311-9543-4

Publisher/Editorial Director: Judy Bass
Copy Editor: James Madru
Compositor: Patricia Wallenburg, TypeWriting
Proofreader: Debbie Anderson
Indexer: Claire Splan

books.industrialpress.com
ebooks.industrialpress.com

# Contents

# Introduction

This guide serves as a workbook to be used in conjunction with an approved course offered by a training partner who is also recognized by the International Council for Machinery Lubrication (ICML) for a Machinery Lubrication Technician (MLT® I or MLT® II) certification exam preparation course. It is required and essential that the course instructor shall have obtained the certification that he or she is teaching as well as several years' experience in the field of lubrication, maintenance, and reliability. The ICML also requires that you take a course from an approved training company and provide a certificate of completion of the course.

## What This Book Is

- A working resource
- A preparation aid to use with a course
- An accompaniment with guidance

The essence of this work is a guided cooperative argumentative dialogue between authors with continuous asking and answering of questions that stimulate critical thinking to draw out underlying presumptions.

We have also done something even more unconventional. We are looking to have you think like an examiner (that person) who actually built the exam!

## What This Book Isn't

- A substitute for a structured, approved exam preparation course
- An answer book

This workbook is *not* intended to be used without guidance from an approved training company that is competent at providing instruction toward mastery of the body of knowledge deemed appropriate for obtaining the MLT® certification. This workbook is a supplement and not a substitute.

## Guide to Using This Book

### Structure

I.  Outline Notes. This is for you to fill in. This is your working notebook section to be used during the course that you will take. It will help you organize your ideas and thoughts.

II. Guided Cooperative Arguments. This is a dialogue between authors concerning the topics. This section will help you understand how to answer questions that are asked of you in your job func-

tion. By mastering this content and understanding the application, you will have a solid working understanding of the content that you are required to know.

III. Statements of Truth and Exam Development. This is from the recommended "Body of Knowledge" with space to develop your multiple-choice questions. This section is vital to complete by developing your critical thinking process as well as making you understand how the exam questions are structured.

IV. Body of Knowledge Outline. This section is provided to work as a reference to help you answer questions that may arise.

V.  Practice Exam. This is a mock exam designed to get you used to taking a multiple-choice test that is similar in structure and content to the one you will take.

VI. Glossary. This is a list of common terms you should have mastery of.

VII. Appendix. This contains useful charts and graphs.

## The Process

After you have registered for your exam preparation course (live online, virtual, or in person), it is suggested that you follow this process in order to get the most out of your learning experience:

- Step 1: Familiarize. Familiarize yourself with this workbook in its entirety. Make sure that you know how the Outline Notes section is structured. Some courses arrange content differently. You may find yourself having to flip through several pages in this section to find the part that the class is reviewing.

Review the Body of Knowledge Outline, Glossary, and Appendix sections to become familiar with where to find information pertaining to the content being taught.

- Step 2: Find Socrates. After each section of the course work is covered during class, review the Guided Cooperative Arguments section to get a better understanding of how the content is delivered and applicable. This section is an example of how instructors and students would engage in questions and answers in order to draw out the truth. This method was perfected by the Greek philosopher Socrates and has been a cornerstone for competency development ever since.

- Step 3: Be the Exam. After you have completed your course work, begin work on the Statements of Truth and Exam Development section. In this section, you will be taught how to think like an examiner. This section has been an influencing reason why so many candidates have obtained their certifications the first time. It is a fantastic way to truly understand the content as well as the logic behind the questions asked. Take your time with this section. It's very important!

- Step 4: Practice Exam. The practice exam questions will get you used to taking the certification exam. It is suggested that you time yourself. ICML will give you three hours to complete the 100 multiple-choice questions. A passing grade is 70% correct.

- Step 5: Explore. Take time to review the Glossary and Appendix sections. There are plenty of useful terms to learn as well as helpful charts and tables that

you will use in any job function regarding lubrication. These valuable resources are found all in one place.

This process has been proven to be exceptionally effective not only for passing the exam but also to direct you to become an expert in the content you are studying.

When the instructions call for you to outline, explain a certain way, actually draw a diagram, or build a graph or chart, please follow those instructions. They are your keys to success. When the day comes and you are about to take the exam, you will be very glad that you followed the recommended steps to completion. Failure to follow the directions most likely will result in a failed grade.

According to the ICML, a person who has achieved Machinery Lubrication Technician Level I certification

must demonstrate skills in the day-to-day activities associated with proper lubrication of machinery. Level I is oriented toward lubrication basics and the proper application and storage of lubricant. Level II is directed toward advanced understanding of lubrication fundamentals, lubricant selection and lubrication schedule and program management.

Level I – Targets in-plant technicians responsible for daily lubrication tasks, including oil changes, top-ups, greasing bearings, lubricant receiving and proper storage and care of lubricants and dispensing devices.

Level II – Targets in-plant technicians or engineers responsible for managing the lube team, selecting lubricants, troubleshooting abnormal lubricant performance and supporting machine design activities.

In order to become a certified MLT® I, ICML states that an individual must meet the following requirements:

■ Education and/or Experience. Candidates must have at least two years education (post-secondary) or on-the-job training in one or more of the following fields: machine lubrication, engineering, mechanical maintenance and/or maintenance trades.

■ Training. Candidates must have received 16 hours of documented formal training in machinery lubrication as outlined in the Body of Knowledge of the MLT® I. For online or recorded training, exercises, practice exams, and review exercises may be included in the training time total but shall not exceed three hours of the required course time. Candidates shall be able to provide a record of this training to ICML that shall include the candidate's name, the name and signature of the instructor, the dates of the training and the number of hours spent in the training.

■ Examination. Each candidate must successfully pass a 100 question, multiple-choice examination that evaluates the candidate's knowledge of the topic. Candidates have three hours to complete the closed-book examination. A score of 70% is required to pass the examination and achieve certification. Contact ICML about the availability of the exam in other languages.

In order to become a certified MLT® II, ICML states that an individual must meet the following requirements:

■ Education and/or Experience. Candidates must have at least three years education (post-secondary) or on-the-job training in one or more of the following fields: machinery lubrication, engineering, mechanical maintenance and/or maintenance trades.

■ Hold Level I Machinery Lubrication Technician (MLT® I) certification.

■ Training. Candidates must have received 16 hours of documented formal training in machinery lubrication as outlined in the Body of Knowledge of the MLT® II. For online or recorded training, exercises, practice exams, and review exercises may be included in the training time total but shall not exceed three hours of the required course time. These 16 hours are in addition to the previous 16 hours of training required for MLT® I, for a total cumulative training of 32 hours. Candidates shall be able to provide a record of this training to ICML that shall include the candidate's name, the name and signature of the instructor, the dates of the training, and the number of hours spent in the training.

■ Examination. Each candidate must successfully pass a 100 question multiple-choice examination that tests the candidate's mastery of the body of knowledge. Candidates have three hours to complete the closed-book examination. A score of 70% is required to pass the examination and achieve certification. Contact ICML about the availability of the exam in other languages.

# MLT® I and II Body of Knowledge

The MLT® I Body of Knowledge is an outline of concepts that one should have in order to pass the exam. References from which exam questions were derived can be found in the Domain of Knowledge from ICML.

I.   **Maintenance Strategy (5%)**
A.   Why machines fail
B.   The impact of poor maintenance on company profits
C.   The role of effective lubrication in failure avoidance

II.  **Lubrication Theory (10%)**
A.   Fundamentals of tribology
B.   Functions of a lubricant
C.   Hydrodynamic lubrication (sliding friction)
D.   Elastohydrodynamic lubrication (rolling friction)
E.   Mixed-film lubrication
F.   Boundary lubrication

III. **Lubricants (15%)**
A.   Base oils
B.   Additives and their functions
C.   Oil lubricant physical, chemical, and performance properties and classifications
D.   Grease lubrication
1.   How grease is made
2.   Thickener types
3.   Thickener compatibility
4.   Grease lubricant physical, chemical, and performance properties and classifications

IV.  **Lubricant Selection (15%)**
A.   Viscosity selection
B.   Base-oil type selection
C.   Additive system selection
D.   Machine-specific lubricant requirements
1.   Hydraulic system
2.   Rolling-element bearings
3.   Journal bearings
4.   Reciprocating engines
5.   Gearing and gearboxes
E.   Application- and environment-related adjustments

V.   **Lubricant Application (25%)**
A.   Basic calculations for determining required lubricant volume
B.   Basic calculations to determine relube and change frequencies
C.   When to select oil; when to select grease
D.   Effective use of manual delivery techniques
E.   Automatic delivery systems
1.   Automated delivery options
a.   Automated grease systems
b.   Oil mist systems
c.   Drip and wick lubricators
2.   Deciding when to employ automated lubricators
3.   Maintenance of automated lubrication systems

VI.  **Preventive and Predictive Maintenance (10%)**
A.   Lube routes and scheduling
B.   Oil analysis and technologies to ensure lubrication effectiveness
C.   Equipment tagging and identification

VII. **Lube Condition Control** (10%)
   A. Filtration and separation technologies
   B. Filter rating
   C. Filtration system design and filter selection

VIII. **Lube Storage and Management** (10%)
   A. Lubricant receiving procedures
   B. Proper storage and inventory management
   C. Lube storage containers
   D. Proper storage of grease guns and other lube application devices
   E. Maintenance of automatic grease systems
   F. Health and safety assurance

The MLT® II Body of Knowledge is an outline of concepts that one should have in order to pass the exam. References from which exam questions were derived can be found in the Domain of Knowledge from ICML.

I. **Maintenance Strategy** (5%)
   A. The impact of lubrication on machine reliability
   B. The impact of lubrication on lubricant life and consumption
   C. Maintenance program strategies for achieving lubrication excellence

II. **Lubrication Theory** (5%)
   A. Friction and tribology
      1. Types of friction and wear
         a. Wear modes and influencing factors
         b. Machine frictional surfaces most at risk for specific wear modes (e.g., abrasion)
      2. Mechanisms of lubrication regimes
         a. Boundary
         b. Mixed film
         c. Hydrodynamic
         d. Elastohydrodynamic
   B. Lubricant categories
      1. Gaseous
      2. Liquid
      3. Cohesive
      4. Solid

III. **Lubricant Formulation** (10%)
   A. Base-oil refining methods and American Petroleum Institute categories
      1. Solvent refined
      2. Hydrotreated
      3. Severely hydrotreated
      4. Hydrocracked
   B. Mineral base oils
      1. Naphthenic
      2. Paraffinic
      3. Aromatic
   C. Vegetable base oils and biolubes
   D. Synthetic lubricant characteristics/applications/compatibility
      1. Synthesized hydrocarbons (e.g., polyalphaolefins)
      2. Dibasic acid esters
      3. Polyol esters
      4. Phosphate esters
      5. Polyalkylene glycol
      6. Silicones
      7. Fluorocarbons
      8. Polyphenyl ethers
   E. Food-grade lubricant classification
   F. Types and functions of additives
   G. Types and functions of solid additives
   H. Modes of additive depletion

IV. **Grease Application and Performance** (5%)
   A. Grease applications requiring high consistency
   B. Grease applications requiring low consistency
   C. Causes of grease separation
   D. Multipurpose greases

P. Rolling Stability of Greases ASTM D1831

Q. Water Washout Test for Greases ASTM D1264

R. Water Spray Test for Greases ASTM D4049

S. Rolling Bearing Rust Test ASTM D1743

T. Koppers centrifugal stability test

U. Oil Separation in Grease Storage ASTM D1742

V. Oxidation Stability – Greases ASTM D942

VII. **Lubricant Application (15%)**

A. Procedures for
  1. Oil drain
  2. Reservoir/system flushing
  3. Disassembling/cleaning reservoirs and sumps
  4. Filling
  5. Top-up
  6. Grease packing
  7. Regreasing
  8. Grease changeout

B. Determine/calculate correct amount for regreasing

C. Determine/calculate correct frequency interval for regreasing

D. Select and manage optimal equipment/systems for lubricant application according to machinery requirements

E. Safety/health requirements for lubricant application

F. Manage proper maintenance of lubrication equipment

G. Manage proper maintenance of automatic lubrication systems

H. Create/update lube survey

I. Record execution of lube program

J. Proactive management and detection of leaks

K. Waste oil/filters management/disposal

L. Writing a lubrication preventive maintenance (PM) plan

VIII. **Preventive and Predictive Maintenance (10%)**

A. Creating and managing lube PM plans and routes

B. Creating and managing a lubrication inspection checklist

C. Used-oil analysis to determine optimal condition–based oil changes

D. Used-oil analysis to troubleshoot abnormal lubricant degradation conditions

E. Used-oil analysis to troubleshoot abnormal wear related to lubricant degradation/contamination

F. Procedures and methods for identifying root cause of lubricant failure

G. Use of technology aids to determine optimal regrease frequency/quantity (ultrasonic, temperature monitoring, shock impulse, etc.)

IX. **Lubricant Condition Control (20%)**

A. Proper sampling procedures

B. Proper sampling locations

C. Proper selection of breathers/vents

D. Proper selection of filters according to cleanliness objectives

E. Filter rating, beta ratio

F. Sump/tank management to reduce
  1. Air entrainment/foam
  2. Particles
  3. Water
  4. Sediments
  5. Heat
  6. Silt/sediments
  7. Unnecessary lubricant volume

G. Proper selection of reconditioning systems for
  1. Water
  2. Air/gas
  3. Particles
  4. Oxidation products
  5. Additive depletion
H. Lube reclamation
  1. Requirements
  2. Feasibility
  3. Procedures for reclaiming/reconditioning
  4. Use of oil analysis to approve reclaimed/reconditioned lubricants

X. **Lube Storage and Management (5%)**
  A. Design of optimal storage room
  B. Defining maximum storage time according to environmental conditions/lubricant type
  C. Safety/health requirements
  D. Proper sampling procedures/locations for sampling stored lubricants
  E. Procedures for reconditioning/filtering stored lubricants

# MLT® I and II Domain of Knowledge

Association of Iron and Steel Engineers (2010). *The Lubrication Engineer's Manual.* Association of Iron and Steel Engineers, Pittsburgh, PA.

Bannister, K. (2007). *Lubrication for Industry.* Industrial Press, New York.

Bloch, H. (2016). *Practical Lubrication for Industrial Facilities.* Marcel Dekker, New York.

Hodges, P. (1996). *Hydraulic Fluids.* Arnold Publishing, London.

Landsdown, A. (1994). *High Temperature Lubrication.* Mechanical Engineering Publications, London.

Landsdown, A. (1996). *Lubrication and Lubricant Selection.* Mechanical Engineering Publication, London.

Ludema, K. (1996). *Friction, Wear, Lubrication: A Textbook in Tribology.* CRC Press, Boca Raton, FL.

National Lubricating Grease Institute (2017). *Lubricating Grease Guide,* 6th ed. National Lubricating Grease Institute, Kansas City, MO.

Pirro, D. M., & Wessol, A. A. (2016). *Lubrication Fundamentals.* Marcel Dekker, New York.

Scott, R., Fitch, J., & Leugner, L. (2012). *The Practical Handbook of Machinery Lubrication.* Noria Publishing, Tulsa, OK.

Troyer, D., & Fitch, J. (2010). *Oil Analysis Basics.* Noria Publishing, Tulsa, OK.

# About the Authors

**Michael D. Holloway** has more than 35 years' experience in industry, holds all eight ICML certifications: Machinery Lubrication Engineer (MLE); Machinery Lubrication Technician® Level I and II (MLT® I, II); Machine Lubricant Analyst Level I, II, and III (MLA I, II, III) (ISO 18436-4, I, II, III), Laboratory Lubricant Analyst Level I and II (LLA I, II) (ISO 18436-5), as well as the Certified Lubrication Specialist (CLS) and Oil Monitoring Analyst (OMA) certifications from the Society of Tribologists and Lubrication Engineers (STLE); Certified Lubricating Grease Specialist, National Lubricating Grease Institute (CLGS); and Certified Reliability Leader, Association of Asset Management Professionals (CRL). He has authored a dozen books as well as holding several university degrees. Michael started 5th Order Industry as the independent content-development vehicle for competency development and certification preparation classes as well as projects ranging from ghost writing, technical marketing, do-it-yourself (DIY) articles, and building learning management systems. 5th Order Industry is now recognized as a world leader in competency development for certification preparation. Michael created the world-leading 5th Order Industry Competency Development Model, a learning system like no other in the industry, based on the neuroscience cognitive pathway development research that is currently under way as well as an implementation of a heuristic practice.

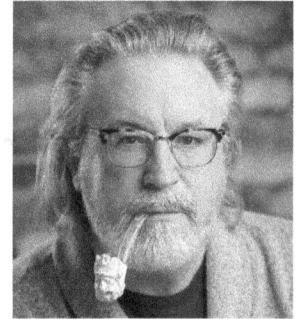

**Sanya Mathura** is founder and CEO of Strategic Reliability Solutions Ltd based in Trinidad & Tobago. Sanya possesses a strong engineering background with a BSc degree in Electrical and Computer Engineering and an MSc degree in Engineering Asset Management. Additionally, she has become the first person (and first woman) to become an ICML-certified Machinery Lubrication Engineer in the Caribbean as well as the first and only female in the world to achieve the ICML VPR (Varnish & Deposits Prevention and Removal) and VIM (Varnish & Deposits Identification and Measurement) badges. She is also the author of *Lubrication Degradation Mechanisms: A Complete Guide* and coauthor of *Lubrication Degradation: Getting into the Root Causes.*

# 1 | Outline Notes

This is your working notebook section to be used during the course. It will help you organize your ideas and thoughts. Some courses structure their content differently. It is suggested that you become familiar with the layout in order to be able to work in the section on which the class is structured. An alternative is to use a blank notebook in which to take notes during class and then transcribe them into this section in a neat and organized manner.

# MLT® I Outline Notes

## I. Maintenance Strategy

*(Explain each of the following items.)*

A. Why machines fail

_____

_____

_____

_____

B. The impact of poor maintenance on company profits

_____

_____

_____

_____

C. The role of effective lubrication in failure avoidance

_____

_____

_____

_____

## II. Lubrication Theory

*(Explain each of the following items.)*

A. Fundamentals of tribology

_____

_____

_____

_____

B. Functions of a lubricant

_____

_____

_____

_____

C. Hydrodynamic lubrication (sliding friction)

_____

_____

_____

_____

D. Elastohydrodynamic lubrication (rolling friction)

_____

_____

_____

_____

E. Mixed-film lubrication

_____

_____

_____

_____

F. Boundary lubrication

_____

_____

_____

_____

## III. Lubricants

*(Explain each of the following items.)*

A. Base oils

_____

_____

_____

_____

B. Additives and their functions

_____

_____

_____

_____

C. Oil lubricant properties and classifications:

1. Physical

_____

_____

_____

_____

2. Chemical

_____

_____

_____

_____

3. Performance

_____

_____

_____

_____

D. Grease lubrication:

1. How grease is made

_____

_____

_____

_____

2. Thickener types

_____

_____

_____

_____

3. Thickener compatibility

_____

_____

_____

_____

4. Grease lubricant properties and classifications:

a. Physical

_____

_____

_____

_____

b. Chemical

_____

_____

_____

_____

c. Performance

_____

_____

_____

_____

## IV. Lubricant Selection

*(Explain each of the following items.)*

A. Viscosity selection

_____

_____

_____

_____

B. Base-oil type selection

_____

_____

_____

_____

C. Additive system selection

_____

_____

_____

_____

D. Machine-specific lubricant requirements
   *(List and sketch the different pump types and provide a schematic of a hydraulic system.)*

    1. Hydraulic system

    2. Rolling-element bearings
   *(Draw a diagram of all the components.)*

    3. Journal bearings
   *(Indicate the use and draw a diagram; include the alloys.)*

    4. Reciprocating engines
   *(Sketch the differences between gasoline engines and the diesel engine.)*

    5. Gearing and gearboxes
   *(List at least five different gear types, and sketch each.)*

E. List of application- and environment-related adjustments

_____

_____

_____

_____

# V. Lubricant Application

*(Explain each of the following items.)*

A. Basic calculations for determining required lubricant volume

_____

_____

_____

_____

B. Basic calculations to determine relube and change frequencies

_____

_____

_____

_____

C. When to select oil; when to select grease

_____

_____

_____

_____

D. Effective use of manual delivery techniques

_____

_____

_____

_____

E. Automatic delivery systems:
   1. Automated delivery options:
      a. Automated grease systems
         *(Draw parallel and progressive systems.)*

b. Oil mist systems

_____

_____

_____

_____

c. Drip and wick lubricators

_____

_____

_____

_____

2. Deciding when to employ automated lubricators
   *(Explain.)*

_____

_____

_____

_____

3. Maintenance of automated lubrication systems
   *(List and draw diagrams.)*

_____

_____

_____

_____

## VI. Preventive and Predictive Maintenance

*(Explain each of the following items.)*

   A. Lube routes and scheduling

_____

_____

_____

_____

   B. Oil analysis and technologies to ensure lubrication effectiveness

_____

_____

_____

_____

   C. Equipment tagging and identification

_____

_____

_____

_____

## VII. Lube Condition Control

*(Explain each of the following items.)*

   A. Filtration and separation technologies

_____

_____

_____

_____

   B. Filter rating
     *(Define and draw the concept of the beta ratio.)*

   C. Filtration system design and filter selection
     *(List and draw the various types.)*

## VIII. Lube Storage and Management

*(Explain each of the following items.)*

   A. Lubricant receiving procedures

_____

_____

_____

_____

   B. Proper storage and inventory management

_____

_____

_____

_____

   C. Lube storage containers

_____

_____

_____

_____

D. Proper storage of grease guns and other lube application devices

_____

_____

_____

_____

E. Maintenance of automatic grease systems

_____

_____

_____

_____

F. Health and safety assurance

_____

_____

_____

_____

# MLT® II Outline Notes

## I. Maintenance Strategy

*(Explain each of the following items.)*

A. The impact of lubrication on machine reliability

_____

_____

_____

B. The impact of lubrication on lubricant life and consumption

_____

_____

_____

_____

C. Maintenance program strategies for achieving lubrication excellence

_____

_____

_____

_____

## II. Lubrication Theory

*(Explain each of the following items.)*

A. Friction and tribology:

1. Types of friction and wear:

   a. Wear modes and influencing factors

   _____

   _____

   _____

   _____

   b. Machine frictional surfaces most at risk for specific wear modes (e.g., abrasion)

   _____

   _____

   _____

   _____

2. Mechanisms of lubrication regimes:

   a. Boundary

   _____

   _____

   _____

   _____

b. Mixed film

_____

_____

_____

_____

c. Hydrodynamic

_____

_____

_____

_____

d. Elastohydrodynamic

_____

_____

_____

_____

B. Lubricant categories:

1. Gaseous

_____

_____

_____

_____

2. Liquid

_____

_____

_____

_____

3. Cohesive

_____

_____

_____

_____

4. Solid

_____

_____

_____

_____

## III. Lubricant Formulation

*(Explain each of the following items.)*

A. Base-oil refining methods and American Petroleum Institute categories:

1. Solvent refined

_____

_____

_____

_____

2. Hydrotreated

_____

_____

_____

_____

3. Severely hydrotreated

_____

_____

_____

_____

4. Hydrocracked

_____

_____

_____

_____

B. Mineral base oils:

1. Naphthenic

_____

_____

_____

_____

2. Paraffinic

_____

_____

_____

_____

3. Aromatic

_____

_____

_____

_____

C. Vegetable base oils and biolubes

_____

_____

_____

_____

D. Synthetic lubricant characteristics/
applications/compatibility:

1. Synthesized hydrocarbons (e.g.,
polyalphaolefins)

_____

_____

_____

_____

2. Dibasic acid esters

_____

_____

_____

_____

3  Polyol esters

_____

_____

_____

_____

4. Phosphate esters

_____

_____

_____

_____

5. Polyalkylene glycol

_____

_____

_____

_____

6. Silicones

_____

_____

_____

_____

7. Fluorocarbons

_____

_____

_____

_____

8. Polyphenyl ethers

_____

_____

_____

_____

E. Food-grade lubricant classification

_____

_____

_____

_____

F. Types and functions of additives

_____

_____

_____

_____

G. Types and functions of solid additives

_____

_____

_____

_____

H. Modes of additive depletion

_____

_____

_____

_____

## IV. Grease Application and Performance

*(Explain each of the following items.)*

A. Grease applications requiring high consistency

_____

_____

_____

B. Grease applications requiring low consistency

_____

_____

_____

C. Causes of grease separation

_____

_____

_____

D. Multipurpose greases

_____

_____

_____

E. Performance and application of specific grease thickeners

_____

_____

_____

_____

F. High-temperature greases

_____

_____

_____

_____

G. Coupling greases

_____

_____

_____

_____

# V. Lubricant Selection

*(Explain each of the following items.)*

A. Viscosity selection/adjustments according to machinery condition/environmental conditions

_____

_____

_____

B. When to use synthetic lubricants

_____

_____

_____

_____

C. When to use biodegradable lubricants

_____

_____

_____

_____

D. Lubricant consolidation

_____

_____

_____

_____

E. Select lubricating oils for:

1. Fire-resistant applications

_____

_____

_____

_____

2. Hydraulics, mobile/industrial

_____

_____

_____

_____

3. Turbines

_____

_____

_____

_____

4. Compressors

_____

_____

_____

_____

5. Bearings

_____

_____

_____

_____

6. Chains/conveyors

_____

_____

_____

_____

7. Mist applications

_____

_____

_____

_____

8. Gears, automotive/industrial

_____

_____

_____

_____

9. Engines, diesel/gas/gasoline

_____

_____

_____

_____

10. Pneumatic tools

_____

_____

_____

_____

11. Spindles

_____

_____

_____

_____

12. Ways/slides

_____

_____

_____

_____

F. Selecting greases for:
  1. Chassis

_____

_____

_____

_____

  2. Couplings

_____

_____

_____

_____

  3. Antifriction bearings

_____

_____

_____

_____

  4. Journal bearings

_____

_____

_____

_____

  5. Automotive bearings

_____

_____

_____

_____

6. Automatic lubrication systems

_____

_____

_____

_____

G. Lubricant selection standards development

_____

_____

_____

_____

H. Procedures for testing and quality assurance of incoming lubricants

_____

_____

_____

_____

I. Procedures for approval of candidate lubricants

_____

_____

_____

_____

## VI. Lubricant Testing and Performance Analysis

*(Explain each of the following items.)*

A. Viscosity

_____

_____

_____

_____

B. Flash/fire point

_____

_____

_____

_____

C. Pour ASTM D97/Cloud Point ASTM D2500

_____

_____

_____

_____

D. Foam ASTM D892

_____

_____

_____

_____

E. Air Release Properties ASTM D3427

_____

_____

_____

_____

F. Neutralization number:

1. Acid Number ASTM D664/D974

_____

_____

_____

_____

2. Base Number ASTM D974/D2896

_____

_____

_____

_____

G. Filterability ISO 13357

_____

_____

_____

_____

H. Oxidation stability:

1. Turbine Oil Oxidation Stability Test ASTM D943

_____

_____

_____

_____

2. Rotary Pressure Vessel Oxidation Test ASTM D2272

_____

_____

_____

_____

I. Rust and corrosion tests:

1. Turbine Oil Rust Test ASTM D665

_____

_____

_____

_____

2. Copper Strip Corrosion ASTM D130

_____

_____

_____

_____

J. Antiwear tests:

1. Four-Ball Wear Test ASTM D2266

_____

_____

_____

_____

2. Vickers Wear Pump Test ASTM D2882

_____

_____

_____

_____

3. SRV Test

_____

_____

_____

_____

K. Extreme pressure:

1. Four-Ball EP Test ASTM D2596

_____

_____

_____

_____

2. Timken Extreme Pressure Test ASTM D2509

_____

_____

_____

_____

3. Falex EP/Wear Test ASTM D2670

_____

_____

_____

_____

4. FZG Four Square Gear Test Rig
   ASTM D5182.97

   _____

   _____

   _____

   _____

L. Demulsibility ASTM D1401

   _____

   _____

   _____

   _____

M. Grease Consistency ASTM D217

   _____

   _____

   _____

   _____

N. Dropping Point of Grease ASTM D2265

   _____

   _____

   _____

   _____

O. Mechanical Stability of Greases ASTM
   D217A

   _____

   _____

   _____

   _____

P. Rolling Stability of Greases ASTM
   D1831

   _____

   _____

   _____

   _____

Q. Water Washout Test for Greases ASTM
   D1264

   _____

   _____

   _____

   _____

R. Water Spray Test for Greases ASTM
   D4049

   _____

   _____

   _____

   _____

S. Rolling Bearing Rust Test ASTM D1743

   _____

   _____

   _____

   _____

T. Koppers centrifugal stability test

   _____

   _____

   _____

   _____

U. Oil Separation in Grease Storage ASTM
   D1742

   _____

   _____

   _____

   _____

V. Oxidation Stability – Greases ASTM
   D942

   _____

   _____

   _____

   _____

## VII. Lubricant Application

*(Explain each of the following items.)*

A. Procedures for:

  1. Oil drain

    _____

    _____

    _____

    _____

  2. Reservoir/system flushing

    _____

    _____

    _____

    _____

  3. Disassembling/cleaning reservoirs and sumps

    _____

    _____

    _____

    _____

  4. Filling

    _____

    _____

    _____

    _____

  5. Top-up

    _____

    _____

    _____

    _____

  6. Grease packing

    _____

    _____

    _____

    _____

  7. Regreasing

    _____

    _____

    _____

    _____

  8. Grease changeout

    _____

    _____

    _____

    _____

B. Determine/calculate correct amount for regreasing

    _____

    _____

    _____

    _____

C. Determine/calculate correct frequency interval for regreasing

    _____

    _____

    _____

    _____

D. Select and manage optimal equipment/ systems for lubricant application according to machinery requirements

    _____

    _____

    _____

    _____

E. Safety/health requirements for lubricant application

_____

_____

_____

_____

F. Manage proper maintenance of lubrication equipment

_____

_____

_____

G. Manage proper maintenance of automatic lubrication systems

_____

_____

_____

H. Create/update lube survey

_____

_____

_____

_____

I. Record execution of lube program

_____

_____

_____

_____

J. Proactive management and detection of leaks

_____

_____

_____

_____

K. Waste oil/filters management/disposal

_____

_____

_____

_____

L. Writing a lubrication preventive maintenance (PM) plan

_____

_____

_____

_____

## VIII. Preventive and Predictive Maintenance

*(Explain each of the following items.)*

A. Creating and managing lube PM plans and routes

_____

_____

_____

B. Creating and managing a lubrication inspection checklist

_____

_____

_____

C. Used-oil analysis to determine optimal condition–based oil changes

_____

_____

_____

_____

D. Used-oil analysis to troubleshoot abnormal lubricant degradation conditions

_____

_____

_____

_____

E. Used-oil analysis to troubleshoot abnormal wear related to lubricant degradation/contamination

_____

_____

_____

_____

F. Procedures and methods for identifying root cause of lubricant failure

_____

_____

_____

_____

G. Use of technology aids to determine optimal regrease frequency/quantity (ultrasonic, temperature monitoring, shock impulse, etc.)

_____

_____

_____

_____

## IX. Lubricant Condition Control

*(Explain each of the following items.)*

A. Proper sampling procedures

_____

_____

_____

_____

B. Proper sampling locations

_____

_____

_____

_____

C. Proper selection of breathers/vents

_____

_____

_____

_____

D. Proper selection of filters according to cleanliness objectives

_____

_____

_____

_____

E. Filter rating, beta ratio

_____

_____

_____

_____

F.  Sump/tank management to reduce:

1.  Air entrainment/foam

_____

_____

_____

_____

2.  Particles

_____

_____

_____

_____

3.  Water

_____

_____

_____

4.  Sediments

_____

_____

_____

_____

5.  Heat

_____

_____

_____

_____

6.  Silt/sediments

_____

_____

_____

_____

7.  Unnecessary lubricant volume

_____

_____

_____

_____

G.  Proper selection of reconditioning systems for:

1.  Water

_____

_____

_____

_____

2.  Air/gas

_____

_____

_____

_____

3.  Particles

_____

_____

_____

_____

4.  Oxidation products

_____

_____

_____

_____

5.  Additive depletion

_____

_____

_____

_____

H. Lube reclamation:

 1. Requirements

    _____

    _____

    _____

    _____

 2. Feasibility

    _____

    _____

    _____

    _____

 3. Procedures for reclaiming/
    reconditioning

    _____

    _____

    _____

    _____

 4. Use of oil analysis to approve
    reclaimed/reconditioned lubricants

    _____

    _____

    _____

    _____

B. Defining maximum storage time
   according to environmental conditions/
   lubricant type

   _____

   _____

   _____

   _____

C. Safety/health requirements

   _____

   _____

   _____

   _____

D. Proper sampling procedures/locations for
   sampling stored lubricants

   _____

   _____

   _____

   _____

E. Procedures for reconditioning/filtering
   stored lubricants

   _____

   _____

   _____

   _____

# X. Lube Storage and Management

*(Explain each of the following items.)*

A. Design of optimal storage room

   _____

   _____

   _____

   _____

# 2 Guided Cooperative Arguments

This section consists of the results of a dialogue between the authors concerning the relevant topics. This section will help you understand how to answer questions that are asked of you in your job function. By mastering this content and understanding the applications, you will have a solid working understanding of what you are required to know.

After each part of the course work is covered during class, review this section to get a better understanding of how the content is delivered and applicable. This section is an example of how instructors and students would engage in questions and answers in order to draw out the truth. The method was perfected by the Greek philosopher Socrates and has been a cornerstone of competency development ever since.

# Machinery Lubrication Technician I (MLT® I)

## I. Maintenance Strategy

*Provide and explain a few reasons for why machines fail.*

Machines fail for two main reasons:

1. Maintenance induced: improper installation, lack of expertise, wrong tools, not enough training, wrong maintenance strategy, no condition monitoring for the assets
2. Non-maintenance induced: design, process, purchasing, cheap spare parts, wrong selection

Machines fail when an anomaly or fault is introduced. An anomaly or fault is the condition when items/components/assemblies degrade or exhibit abnormal behavior, which may lead to failure of the machine (ISO 13372). Anomalies or faults may be introduced during design, manufacture, and/or operation.

Machines consist of parts or items, each of which has a finite designed life. Once the designed end of life has been reached, the machine may fail. If the operation and loading of the machine, including the duty cycle and factors such as contamination, are not managed according to the machine's design intent, the machine's dependability will be affected. Machines need to be maintained as per the original equipment manufacturer's (and therefore the designer's) recommendations. If maintenance is not executed or not executed effectively and correctly and the correct equipment and materials/parts/items are not used when maintaining, anomalies and faults may be introduced, which may lead to failure.

*Describe the impact of poor maintenance on company profits.*

Higher repair costs, longer downtimes, lots of breakdowns and secondary damage, lower profitability, bad return on investments. Anomalies and faults may be introduced by poor maintenance, and this may lead to failures, which will lead to costly downtimes. The repair cost will increase if the failures are unplanned because unplanned personnel will be required for repair. Poor maintenance therefore leads to increased costs and reduced profits.

## II. Lubrication Theory

*What is the role of effective lubrication in failure avoidance?*

Effective lubrication is similar to any other monitoring tools; right lubrication program means right lubricant selection, sampling frequency, analysis, training, and so on. This will lead to failure avoidance by control of wear, corrosion, rust, and contamination and protect from sudden failure, lower lubricant lifetime, and more—all these factors lead to more machine health and lifetime. Effective lubrication will ensure that the correct lubricant is used and that the lubricant is kept clean, dry, and cool. Effective lubrication will ensure that the lubrication function is maintained correctly and that friction and wear, which may lead to failure, are therefore reduced or prevented.

*What are the fundamental concepts of tribology?*

Tribology is the study of friction, wear, and lubrication and how to control friction, wear, corrosion, temperature, and contamination.

*What are the functions of a lubricant?*

Control of friction, wear, corrosion, temperature, contamination and transmission of power.

*Explain boundary lubrication.*

It is friction from viscous drag and mechanical contact; 90% of the load rests on surface peaks; protection depends on boundary anti-wear film; found during slow-moving conditions and on shock loads—surfaces separated only by molecules.

Boundary lubrication exists during the following conditions: when the component's surface roughness significantly exceeds the oil-film thickness; when the equipment is subjected to frequent starts and stops, shock-load conditions, high static loads, or slow speeds; and where operational requirements dictate the use of low-viscosity oil, compromising friction control of some system components.

*Now explain mixed-film lubrication.*

The majority of the load sets on the surface peak, incomplete separation, and both the bulk lubricant and boundary film play a role. Mixed-film lubrication exists where the component's surface thickness exceeds the oil-film thickness to some extent but not as much as in the case of boundary lubrication.

*Now explain hydrodynamic lubrication (sliding friction).*

Pressure builds up internally by the relative motion between the moving and rigid surfaces.

- Increase the oil-film thickness: speed up or provide oil cooling or load reduction.
- Decrease the oil-film thickness: reduce speed, heat the oil, or increase the load.

- Increase contact pressure from 100 to 300 lb/ft$^2$ (psi).
- Example: journal bearings.

Hydrodynamic lubrication is also called *full-film* or *full-fluid-film lubrication*. Moving surfaces are fully separated by a hydrodynamically formed wedge. No contact is expected between the asperities of the two surfaces in relative motion. Formation of the oil wedge depends on surface geometry, speed, load, and oil viscosity. Hydrodynamic lubrication is not achieved during startup and shutdown conditions unless there is pressure on the system in order to get full component separation. This helps reduce wear and excessive power requirements during startup. This is called *hydrostatic lubrication* or *preload lubrication*. This is done with turbine journal bearings and even modern engines with the automatic shutoff.

*Now there is another form of lubrication film development—a different sort of regime, if you will—explain elastohydrodynamic lubrication (rolling friction).*

Pressure builds up because of elastic surface deformation; highly concentrated loads create a wave of metal ahead of the load zone. Elastohydrodynamic lubrication is formed at rolling contacts such as rolling-element bearings, where surfaces in relative motion converge on a point (ball bearings) or line (roller bearings). Because of the elasticity of the bearing material, it can deform elastically in the small area of contact. A hydrodynamic oil film is formed inside the small volume of deformation. The high pressure inside the deformation results in the oil momentarily changing into a solid in the small volume of deformation. Once the pressure is released, the lubricant returns to a liquid state again.

## III. Lubricants

*What are the classes and examples of base oils?*

Mineral base oil and synthetic base oil. There are five American Petroleum Industry (API) groups of oil stocks based on sulfur content, saturation, and viscosity index.

*Provide examples of additives and their functions.*

In general, additives are used to enhance existing desirable base-oil properties, suppress undesirable base-oil properties, and impart new desirable properties to base oil.

- Antioxidants. Zinc dialkyldithiophosphate (ZDDP). Antioxidants delay the formation of acids, varnish, sludge, and the high viscosity that normally results from oxidation.
- Antifoam agents. Antifoam agents suspended microglobule silicon, which weaken air bubbles so that they break. Note that at high viscosity, foam is formed from small bubbles that are hard to break.
- Demulsifying agents. These agents separate oil and water.
- Detergents. Detergents work at higher temperatures than dispersants, removing soot, and they naturalize acids too
- Viscosity index (VI) improvers. These agents increase the VI for multisession viscosity/temperature performance.
- Corrosion inhibitors. ZDDP. These agents neutralize acids and sealing surfaces from contact with water and corrosive acids.
- Rust inhibitors. Rust and oxidation (R&O) inhibitors form a polar film that adheres to steel and cast-iron surfaces. The film repels water, which inhibits rust formation.

- Pour-point depressants. These agents prevent congelation of oil at low temperatures.
- Dispersants. These agents disperse soot particles to prevent clustering, settling, and deposits at low to moderate temperatures.

*Provide lubricant physical, chemical, and performance properties and their classifications.*

- Chemical properties. Viscosity, VI, pour point, flash point, oxidation stability, color, additive response, and thermal stability
- Physical properties. Sealing, operating temperature, food compatibility, speed factor, and installation

*How is grease made?*

Grease is a solid to semisolid fluid material product of a thickener agent (soap) with liquid lubricant; other additives may be included. Grease is manufactured via some or all of the following steps:

- Saponification
- Dehydration
- Cutback
- Milling
- Deaeration

Greases are usually defined as solid or semifluid products of a thickening agent in a liquid lubricant. They can contain additional ingredients that impart special properties. Typically, a grease is composed of 65%–95% base oils, 5%–35% thickeners, and 0%–10% additives.[1]

*What are the grease thickener types?*

■ Soap thickeners
■ Simple metallic soap thickeners. Aluminum soap, sodium soap, and anhydrous calcium or lithium soaps.
■ Complex metallic soap thickener. Aluminum complex soap, lithium complex soap, lithium and calcium complex blend soaps.
■ Nonsoap thickener. Organoclay, polyurea, calcium sulfonate as well as fluorocarbon polymers like Teflon®

*Are there different types of thickeners?*

Yes, there are many different types of thickeners. Each thickener provides particular properties. Generally, they are classified into simple soaps, complex soaps, organic, inorganic, and miscellaneous.

■ Simple soaps include lithium, calcium, sodium, aluminum, and barium.
■ Complex soaps include lithium complex, calcium complex, calcium sulfonate complex, aluminum complex, and sodium complex soaps.
■ Organic thickeners *(nonionic)* include oligourea (commonly called *polyurea*) and polytetrafluoroethylene (PTFE).
■ Organic thickeners *(ionic)* include sodium and calcium salts of stearylamidoterphthalic acid.
■ Inorganic thickeners include clays (bentonite-type aluminosilicates, mainly smectites, montmorillonite, and hectorite) and highly dispersed silicic acid.
■ Miscellaneous thickeners include temporarily thickened fluids such as magnetic fluids, electrorheologic fluids, and liquid crystals.[2]

*Explain grease thickener compatibility and incompatibility.*

Incompatibility causes lowering of mechanical stability and lowering of the National Lubricating Grease Institute (NLGI) number, which cause a change in the dropping point, for example, so many tests are required to check grease compatibility before mixing two different grease types.

*Provide examples of grease lubricant physical, chemical, and performance properties and include the NLGI classifications where applicable.*

■ Chemical properties. Dropping point, extreme pressure, grease hardness, and mechanical stability (NLGI number describes grease hardness).
■ Physical properties. Sealing, operating temperature, food compatibility, speed factor, and installation.

Compatibility of thickeners with other thickeners varies. Greases should not be mixed if the thickeners are incompatible. Check grease consistency, which describes the degree of stiffness of the grease (a measure of the ability of the grease to resist change of form or shape or of the ability of the grease to resist deformation by an applied mechanical force exerted by a penetrating cone of known mass).

*Are there classifications for the physical, chemical, and performance properties of greases?*

The physical properties of a grease are classified by its consistency (cone penetration and NLGI grades), apparent viscosity, and dropping point. The NLGI grades, which determine grease consistency, are grouped from 000 (fluid) to 6 (solid) grease. The chemical properties are classified based on the type of base oil, additives, and thickeners used.

There are 12 phenomena that result in 24 properties of grease performance:

- High temperature (maximum thermal stability, minimum evaporation loss, minimum viscosity)
- Low temperature (no regular crystallization, minimum viscosity)
- Aging (maximum oxidation resistance, resistance to changes in structure)
- Compatibility (no reaction with non-ferrous metals, maximum corrosion inhibition, maximum polymer compatibility, immiscibility with foreign liquids, deflection of solid matter)
- Oil loss (optimum oil loss)
- Toxicity (no toxicity, biodegradability)
- Tackiness (optimum tackiness)
- Flowability (optimum relaxation, maximum pumpability)
- Load (optimum elasticity, maximum lubricating film thickness, maximum emergency running properties)
- Shear (maximum mechanical stability or optimum relaxation time)
- Friction (minimum or optimum friction)
- Wear (minimum wear)

## IV. Lubricant Selection

*Explain viscosity selection and its importance.*

Viscosity is the most important physical property of a fluid lubricant. Optimal viscosity selection maximizes reduced friction by creating and maintaining a proper fluid film that separates moving parts or minimizes metal contact depending on design and application. The selection is done according to speed, operating temperature, and geometric shape.

There is a relationship between viscosity, oil temperature, and the load-carrying capacity of the oil. Proper viscosity selection is necessary to ensure that the oil performs as required according to the specific application, process, and load requirements. Several international standards guide viscosity and its classification:

- ISO3448 (ISO VG)
- SAE J300 Automotive Engine Oil
- AGMA9005-D94 Gear Oil
- SAE J306 Automotive Gear Oil
- ASTM D2270 Viscosity Index

*Explain base-oil type selection and its importance.*

The three most important base-oil properties are:

- Viscosity
- Viscosity Index (VI)
- Service life

The general base-oil properties are:

- Viscosity. Viscosity grade at a temperature
- Viscosity Index. Viscosity-temperature relationship
- Specific gravity. Density of oil compared to water·
- Flash point. High-temperature volatility
- Pour point. Low-temperature fluidity

These are tested according to international standards. Before all the additives are added to the mixture, lubricating oils begin as one or more of five groups from the API base oil categories (groups) based on percent sulfur, percent saturates, and VI:

- Group I. >0.03% sulfur, <90% saturation, VI 80–120
- Group II. <0.03% sulfur, >90% saturation, VI 80–120
- Group III. <0.03% sulfur, >90% saturation, VI >120
- Group IV. All polyalfaolefins, synthetic

- Group V. All others including nathanics, silicon, and biobased, as well as synthetic esters

Mineral base-oil properties include:

- VI
- Pour point
- Flash point
- Additive solvency

Applications include:

- Bearing and turbine oils
- Hydraulic fluids
- Gear oils
- Worm-gear oils
- Internal combustion engines
- Percent paraffinic content
- Percent naphthenic content
- Percent aromatic content

Properties of a lubricant differ based on the base oil used in its creation. In general, synthetic lubricants have superior properties over mineral-oil lubricants. Synthetics also have some disadvantages when compared with mineral lubricants. Synthetic lubricant advantages include:

- Higher VI
- Better thermal stability
- Results in less friction
- Higher shear strength
- Higher oxidation stability
- Better fire resistance
- Higher flash point
- Lower pour point
- Better detergency

Synthetic lubricant disadvantages include:

- Higher cost
- Less mixable; less compatibility with other fluids and lubricants

- Less compatible with different seal types
- Less soluble in terms of additives
- More toxic
- Higher disposal cost
- Less hydrolytic stability; less resistant to the effects of water contamination

### When should additives be used?

Additives usually have one of three purposes: to enhance, to impart new properties, or to reduce unwanted properties of a base oil. Typically, additives must be added to the lubricant such that the final product is balanced. If additives are added in an effort to replenish reduced additive packages from a finished lubricant, this can lead to detrimental effects because the finished lubricant is no longer balanced.

Here are the most popular additives used in lubricants and their purposes:

1. Pour-point depressants. These are used to lower the temperature at which wax crystals form, thereby lowering the actual pour point of a lubricant.
2. VI improvers. These are placed in an oil to allow its viscosity to increase at higher temperatures. VI improvers are used in engine oils, automatic transmission fluids (ATFs), multipurpose tractor fluids, hydraulic fluids, and even automotive gear lubricants.
3. Defoamants. These have the ability to attach themselves to air bubbles and allow them to rise to the surface. Usually, when they rise to the surface, they may form foam, but because of the nature of the defoamant, they cause unstable bridges that help in collapsing the air bubbles, preventing foam.
4. Oxidation inhibitors. These have the expressed purpose of either reacting with the initiators (which begin the oxidation process) to form an inactive

compound or decomposing the materials to form less reactive compounds and thereby reduce the rate of oxidation.

5. **Rust and corrosion inhibitors.** Corrosion inhibitors form a protective film on the surface of the metal to prevent its attack from acids. In contrast, rust inhibitors have a high polar attraction to the metal surfaces and also form a film to protect the metal surfaces from any water that may reach them.

6. **Detergents and dispersants.** Detergents are responsible for neutralizing deposit precursors that tend to form under high-temperature conditions (particularly acid blow by combustion products in engines). These detergents also have the ability to disperse and suspend contaminants. In contrast, dispersants suspend the potential varnish- or sludge-forming materials. Dispersants also can prevent the formation of high-temperature deposits.

7. **Antiwear additives.** These adhere to the surfaces of the materials to reduce friction and wear scuffing and scoring during the boundary lubrication phase.

8. **Extreme-pressure additives.** These begin to work at high temperatures or under heavy loads where severe sliding friction exists. These additives are required to reduce friction, control wear, and prevent surface damage.[3]

*Explain the additive system selections for gear oil, hydraulic oil, turbine oil, ATF, and engine oil.*

■ Gear oil. Antifoam, antioxidant, corrosion and rust inhibitor, demulsifier, pour-point depressant, antiwear, VI improver, and extreme pressure and solid-film-strength additive.

■ Hydraulic oil. Antifoam, antioxidant, corrosion and rust inhibitor, demulsifier, pour-point depressant, antiwear, and VI improver.

■ Turbine oil. Antifoam, antioxidant, corrosion and rust inhibitor, demulsifier, and pour-point depressant. Steam, gas, and hydroelectric turbines operate on a family of lubricating oils known as *rust- and oxidation-inhibited oils* (R&O oils). Oxidation stability is a requirement because large quantities of turbine oils with a low makeup rate are used for each turbine. Demulsibility improvers improve the rate at which water separates from the oil, from being emulsified in the oil to being separated from the oil. Hydrolytic stability is required for hydroelectric turbines. Some gas turbine oils are required to have very special thermal and oxidation stability because of the high temperatures they experience. Antifoam additives also may be required for some turbine oil applications.

■ Engine oil. Antifoam, antioxidant, corrosion and rust inhibitor, pour-point depressant, antiwear, VI improver, friction modifier, and metal deactivator.

■ Machine-specific lubricant requirements. Viscosity, temperature, speed range, application, and compatibility.

*What are the lubricant requirements of hydraulic systems?*

The main consideration when selecting hydraulic oils is the pump design and required corresponding viscosity grade. Selection of the additive package is based on the application as per below:

■ High film strength to resist wear (antiwear capability)
■ High demulsibility
■ Noncompressible

- Oxidation stability
- Thermal stability
- Corrosion protection
- Applicable viscosity
- Applicable VI
- Optimal viscosity for different temperature ranges
- Incompressible
- Reduced wear rate and boundary contact points
- Ultrafine filters
- Foam resistance
- Hydrolytic stability
- Seal compatibility

Detergents are normally required in mobile hydraulic machines, whereas nondetergent additives are required for industrial hydraulic machines. Antiwear additives are used in high-pressure and high-performance hydraulic systems.

### What are the lubricant requirements of journal bearings?

Journal bearings are designed to operate on the full-film or hydrodynamic lubrication regime in most cases, although some low-speed pins and bushings are designed for boundary lubrication. Dry and semi-lubricated journal bearings also exist. Journal bearings are normally lubricated with oil, but grease lubrication also may be used in some applications.

One of the most important lubrication characteristics of journal bearing oil is the viscosity grade:

- Temperature operating range should be correct.
- Rust and oxidation inhibitors may be required.
- It is necessary to have the correct flash point, pour point, and acid number.

Speed, load, temperature, type of bearing, and how the lubricant is applied are important. Journal bearing lubricant requirements include:

- Maintenance of full fluid film
- Prevention of excessive friction
- Heat conductance
- Maintenance of stability under severe operating conditions

The main requirements are the correct:

- Consistency
- Temperature range
- Rust inhibition
- Load-carrying capability
- Miscibility (i.e., how well it mixes at any concentration)

### What are the lubricant requirements of rolling-element bearings?

For most applications, rolling-element bearings are lubricated with grease. In some applications, rolling-element bearings may be lubricated with oil. In special cases, even solid lubricants may be used.

The lubricant should be suitable for bearing operating temperature, load, speed, and environmental conditions. The following factors need to be taken into consideration when selecting solid lubricants:

- Coefficient of friction
- Load-carrying capacity
- Corrosion resistance (i.e., susceptibility to galvanic corrosion)
- Electrical conductivity
- Environmental factors such as temperature, pressure, humidity, oxygen content, and radiation

*What application- and environment-related adjustments would one make for a gearbox?*

To handle increased demands, today's industrial gear oils must contain high-performance additive chemistry. The goal is to keep the lubricant thermally stable and robust enough to ensure that it lasts longer, protects better, and performs more efficiently while at the same time keeping the system clean and carrying away heat and contaminants. Gear oils need to be kept clean, and the viscosity needs to suit the application. The correct additive package needs to be selected.

Typical additives for gear oils include extreme pressure and fluid demulsibility. However, there are two types of industrial gear lubricants: universal and dedicated gear oils. Dedicated gear oils contain specific additive packages for specific industrial applications.

Universal gear oils typically are used in automotive applications and may be unsuitable for industrial applications.

A challenge in gear oil additives is that extreme-pressure additives can be prone to thermal instability, resulting in sludge formation. Thermal stability and extreme-pressure protection need to be carefully balanced for gear oil applications.

ATFs have different characteristic colors, achieved by adding a dye to the oil. Because automatic transmissions are normally sealed units, dispersant additives have limited use in ATFs:

■ Corrosion inhibitors are required for ATFs.
■ VI improvers are used.
■ Seal-swell agents are used to ensure effective sealing.
■ Pour-point and cloud-point suppressors are required to ensure proper functioning at lower temperatures.
■ Foam inhibitors are required.

Also required are antiwear and antioxidant additives. Proper viscosity is selected based on load and operating temperature. Some adjustments include:

■ Correct amount of extreme-pressure additives based on the load, speed, and design
■ Correct corrosion control based on the environment in terms of water contamination and temperature variations
■ Adjustments for size
■ Adjustments for load and speed
■ Adjustment for contamination from the environment
■ Adjustments for operational temperature, including ambient temperature variations

*What are the lubricant requirements for engines?*

Speed, load, temperature, type of bearing, and how the lubricant is applied have to be considered. Additive packages for gasoline engines and diesel engines differ. One reason is that the catalytic converters differ. Diesel engine oils have a higher loading of antiwear additives in the form of zincdialkyldithiophosphate (ZDDP), and diesel engine catalytic converters can handle the ZDDP, whereas catalytic converters for gasoline engines cannot. Diesel engine oils have more additives per volume. Detergents are used in engine oils; too much may cause loss of compression.

Engine lubricants must have the correct:

■ Viscosity grade
■ VI
■ Flash point
■ Pour point
■ Cloud point
■ Oxidation resistance
■ Ashless dispersant additives to minimize deposit formation

Additives used in engine oils in general include:

- Detergents
- Dispersants
- VI improvers
- Pour-point depressants
- Foam depressants/air-release agents
- Antiwear additives
- Oxidation inhibitors
- Friction modifiers
- Extreme-pressure additives
- R&O inhibitors

## V. Lubricant Application

*Give examples of the basic calculations for determining required lubricant volume in bearings.*

Bearings should be 100% filled with lubricant with the housing partly filled (30%–50%). For low-speed bearings, the housing should be filled up to 90%, and the seals should be packed with grease. To determine grease quantity:

$$G = 0.005DB$$

where $G$ = grease quantity in grams (g)

$D$ = bearing outside diameter in millimeters (mm)

$B$ = total bearing width in millimeters (mm) or henrys (H) for thrust bearings

*Provide basic calculations to determine relube and change frequencies for lubricated systems.*

$$A \text{ (speed factor)} = n \times dm$$

where $n$ = speed in revolutions per minute (rpm)
$dm = 0.5(d + D)\text{mm}$

Load ratio $C/P$ is obtained from a chart, and the bearing factor $bf$ is given.

*Explain how to select and change oil and grease.*

Oil and grease are selected based on the design and operational requirements of the machine. Base oil types, oil properties (including viscosity and VI), and the additive package are taken into account when considering the design and operational requirements. These include considering the use of fluid oil, semifluid grease, or a solid lubricant.

*Explain the basics on effective use of manual delivery techniques, and list automatic delivery systems for oils and for greases.*

Oil mist, circulating oil, single-line grease, multiple-line grease, electroluber, and spray automated deliver options:

- Spring feed
- Electromechanical, up to 350 psi
- Electrochemical, up to 75 psi

*Explain the different designs, applications, and advantages of automated grease systems.*

A manual delivery system may be subject to many potential errors. Quality control processes must be used to ensure that the correct lubricant is used, that the correct quantity of lubricant is applied, that the lubricant is applied at the most effective intervals, and that contamination is controlled.

For a manual lubrication technique to be effective, it is not enough to just have the same lubrication frequency and volume for all equipment (e.g., once a month or once every 3 months all points are lubricated with the same amount of lubricant). Each lubrication point should have its own lubrication frequency and volume based on the equipment, its design, load, and environment.

### And the design?

Automatic lubrication systems apply lubrication automatically, not manually. Automatic lubrication includes single-point and centralized (piped) distributed systems. Automatic lubrication systems could be a fixed multipoint piped system or single-point applicators. Piped systems could be single or multiline. Piped systems could be actuated electrically or hydraulically via gas or with mechanical springs. Sometimes pipe systems are actuated manually (not classed as automatic).

An automatic system could consist of:

- Controllers
- Pumps
- Valves
- Reservoirs
- Heat exchangers
- Nozzles
- Injectors
- Pipes
- Filters
- Switches
- Instrumentation
- Regulators
- An energy supply

### The application?

Grease is more susceptible to particulate contamination. Automatic grease lubrication systems reduce the risk of contamination during lubrication. Centralized applicators are ideal for equipment with many lubrication points in close proximity, such as a mill system. Centralized applicators may carry sufficient reservoirs of lubrication supplied through a supply pipeline under sufficient pressure. Single-point applicators are more suited to equipment with many or widely separated lubrication points, such as a conveyor-belt system.

### Explain the design, application, and advantages of oil mist systems.

Design. This is a centralized system that provides an efficient lubrication technology to different classes of rotating equipment widely dispersed at a physical location. Instrument air is used to produce and convey oil mist through a piping distribution system and then inject the oil mist into bearing cavities for lubrication. This equipment may be operating, on standby, or in storage condition. Some conditions that affect system design are ambient temperature, area classification, project specifications, and/or location and arrangement of the equipment.

Applications. Petrochemical, power, steel, and pulp and paper. Most antifriction bearings can be pure mist applications, sleeve-type bearings, and gearboxes.

Advantages. Very good filtration, 1–3 μm compared with conventional liquid/sump lubrication; oil mist systems decrease oil usage by as much as 40% while completely eliminating the need for oil changes. Reduced oil consumption lowers operating costs and helps protect the environment by eliminating waste. Advantages also include the following:

- Bearings run cooler by 8–19°C.
- Bearing service life is extended.
- Reduce failures are reduced by as much as 90%.
- Mechanical seal service life is extended.
- Mechanical seal failures are reduced by as much as 30%.
- Oil consumption is decreased by as much as 40%.
- Overall maintenance costs are reduced.
- The impact of used oil on the environment is reduced (less oil is used).

*Explain the design, application, and advantages of drip and wick lubricators.*

A wick is used to absorb oil from an oil reservoir by capillary action. The oil is then allowed to drip from the wick due to gravity, ensuring a uniform distribution of oil on the lubricated surface over time.

*What decision criteria are employed for automated lubricators?*

First is criticality, but you have to do the "5 rights":

- The right lubricant
- In the right quantity
- At the right time
- At the right point
- With the right method

Based on machine criticality, the cost of applying manual lubrication to the plant is considered. The expected failure rate, cost to repair, and cost of lost production due to inadequate manual lubrication are considered, and a risk rating is calculated. The cost of automated lubricators is considered. The expected reductions in failure rate, cost to repair, and production are considered, and a risk rating is calculated.

A thorough criticality analysis of each asset will illustrate the impact of a failure in terms of:

- Overall production cost
- Overall maintenance cost
- Environmental impact
- Health and safety of personnel

The most critical assets are commonly the first targets of automatic lubrication.

*Explain the basics of the maintenance of automated lubrication systems.*

- Check the level in the reservoir
- Check the delivery tubing
- Check the grease expelled from bearings
- Check for piping damage
- Check the filters

Automatic lubricators must be on a formal inspection and maintenance schedule as determined by the condition-based predictive maintenance strategy. Formal and regular inspections are of specific importance for single-point automatic applicators. Visual inspections could result in anomalies being detected early.

Check and repair the following if required:

- Energy- and pressure-generating components:
  - Electric pumps
  - Pneumatic pumps
  - Hydraulic pumps
  - Mechanical pumps
  - Manually operated pumps
- Control components:
  - Timers
  - Reversing valves
  - Pressure regulators
  - Signals (lights or warning devices)
- Distribution network:
  - Primary header piping
  - Secondary piping (header to metering device)
  - Final piping (metering device to lube point)
  - Valves, etc.
  - Metering devices
  - Pressure-sensing devices

## VI. Preventive and Predictive Maintenance

*List the basics in establishing a lube route and scheduling practice.*

Choose right lubricant, right time, right sampling producer, right sampling time, right machine, and right amount of lubricant and arrange with all other preventive maintenance techniques.

■ Identify lubrication points.
■ Identify type of lubricant to be used for each lubrication point.
■ Identify the quantity of lubricant to be applied at each lubrication point.
■ Identify lubrication tasks and inspection tasks.
■ Identify lubricant application method for each lubrication point.
■ Identify inspection points and certify which points are for lubrication condition and which are for machine condition.
■ Establish expected (healthy) and unexpected (unhealthy) inspection findings and their characteristics.
■ Identify reporting requirements.

## VII. Lube Condition Control

*List the basic types of lubricant filtration systems and the three different types of separation technologies.*

Filtration is the process of removing unwanted solid particles from fluids by allowing the fluid to flow through a medium through which the fluid can pass freely while restricting solid particles larger than a specified size. Filter types could be surface filtration or depth filtration.

Filtration systems are classified in terms of the number of times the lubricant passes through the filter and in terms of the position of the filtration system in the lubrication circuit (in line or off line). Some include:

■ Single- or multipass systems
■ Draw-down systems
■ In- or off-line systems
■ Fixed or portable systems
■ Some combination of the above

Separation:

■ Settling and heating
■ Centrifuging
■ Vacuum dehydration

*Explain what filter ratings are.*

Filter ratings are based on the diameter of the largest hard spherical particle that will pass through a filter under specified test conditions. This is an indication of the largest opening in the filter elements.

Filters are rated in microns. The micron (micrometer) rating refers to the smallest particle size to be captured or filtered out.

■ Nominal micron rating. This expresses the ability of the filter to capture particles of a specified size at an efficiency between 50% and 98.6%.
■ Absolute micron rating. This means that the filter is capable of removing at least 98.7% of a specific size particle. This rating is more informative than the nominal micron rating.

The *beta ratio* is the ratio between the quantity of particles larger than the micron rating upstream of the filter and the quantity of particles larger than the micron rating downstream of the filter. *Filter efficiency* can be calculated from the beta rating:

$$\text{Filter efficiency} = (\beta - 1)/\beta \times 100\%$$

*Provide examples of a filtration system design and the rationale for the filter selection.*

- Surface filter
- Depth filter
- Electrostatic filter
- Centrifugal filter
- Vacuum filter

- Off-line filtration system. Periodic off-line filtering of machine oil for applications requiring additional filtration because of conditions and/or load could be done with an electrical filter cart.
- In-line high-pressure filtration system. High-pressure filters keep the oil that comes directly from the pump clean so that the more expensive downstream components (such as valves and actuators) are protected. Pressure-line filters offer protection from catastrophic pump failure.
- Draw-down filtration system. Common in off-line filtration cleanup and new fluid cleanup, these systems are similar to a multipass system without continuous ingression. A constant volume of fluid is circulated through a filter system with no additional contaminant added to the fluid.

## VIII. Lube Storage and Management

*What are the steps for lubricant receiving procedures?*

- A proper lubrication storage room that is closed and air conditioned should be available to keep lubricants clean, dry, and cool in storage.
- Enough space needs to be available inside the storage room to provide for all necessary quantities of lubricants to be stored properly.

- Lubricants need to be identified upon arrival and stored with identification. This could include color coding and signage.
- Transfer of lubricant from delivery vehicle to storage room needs to be controlled.
- Dedicated lubricant transfer containers must be available to prevent cross-contamination.
- The room should be fitted with shelves and hangers to allow for the storage of grease pails, grease guns, filters, and oil samples and oil sample packaging.
- Main oil receiving tanks must have sampling and level-monitoring facilities to ensure oil quality and cleanliness as received.
- Piped delivery from the main oil receiving tanks to individual tanks will prevent contamination.

*Explain the basics requirements for proper storage and inventory management.*

- Right lube room design and requirements. A properly designed lube room must be functional, safe, and expandable and provide all the necessary storage and handling requirements for the facility.
- Bulk oil storage. Determine lubricant consumption rate, lubricant storage capacity, and lubricant supplier turnaround time.
- Quality control. Lubricants delivered from lube suppliers must be verified to ensure that the correct product is being delivered and that the cleanliness of the delivered lubricant is up to current target particle and moisture cleanliness levels.
- Proper top-up container and grease gun storage.
- Labeling and identification.

- Lubricants must be identified correctly when stored.
- Lubricants must be identified correctly when dispensed.
- Lubricants must be kept clean.
- Lubricants must be kept dry.
- Lubricants must be kept cool.

*Give different examples of lubricant storage containers.*

- Oil pails
- Oil drums
- Oil totes
- Bulk storage
- Tankers

*Explain the proper storage of grease guns and other lube application devices.*

Grease guns should be stored in a clean, dry, and controlled environment. They are precision tools that must be taken care of in order for them to provide the maximum degree of accuracy and reliability. Grease guns should be cleaned and inspected regularly for proper function, and an annual calibration should be performed.

*Explain the proper maintenance of automatic grease systems.*

Calibration will ensure that the same volume of grease is still being dispensed with one shot as when the gun was new. The best method for grease gun calibration is to use a postal scale to measure how much grease is dispensed with one pump. Maintenance of automatic lubrication systems needs to be part of the condition-based predictive maintenance strategy.

Automatic lubricators must be on a formal inspection and maintenance schedule. Formal and regular inspection is of specific importance for single-point automatic applicators. Visual inspections could allow anomalies to be detected early.

*What measures must be taken to ensure proper health and safety assurance when using and handling lubricants?*

Proper receiving techniques should include filtration of incoming oils. Frequently, new oils are dirtier than the defined particle target cleanliness level. This means that if you define your particle target cleanliness level and spend time, money, manpower, and so on to achieve these levels of in-service lubricant cleanliness, the last thing you want is to contaminate your system with "dirty" new oils. Such a program relies on:

- Quality of base stocks
- Additive quality and concentration
- Lubricant performance properties
- Thickener performance properties (grease)

Larger storage containers such as 200-liter drums are heavy. Normal care should be taken when handling these drums to ensure safety of personnel:

- Use two people to move one drum.
- Bend the knees when lifting.
- Use appropriate lifting gear.
- Do not work under suspended loads.
- For smaller cans, use a safety locker with ventilation.
- Have material safety data sheets available for stored product, and understand the content.
- Fire extinguishers (inspected and maintained) should be available in storage areas.
- If solvent containers are stored with oil containers, these should be grounded to prevent sparks from static electricity.
- Avoid allowing oil to contact the skin.
- In case of contact with skin, clean with warm water and soap, not with solvents.
- Preferably use barrier cream and gloves.

- Avoid breathing oil mist. Of course, no smoking should be allowed in or near oil storage areas; signs should be posted.
- Spills should be cleaned up immediately to reduce slipping hazard.
- Oil storage areas should be kept clean; no loose, oil-soaked rags and other materials should be allowed.
- Use appropriate personal protective equipment (PPE), with safety glasses being compulsory.
- Some dispensing systems, especially manual and automatic grease application systems, can generate very high pressures. At these high pressures, leaks could cause fluid injection into the body.
- For air-pressurized systems, care must be taken, and such systems should be depressurized before attempting to work on them.
- Follow lock-out and tag-out procedures when working on grease systems.
- Care should be taken when lubricating rotating or moving equipment.

# Machinery Lubrication Technician II (MLT® II)

## I. Maintenance Strategy

*What is the impact of lubrication on machine reliability?*

Effective lubrication contributes to a dependable plant. The main function of lubrication is to protect surfaces in relative motion from wear damage caused by friction and a reduction in damage due to friction and wear. Any lubrication failure will result in a secondary failure linked to its function with reduced unplanned failures and increased repair costs.

*Does lubrication affect machine reliability?*

Yes, lubrication provides the basic functions of reducing friction, distributing heat, removing contaminants, improving efficiency, and minimizing wear.[4] Thus, if the lubricant does not perform these functions, then the machine can fail, reducing its reliability.

As J. B. P. Williamson explains:

All surfaces are rough. The world of the engineer is made of solids whose surfaces acquire their texture as the result of a great variety of processes. In some cases it is merely a by-product of forming the bulk shape, for example, in casting, molding, or cutting. More often a separate process affecting only the surface layers is applied after the part has been formed to its bulk dimensions. . . . [Furthermore] whenever two solids are brought together, they touch first where hills on one contact the surface of the other. As the hills flatten, contact areas grow and the pressure falls until it becomes too low to cause further deformation. Contact is thus limited to a relatively small area, and the rest of the surfaces are held apart. The interfacial gap formed is usually continuous, permitting gaseous and liquid access to the whole interface. Lubricants provide a film to separate the surfaces.

A lubricant produces the separation by establishing a film.

*What is the impact of lubrication on lubricant life and consumption?*

Lubricant life and consumption are influenced by the condition of the lubricant, which may be influenced by machine condition. Temperature, oxidation, and contamination all reduce life.

One of the greatest examples of lubricant life as well as consumption is in diesel engines. Diesel engines present an exceptionally aggressive environment as well as controlled consumption of the engine oil over time.

*Explain.*

One area that is especially aggressive on the engine oil in a diesel engine is the combustion chamber. As the piston moves, so do the various rings that act as partial seals between the chamber wall and the pistons. A thin film of engine oil coats the chamber and acts as a dynamic seal that travels the length of the piston throw.

The oil film is burned off and a new film is formed during the stroke of the piston. Replacing oil consumed by the engine is an added maintenance item that increases the cost of engine service. In some cases, excessive oil consumption can lead to shortened oil change intervals.

Lubricating oil consumption also can affect the rate of formation of deposits in the engine. Lubricating oil can contribute to increased engine-out emissions. In particular, lubricating oil can be a significant source of hydrocarbon and carbon monoxide emissions, especially in smaller engines, and total particulate matter mass and particle number emissions. The ash accumulation can have a significant impact on the diesel particulate filters (DPFs). Most ash accumulating in the DPF is attributed to lubricating oil consumed in the engine. Keeping lubricating oil consumption low allows for smaller DPFs to be fitted to the engine because of lower ash accumulation rates, less frequent regeneration (and fuel economy penalty) to avoid excessive pressure drop, and less deterioration of DPF ceramic substrates.

Excessive accumulation of oil-derived hydrocarbons in the DPF also can lead to uncontrolled regeneration and subsequent DPF damage. All emission-control catalysts must be designed—in terms of sizing and precious metal loading—to account for catalyst activity loss due to exposure to oil-derived catalyst poisons and, in the case of selective catalytic reduction (SCR) catalysts, oil-derived hydrocarbons.

Diesel Engine Example. A diesel engine provides an excellent example of all the various tribologic regimes as well as a full palette of lubrication functions.

1. Permit easy starting. The lubricant must have a low viscosity at low temperatures and be pumpable so as to instantaneously reach the engine parts that need lubrication. This is an important attribute because most of the engine wear occurs during the startup, primarily due to lubricant starvation.

2. Maintain adequate viscosity at high temperatures. This is important because most oils experience a decrease in viscosity at high temperatures, such as those in and around the combustion engine. If the viscosity of the oil drops too far, the lubricant loses its ability to form the lubricating film of the appropriate thickness, which will permit metal-to-metal contact and wear.

3. Lubricate and prevent wear. This translates into the oil forming a lubricating film of appropriate thickness to prevent metal surfaces from contacting each other and experiencing wear. For most engine parts, the surfaces are well separated, which makes lubrication easier. However, there are parts such as the piston rings and cam lobes that are designed to have metal-to-metal contact, and the function of the lubricant is to minimize wear by making chemical surface films.

4. Reduce friction. The formation of a lubricant film of proper thickness on surfaces and its maintenance will reduce friction and the accompanying wear. This is especially true during startup and idle, when the lubrication is inadequate and frictional losses occur. Therefore,

controlling friction will improve fuel economy.

5. Protect against rust and corrosion. Water resulting from the fuel combustion, while meant to escape through the exhaust, can condense on the cylinder walls or travel past piston rings as part of the blow-by and enter the crankcase. This typically occurs in cold weather or during short-distance driving because the engine and the lubricant are not hot enough for water to be removed via evaporation. Water can initiate rust and, in the presence of the acidic materials resulting from lubricant oxidation and additive decomposition, can cause corrosion.

6. Keep engine parts clean. Partial fuel combustion products, such as free radicals, soot, sulfur, and nitrogen oxides, enter the crankcase as the blow-by and react/interact with the lubricant to form highly polar deposit precursors and corrosive materials. These species have the tendency to separate on the hot surfaces to form deposits and lead to corrosion. Engine lubricants are designed to prevent the formation of these species and/or keep them from separating on the surfaces by suspending them in the bulk lubricant.

7. Cool engine parts. Cooling of the engine parts is crucial to trouble-free operation. Parts that must be cooled include cylinder heads, cylinder walls, valves, crankshaft, main and connecting rod bearings, timing gears, pistons, and others. Certain parts of the engine can be cooled by the use of a coolant, which is typically a mixture of water and ethylene glycol. Other parts cannot be cooled effectively by a coolant either because of their proximity or because

the part temperature is extremely high, which leads to the rapid evaporation of water. In such situations, the lubricant acts as a coolant.

8. Seal combustion pressures. Surfaces of piston rings, ring grooves, and cylinder walls do not have an ideal fit primarily because of machining limitations. It is important that these parts act as a good seal to prevent the loss of the high combustion and compression pressures, which are needed for efficient engine operation. A loss into the low-pressure area of the crankcase would result in a reduction in engine power and efficiency. Engine oils therefore improve the seal by filling spaces in the above-listed parts. Typically, the oil film that acts as a seal is only 0.025 mm thick; hence it is ineffective in filling spaces that are larger because of intensive wear. Incidentally, the oil consumption in a new engine is high until the surfaces in these parts become smoother due to wear for the oil to form a better seal.

9. Control foam. Foaming of engine oil due to air entrainment occurs because of the rapidly moving engine parts that create turbulence. The result is the formation of air bubbles, which normally rise to the surface of the oil and break. However, the presence of water and additives, many of which have surfactant properties, slows down this process. Foam in the engine oil is undesirable because of its poor cooling ability and noncontiguous film formation, which will result in excessive engine wear.

*How do you go about the practice of maintaining your assets?*

A person in this position would first fix what is broken, perhaps look to prevent something from breaking, predict whether something will break, and establish a sound performance specification so that the next time an asset is needed it will be better than before.

*And if you were to define each method, what would they be?*

Reactive, preventive, predictive, and reliability centered

*What are the fundamental aspects of reliability-centered maintenance?*

Reliability-centered maintenance (RCM) is a corporate-level maintenance strategy that is implemented to optimize the maintenance program of a company or facility. The final result of an RCM program is the implementation of a specific maintenance strategy on each of the assets of the facility.

*What are the fundamental aspects of condition-based maintenance?*

Condition-based maintenance (CBM) is a maintenance strategy that monitors the actual condition of the asset to decide what maintenance needs to be done. CBM dictates that maintenance should only be performed when certain indicators show signs of decreasing performance or upcoming failure.

## II. Lubrication Theory

*List the types of friction, and define tribology.*

Friction is the force that opposes motion between any surfaces that are in contact. There are four types of friction: static, sliding, rolling, and fluid. Static, sliding, and rolling friction occur between solid surfaces. Fluid friction occurs in liquids and gases.

Tribology is the science and technology of interacting surfaces in relative motion. It includes the study of friction, wear, and lubrication.

*Are there limitations on the body types or surfaces?*

Aerodynamic friction exists when a gas and solid move over each other, whereas liquid friction occurs when a liquid and solid move against each other. Internal friction, in contrast, occurs when internal energy-dissipation processes occur within one body. Generally, in solid friction, the two surfaces will have asperities (microscopic peaks) that will touch and rub against each other.

There are four laws that govern friction:

1. Static friction may be greater than kinetic friction.
2. Friction is independent of sliding velocity.
3. Friction force is proportional to applied load.
4. Friction force is independent of contact area.[5]

*What types of friction contribute to wear?*

Static, sliding, and rolling

*Is there a defining formula for friction?*

The coefficient of friction is the ratio between force and load.

*What would be the typical ranges for the coefficient of friction in regular applications?*

Typically, a well-lubricated bearing has a coefficient of friction of 0.03, whereas dry sliding has a range of 0.5 to 0.7. In contrast, clean metal surfaces in a vacuum can have a coefficient of friction 5 or more.[6]

### What is common among all wear types?

With all types of wear, there must be the removal of a solid material through the rubbing of surfaces. There are six main forms of wear: adhesive, abrasive, fatigue, impact by erosion and percussion, chemical (or corrosive), and electrical arc induced.

### What influences these types of wear?

Adhesive wear usually occurs when two bodies slide over each other. The asperities on both surfaces form a weld and can be sheared off as a result of the sliding motion. During this motion, some of the fragments are either transferred to the other material or become loose. Severe types of adhesive wear can include galling, scuffing, welding, and smearing.

Abrasive wear occurs when there is a difference in the hardness of materials as they pass over each other. The harder particles typically slide over the softer particles and damage the softer particles through plastic deformation or fracture. There are two types of abrasive wear: two-body and three-body wear. In two-body wear, one of the materials is harder than the other, causing the softer one to deform. This is usually seen in mechanical operations such as cutting, grinding, or machining. However, in three-body wear, the harder material is usually in the form of a particle that damages either of the two surfaces with which it comes into contact. Typically, this type of wear starts as adhesive wear, but after the weld is broken and particles are broken off, these particles can form the hard surface that, in turn, damages the other surfaces. Abrasive wear is also associated with scratching, scoring, and gouging. Plowing, wedge formation, and cutting are also wear modes that can occur during plastic deformation in the abrasion process.

Fatigue wear is introduced with the advent of repeated rolling and sliding. The cycles of loading and unloading eventually lead to the breakup of the surface, leaving large fragments of material displaced. These displacements resemble pits, and this form of wear is also linked to pitting. Spalling is also observed when a crack forms below the surface and the material is removed afterward. Static fatigue, in contrast, occurs as a result of chemically enhanced cracks. This type of deformation on the surface can result in increased wear on the surface layers with static and dynamic (rolling) conditions.

Impact wear can be classified into two regimes: erosion and percussion. Erosion usually has kinetic energy associated with the solid particles, liquid droplets, or implosion of air bubbles depending on the medium of transport. In contrast, percussion occurs as a result of repetitive solid-body impacts that result in a loss of material. Usually this is a result of sliding and can combine several wear mechanisms, including adhesive, abrasive, surface fatigue, fracture, and tribochecmical wear. Erosion also can be classified as a form of abrasion where particle velocity, impact angle, and size of the abrasion give rise to the kinetic energy associated with the impact velocity. The wear debris that is formed as a result of erosion is created by repeated impacts.

Corrosive wear is also called *chemical wear*, and it occurs when sliding occurs in a corrosive environment. Typically, the action of sliding removes any protective layer on the surface and allows the corrosive chemical to attack. Hence corrosion can only occur in the presence of both sliding and corrosive chemicals.

Electrical-arc-induced wear occurs when an arc is formed due to dielectric breakdown. This breakdown is initiated if a high potential is present over a thin film of air during the sliding process. When the arc is formed, large craters can be produced. If these are subjected to sliding or oscillation, fractures can occur that lead to three-body abrasion, corrosion, surface fatigue, and even fretting in some instances.

Fretting usually occurs when two surfaces are at rest but experience low-amplitude oscillatory motion in a tangential direction. Typically, fretting is as a result of abrasion or adhesive wear. Adhesion occurs between the asperities that come into contact with each other under the normal load; then the oscillatory movement disrupts this bond, and wear debris is formed that can be used in an abrasive manner.[7]

*What are the types of wear modes and their influencing factors?*

Wear is damage to a solid surface, generally involving progressive loss of material, due to relative motion between that surface and a contacting substance or substances. The type of relative motion is often used to define the wear that is generated. Because of its complexity, a number of wear modes have been recognized.

- Abrasive wear. Wear due to hard particles or hard protuberances forced against and moving along a solid surface. These hard particles might be commercial abrasives such as silicon carbide and aluminum oxide or naturally occurring contaminates such as dust particles and sand (crystalline silica [quartz]). If the abrasive particles are allowed to roll, rolling abrasion or three-body abrasion occurs.
- Adhesive wear. Wear due to localized bonding between contacting solid surfaces leading to material trans-

fer between the two surfaces or loss from either surface. Adhesive wear is not as prevalent as abrasive wear and is induced when like materials slide against each other with no lubrication. This type of wear involves the formation of local cold welds between surfaces coming into contact under a load and tangential shearing or plowing of the junctions. Material can be transferred from one surface to the other during this process.

- Corrosive wear. Wear in which a chemical or electrochemical reaction with the environment is significant.
- Cutting wear. In solid impingement erosion, the erosive wear is associated with the dissipation of the kinetic energy of impact arising from the tangential component of the velocity of the impacting particles.
- Deformation wear. In solid impingement erosion, the erosive wear of a material is associated with dissipation of the kinetic energy of impact arising from the normal component of the velocity of the impacting particles. It is therefore the sole component of wear for particles impacting at a 90-degree angle of attack.
- Erosive wear. Progressive loss of original material from a solid surface as a result of mechanical interaction between that surface and a multicomponent fluid or impinging liquid or solid particles (erosion—damage caused by particulates in gases or liquids striking a surface).
- Fatigue wear. Wear of a solid surface caused by fracture arising from material fatigue.
- Fretting wear. Wear arising as a result of fretting (where fretting, in tribology, involves a small amplitude oscillatory motion, usually tangential, between two solid surfaces in contact).

- **Frosting.** A change in color in a limited area of fabric caused by abrasive wear.
- **Impact wear.** Wear due to collisions between two solid bodies where some component of the motion is perpendicular to the tangential plane of contact.
- **Pitting.** A form of wear characterized by the presence of surface cavities, the formation of which is attributed to processes such as fatigue, local adhesion, and cavitation.
- **Rolling wear.** Wear due to the relative motion between two nonconforming solid bodies whose surface velocities in the nominal contact location are identical in magnitude, direction, and sense.
- **Rolling abrasion.** A form of abrasion that occurs when abrasive particles or debris are allowed to "roll" between the surface and a contacting substance (also see three-body wear).
- **Scoring.** A severe form of wearing characterized by the formation of extensive grooves and scratches in the direction of sliding.
- **Scratching.** The mechanical removal or displacement, or both, of materials from a surface by the action of abrasive particles or protuberances sliding across the surfaces—typically in the form of a line caused by the relative movement of an object across and in contact with the surface.
- **Scuffing.** A form of wear occurring in inadequately lubricated tribologic systems that is characterized by macroscopically observed changes in surface texture with features related to the direction of relative motion.

- **Sliding.** Wear due to the relative motion in the tangential plane of contact between two solid bodies—typically recognized by linear grooves that are generated from a reciprocating or unidirectional contact.
- **Three-body wear.** A form of abrasive wear in which wear is produced by loose particles introduced or generated between the contacting surfaces.
- **Two-body abrasive wear.** A form of abrasive wear in which the hard particles or protuberances that produce the wear of one body are fixed on the surface of the opposing body.

## Mechanisms of Lubrication Regimes

*Explain boundary lubrication and the friction and wear sources attributed to it.*

Boundary lubrication exists during the following conditions: when the component's surface roughness significantly exceeds the oil-film thickness; when the equipment is subjected to frequent starts and stops, shock-load conditions, high static loads, or slow speeds; and when operational requirements dictate the use of low-viscosity oil, compromising friction control of some system components.

*What would be characteristic of boundary lubrication?*

Boundary lubrication usually occurs at startup and shutdown of the equipment. It is the phase during which the asperities of the two surfaces still make contact even though there is the presence of the lubricant. Typically, this occurs when the lubricant is forming the film between the two surfaces. During this type of lubrication, one usually observes adhesive and chemical/corrosion wear.[8]

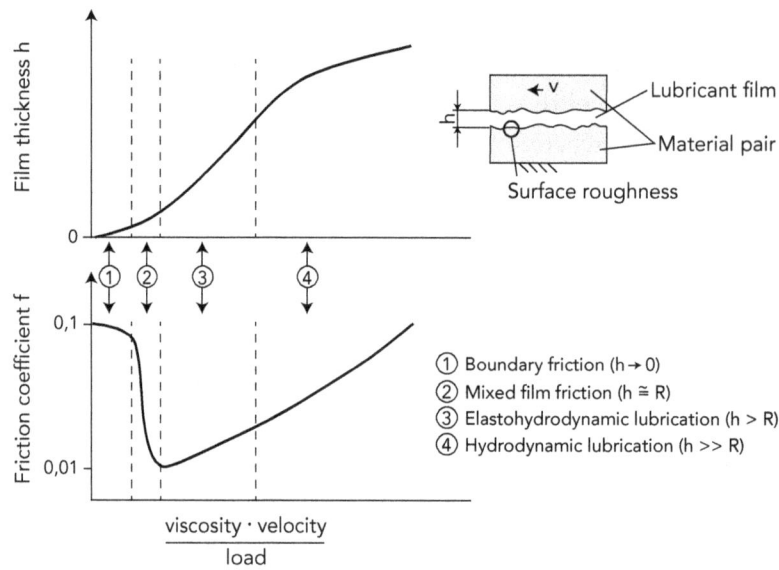

① Boundary friction (h → 0)
② Mixed film friction (h ≅ R)
③ Elastohydrodynamic lubrication (h > R)
④ Hydrodynamic lubrication (h >> R)

### How does this relate to the Stribeck curve?

In the boundary lubrication regime, there is an increase in load as the machine starts up and a decrease in speed. This leads to an increase in the coefficient of friction to levels of 0.1 or higher as shown in the diagram above.[9]

### Explain mixed-film lubrication and the friction and wear sources attributed to it.

Mixed-film lubrication exists where the component's surface roughness exceeds the oil-film thickness to some extent but not as much as in the case of boundary lubrication.

- Friction source. Sliding friction
- Wear source:
  - Adhesive
  - Scuffing
  - Sliding
  - Scratching

### Does boundary lubrication differ from mixed-film lubrication?

Mixed-film lubrication is exactly that, a mixed type of regime. This mixed regime can include boundary lubrication and hydrodynamic lubrication. In this regime, the surfaces undergo a transition from boundary lubrication to hydrodynamic or even elastohydrodynamic lubrication. As such, parts of the surfaces experience areas of contact between their asperities, whereas other parts are fully supported on a film of lubricant.

### What are the typical wear patterns associated with mixed-film lubrication?

Usually, adhesion can occur in the areas of boundary lubrication where there is contact between the two surfaces. This can eventually lead to metal transfer, wear particle formation, or seizure between the surfaces.

### Explain hydrodynamic lubrication and the friction and wear sources attributed to it.

Hydrodynamic lubrication is also called *full-film* or *full-fluid-film lubrication*. Moving surfaces are fully separated by a hydrodynamically formed wedge. No contact is expected between the asperities of the two surfaces in relative motion. Formation of the oil wedge depends on surface geometry, speed, load, and oil viscosity. Hydrodynamic lubrication is not achieved during startup and shutdown conditions.

■ Friction source. Fluid friction

■ Wear source. None. Surfaces are separated; assumed clean lubricant and no shaft voltages.

### What makes hydrodynamic lubrication the most sought-after regime?

Hydrodynamic lubrication is usually achieved when the lubricant forms a film between the contact surface areas that adequately separate the asperities, allowing the surfaces to move over each other with minimal friction. Typically, the coefficient of friction in this regime can be as small as 0.001 and conforms to sliding friction.[10]

### Will any wear take place in this regime?

Yes, wear can take place, but this is usually in the form of adhesion during start and stop operations, or it can also be present in the form of corrosive or chemical wear. With corrosive or chemical wear, the surfaces can become corroded as a result of the interaction of the lubricant on the surface.

### Explain elastohydrodynamic lubrication and the friction and wear sources attributed to it.

Elastohydrodynamic lubrication is formed at rolling contacts such as rolling-element bearings, where surfaces in relative motion converge on a point (ball bearings) or line (roller bearings). Because of the elasticity of the bearing material, it can deform elastically in the small area of contact. A hydrodynamic oil film is formed inside the small volume of deformation. The high pressure inside the deformation results in the oil momentarily changing into a solid in the small volume of deformation. Once the pressure is released, the lubricant returns to its liquid state.

■ Friction source. Sliding, rolling, and fluid friction (drag in grease)

■ Wear source:
  • Fatigue wear
  • Impact wear
  • Rolling wear
  • Rolling abrasion
  • Sliding wear

Assume clean lubricant and no shaft voltage.

### How does elastohydrodynamic lubrication differ from hydrodynamic lubrication?

Elastohydrodynamic lubrication usually occurs when the surfaces experience hydrodynamic lubrication. When elastohydrodynamic lubrication occurs, there is an elastic deformation of the contact surfaces. This is seen with heavily loaded contacts when the load becomes concentrated at a small contact area such that the contact area becomes deformed to allow the film of fluid to continue its separation of the surfaces. Typically, this concentration of load is achieved in rolling friction unlike in hydrodynamic lubrication, which uses sliding friction.[11]

### If the contact surface deforms to allow the film to remain, will wear take place in this regime?

In elastohydrodynamic lubrication, wear can still occur in the start and stop phases or, more commonly, through fatigue given the high loads being concentrated at smaller contact areas.

## Lubricant Categories

*Provide examples of gaseous lubricants.*

A gas bearing may be hydrodynamic or hydrostatic. In hydrodynamic bearings, the gas is introduced into the bearing surfaces by the action of the bearing. In hydrostatic bearings, the gas is introduced under pressure from an external source. Air bearings are also widely used to allow frictionless movement of large, heavy items over flat surfaces using air pads. Air spindles are fitted with water cooling systems (1) to maintain a constant, near-ambient operating temperature for minimizing thermal growth and (2) in less critical applications simply to remove excessive heat that might damage the motor or reduce bearing operating clearances.

*Provide examples of cohesive lubricants.*

Oil used in grease to lubricate rolling-element bearings is an example of cohesive lubrication. Cohesion is the attraction between the molecules of a substance that tends to hold the substance together. *Adhesion* is the property of a lubricant that, in liquid form, causes the lubricant to stick (adhere) to the parts being lubricated. *Cohesion* is the property that holds the lubricant together and enables it to resist breakdown under extreme pressure.

*Provide examples of solid lubricants.*

Solid lubricants are used primarily in extreme-pressure and extreme-wear environments:

- As an additive in grease, molybdenum disufide and graphite are used for extreme pressure—typical applications in pins and bushings.
- Graphite is particularly useful where moisture is present.
- As an additive in oil, molybdenum and polytetrafluoroethylene (PTFE) are used as antiwear additives in engine oils.

- Solid lubricants are bonded or impregnated into the surfaces of machine components such as cylinder liners.
- Solid lubricants may be present in the friction area in forms of either dispersed particles or surface films.
- A coating (film) of a solid lubricant may be applied on a part surface.
- A composite coating consisting of particles of a solid lubricant dispersed throughout a matrix may be used.
- Particles of a solid lubricant may be dispersed throughout the bulk of the part material (composite material).
- Powder of a solid lubricant may be delivered to the rubbing area (dry lubrication).
- Solid lubricants may be used as additives in lubricating oils or greases.

The most commonly used solid lubricants are:

- Graphite. Used in air compressors, the food industry, railway track joints, brass instrument valves, piano actions, open gears, ball bearings, and machine-shop works. It is also used very commonly for lubricating locks because a liquid lubricant allows particles to get stuck in the lock, worsening the problem. It is also often used to lubricate the internal moving parts of firearms in sandy environments.
- Molybdenum disulfide ($MoS_2$). Used in constant-velocity (CV) joints and space vehicles. Lubricates in vacuum.
- Hexagonal boron nitride. Used in space vehicles; also called *white graphite*.
- Tungsten disulfide. Similar usage as molybdenum disulfide, but because of the high cost, only found in some dry lubricated bearings.

Graphite and molybdenum disulfide are the predominant materials used as dry lubricants.

## III. Lubricant Formulation

*The API classification system is determined by the base stock characteristics and not the refining process. Are there refining processes that produce these characteristics?*

Typically, solvent refining and dewaxing can produce Group I base oils, whereas hydro processing is used for Group II and III base stocks.

*What is the process of solvent refining?*

For Group I base oils, the process of solvent refining begins with solvent extraction. A solvent is used to extract the sulfur, aromatic, and nitrogen compounds from the vacuum gas oil (heavy refinery feedstock from crude distillation). Afterward, another light solvent is added to the mixture to remove paraffins. This final mixture is the base oil.[12]

*Explain solvent refining.*

This is the term for the removal of most of the aromatics and undesirable constituents of oil distillates by liquid extraction. Common and suitable extractants are phenol, furfural, and sulfur dioxide. Furfural is used extensively for the refining of paraffinic oils. The resulting base stocks are raffinates (referred to as *neutral oils*) and an extract fluid that is rich in aromatic content, which is highly sought after as a process oil and fuel oil.

After solvent extraction, the raffinates are dewaxed to improve low-temperature fluidity and then hydro finished to improve color and stability. The final quality of the base oil is determined by the severity of the application of temperatures and pressures in the hydro finishing process. The base oils are now ready to be selectively blended with the appropriate additives.

*How does hydrogen treatment vary from solvent refining?*

While the same vacuum gas oil feedstock is used (as for Group I base oils), the feedstock is instead passed through a hydrogen cracker at high pressures (mostly above 2000 psi). This allows the reshaping of the molecules and saturation of the aromatic compounds with hydrogen, which improves their stability. The final output contains more than 90% saturated compounds. This is then passed to a second reactor where hydrogen isomerization takes place. This occurs when the wax molecules are converted to isoparaffins, and more saturation takes place. Eventually, the solution goes through a lower temperature but high pressure to let the remaining residual aromatic compounds form the Group II base oils.[13]

*Explain hydrogen treatment.*

Also called *catalytic hydrogenation*, this refining process subjects the distillates to a chemical reaction with hydrogen in the presence of a catalyst at temperatures as high as 420°C and pressures up to 3000 psi. Hydrotreating processes are the favored methods used by many base-oil refiners and lubricant manufacturers. This is due to the small material losses involved. When hydrotreating is performed, more than 90% of the aromatic content is converted to hydrocarbons.

The principle of all hydrotreating processes is quite similar. The crude distillate is preheated to temperatures between 150 and 420°C and then brought together with hydrogen or a hydrogen-enriched gas through a fixed-bed reactor. The oil reacts with the hydrogen in the presence of a catalyst to control and speed up the reactions. Hydrotreated base oils cannot retain their solubility for some chemicals, and thus additive retention may be seriously affected.

*Explain severe hydrogen treatment.*

Severely hydrotreated base oil contains almost no aromatics; these oils must be fortified with seal swell agents in the additive package. Solvent-refined base oils, in contrast, retain some aromatics, which are "natural" seal swell agents.

*Is there a difference between the methods used for hydrotreatment and severe hydrotreatment?*

Severely hydrotreated products undergo the same hydrogen treatment process but with more severe operating conditions. This generally causes chemical changes to the molecular composition that remove nitrogen and sulfur while also reducing the aromatics present.[14]

*Explain hydrocracking.*

Hydrocracked lubricants are formulated via hydrocracking, in which crude oil molecules are chemically altered and rearranged into building blocks with favorable characteristics. The resulting base oil is purer (99%+) than solvent-refined base stocks, making it more resistant to oxidation, more responsive to additives, more pumpable at low temperatures, and lighter in color. Hydrocracked base oils offer superior product performance, resulting in greater oxidation and thermal stability, soot dispersancy in heavy duty motor oils, and low-temperature performance. In addition to these benefits, hydrocracked base oils also have a high VI and low volatility. Hydrocracked base oils can be formulated with a wide variety of additives either to achieve the latest industry specifications or simply to produce premium performance lubricants. Hydrocracked base oils can help to meet challenging future lubricant specifications cost-effectively, whereas Group I oils often cannot. Hydrocracked base oils are especially valuable where they can replace traditional synthetic base oils to achieve better performance.

*Is there any significant difference between hydrocracked and severely hydrotreated lubricants?*

When a product undergoes hydrocracking, it is subjected to a highly exothermic reaction with severe operating conditions. These conditions give rise to complete destruction of any aromatics and make the base oil very paraffinic with a higher VI and lower volatility.[15]

## Mineral Base Oils

*What is the composition of a base oil?*

Base oils are refined from crude oils. These crude oils usually contain naphthenes, paraffins, aromatics, resins, and asphaltenes.[16]

*Explain what a naphthenic oil is, list its properties, and provide application examples.*

Carbon atoms are connected with cyclic single bonds, and the other carbon bonds are saturated with hydrogen or connected to paraffinic chains.

- Lower VI (less stable viscosity behavior during temperature variations)
- Oxidation stability not as good as that of paraffinic base oils

Used in the metalworking fluids market—its better solubility and ability to form stable emulsions provide clear advantages over paraffinic base oils.

*What makes naphthenic base oils unique?*

Naphthenic base oils essentially have no wax and are high in hydrocarbons containing ring structures. They also have a very low pour point (–50°F, –46°C). Naphthenic base oils have a low VI, which means that the viscosity changes significantly with temperature. Hence these oils are used in applications with very little temperature variations.[17]

*Explain what paraffinic oil is, list its properties, and provide application examples.*

Most base-stock oils are paraffinic. Carbon atoms are connected by a single bond, and the other carbon bonds are saturated with hydrogen atoms.

- Relatively high VI (stable viscosity behavior during temperature variations)
- Relatively good oxidation stability

Paraffinic oils continue to be the preferred option for high-temperature applications and when longer lubricant life is required.

*How do paraffinic base oils differ from naphthenic base oils?*

Paraffinic base oils need to be dewaxed so that their pour point can be reduced from 80 to 0°F (27 to −18°C). Paraffinic base oils have a higher VI, thus making them desirable for applications with wide variations in temperature.[18]

*Explain what an aromatic oil is, list its properties, and provide application examples.*

Carbon atoms are connected with cyclic single bonds and some double bonds, and the other carbon bonds are saturated with hydrogen or connected to paraffinic chains. Aromatics in oil are undesirable. Group I oils are solvent refined and may contain significant proportions of aromatics, whereas Group II (hydrotreated) and Group III (hydrocracked) oils contain very small amounts of aromatics.

*Are the properties of aromatic base oils desirable?*

Aromatic base oils are highly unsaturated and have a high level of reactivity. However, they do have good solubility for additives, whereas their oxidation stability is poor. Additionally, their melting points are low, and they have poor vis-

cosity/temperature characteristics. They are usually used in refrigeration oils.[19]

*Explain the different vegetable base oils and biobased lubes, list their properties, and provide application examples.*

Vegetable base oils, which are derived from plant oils, represent a very small percentage of lubricants and are used primarily for renewable and environmental interests. They are used in food-processing and in biodegradable applications. They have:

- Poor oxidation stability
- Renewable source
- Expensive

By using commercially available palm oil and Jatropha oil, biobased lubricants may be produced through two stages of transesterification. The first stage is the process of using methanol in the presence of potassium hydroxide to produce biodiesel. The second stage is the reaction of biodiesel with trimethylolpropane using sodium methoxide as catalyst to yield palm or Jatropha oil-base trimethylolpropane esters (biobased lubricants).

*What are the properties of vegetable oils, and can they be used?*

Vegetable oils are naturally occurring, and hence they are more environmentally friendly. Usually, they are composed of fatty acid content, which helps determine their characteristics. They have high flash points (above 300°C) and may be prone to hydrolysis (trimester-based oils). The main disadvantage of vegetable oils is their performance at high or low temperatures. At low temperatures they solidify, whereas at high temperatures they can be easily oxidized.

Biolubricants are environmentally acceptable fluids that satisfy biodegradation and bioaccumulation standards.[20]

*List synthetic lubricant characteristics, applications, and compatibility.*

**Advantages:**

- Higher VI
- Better thermal stability
- Less friction
- Higher shear strength
- Higher oxidation stability
- Better fire resistance
- Higher flash point
- Lower pour point
- Better detergency

**Disadvantages:**

- Higher cost
- Less mixable (i.e., less compatibility with other fluids and lubricants)
- Less compatible with different seal types
- Less soluble in terms of additives
- More toxic
- Higher disposal cost

- Less hydrolytic stability (i.e., less resistant to the effects of water contamination)

In applications with extremely high or extremely low temperatures or very heavy loads, synthetic lubricants may overcome the challenges better than nonsynthetic lubricants. For example, a synthetic gear oil would provide exceptional protection in heavily loaded gearboxes exposed to high- and low-temperature extremes. Another example is a synthetic engine oil or synthetic motor oil, which would provide better viscosity stability in extreme temperature conditions.

The performance advantages offered by synthetic oils and synthetic greases vary by product, but they include improved thermal stability, improved oxidation resistance, a high VI, improved low-temperature properties, lower evaporation losses, reduced flammability, and a lower tendency to form residues as shown in Table 2.1.

**Table 2.1** Comparison of Fully Formulated Environmentally Acceptable Lubricants with Various Bases

| Properties | Mineral | Vegetable | PAOs | Diesters | Polyol esters | PAGs |
|---|---|---|---|---|---|---|
| Viscosity temperature characteristics | Fair | Good | Good | Fair | Very good | Very good |
| Low-temperature properties | Poor | Poor | Very good | Good | Good | Good |
| Oxidation stability | Fair | Poor | Very good | Good | Good | Good |
| Compatibility with mineral oils | Excellent | Excellent | Excellent | Good | Fair | Poor |
| Low volatility | Fair | Good | Excellent | Excellent | Excellent | Good |
| Varnish and paint compatibility | Excellent | Very good | Excellent | Poor | Poor | Poor |
| Seal swell (NBR) | Excellent | Excellent | Very good | Fair | Fair | Good |
| Lubrication properties | Good | Very good | Good | Very good | Very good | Good |
| Hydrolytic stability | Excellent | Poor | Excellent | Fair | Fair | Very good |
| Thermal stability | Fair | Fair | Fair | Good | Good | Good |
| Additive solubility | Excellent | Excellent | Fair | Very good | Very good | Fair |

*These ratings are generalizations. Specific manufacturers of products should be consulted for current data.*
Source: D. M. Pirro and A. A. Wessol, *Lubrication Fundamentals*, 2nd ed. New York: Marcel Dekker, 2001, p. 116.

*Explain what synthesized hydrocarbons (e.g., polyalphaolefins) are, list their properties, and provide application examples.*

Synthetic base-stock oils are made from synthesized hydrocarbons such as polyalphaolefins (PAOs) and diesters. These are special hydrocarbons with a double bond between the primary carbon atoms. They are:

- Most stable oils available
- Really good VIs
- Really good oxidation stability (long service life)
- Expensive

Application would be in extreme and varying operating temperatures and where the load is extremely high. These oils could be applied as lubrication in high-performance car engines and aircraft jet engines.

*What are the characteristics, applications, and compatibility of synthesized hydrocarbons?*

Synthesized hydrocarbons have high VIs (>135), excellent low-temperature fluidity, and low pour points. They also have excellent shear stability and hydrolytic stability and good thermal stability. However, they have poor solvency for various additive types, and the low-viscosity grades tend to cause seal shrinkage. They are widely used in automotive and industrial lubricants.[21]

*Explain what dibasic acid esters are, list their properties, and provide application examples.*

In chemistry, a dibasic ester (DBE) is an ester of a dicarboxylic acid. Depending on the application, the alcohol may be methanol or higher-molecular-weight monoalcohols. Mixtures of different methyl dibasic esters are commercially produced from short-chain acids such as adipic acid, glutaric acid, and succinic acid. They are nonflammable, readily biodegradable, noncorrosive, and have a mild, fruity odor.

Dibasic esters of phthalates, adipates, and azelates with C8–C10 alcohols have found commercial use as lubricants, spin finishes, and additives. The first qualified synthetic crankcase motor oils were based entirely on ester formulations, and these products were quite successful when properly formulated. Esters have given way to polyalphaolefins (PAOs) in this application because of lower cost and their formulating similarities to mineral oil. Nevertheless, esters are often used in combination with PAOs in full synthetic motor oils in order to balance the effect on seals, solubilize additives, reduce volatility, and improve energy efficiency through higher lubricity. The percentage of ester used can vary anywhere from 5% to 25% depending on the desired properties and type of ester employed.

*What are the characteristics, applications, and compatibility of dibasic acid esters?*

Dibasic acid esters are organic acids that exhibit good metal wetting ability, high film strength, high oxidation and thermal stability, and good shear stability. They can dissolve system deposits, but their hydrolytic stability and antirust properties are fair. These esters are compatible with mineral oils but not all elastomers and paints. Dibasic acid esters were used for old military jet engines but can also be used in automotive engine oils and air compressor lubricants and as the base for some aircraft greases.[22]

*Explain what polyolesters are, list their properties, and provide application examples.*

The beta hydrogen in polyolesters is very reactive toward oxygen, so esters with no beta hydrogen are more thermally stable. These are known as *neopolyolesters*, with their name derived from their structural similarity to neopentane. Neopolyol is shortened to polyolesters and abbreviated as POE. All POEs have good oxidative stability because they have no beta hydrogens.

Polyolester oil (POE oil) is a type of synthetic oil used in refrigeration compressors that is compatible with the refrigerants R-134a, R-410A, and R-12. It is recommended by experts as a replacement for hydrofluorocarbons (HFCs). Along with R-134a, POE oil is recommended as a replacement for R12 mineral oil because R134a does not mix well with mineral oil. These wax-free oils are suggested for use with chlorine-free HFC systems because they provide better lubrication and stability and are more miscible with HFC refrigerants. They can meet the lubricity requirements of the mineral oils used with chlorofluorocarbons (CFCs) and HFCs. They are compatible with most lubricants in the market. It is noted that the viscosity of the oil decreases with temperature. The dispersion behavior of this oil also has been the subject of a lot of study. It is also considered by some to be a good additive to engine oil.

### What are the characteristics, applications, and compatibility of polyolesters?

Polyolesters have better high-temperature stability than diesters with a lower VI. Their volatility is equal to or lower than that of diesters, but they can cause swelling of elastomers and have an effect on paints. These esters are used primarily in Type II jet engine oils, air compressor oils, and refrigeration lubricants. They are environmentally sensitive and can be used in hydraulic oils that need to be biodegradable.[23]

### Explain what phosphate esters are, list their properties, and provide application examples.

Since the discovery of their excellent antiwear and fire-resistance properties in the 1940s, phosphate ester use by industry has increased steadily. Phosphate esters are used primarily as fire-resistant base stocks in several applications, including hydraulic systems, turbines, and compressors. The first commercial products were synthesized from coal-tar derivatives and were consequently composed of a mixture of various isomeric aryl phosphates, including the neurotoxic orthodoxy phosphate. Today, other raw materials are used for the synthesis of phosphate esters. Modern technology allows better control of the manufacturing process, and the toxicity of the final product is normally low.

### What are the characteristics, applications, and compatibility of phosphate esters?

Phosphate esters are known for their fire-resistance properties. However, they have poor viscosity/temperature characteristics, low pour points, and low volatility. They may cause the swelling of seals and can have effects on paints and finishes. Typically, they have a good compatibility with mineral oils, but this depends on the ester used. Their specific gravities are greater than 1; hence water floats rather than settles to the bottom.

These phosphate esters are used mainly for fire-resistant fluids in commercial aircraft or industrial hydraulic fluids. They also can be used in turbines and compressors. Some phosphate esters are used in greases and mineral oils as wear- and friction-reducing additives.[24]

### Explain what polyalkylene glygols are, list their properties, and provide application examples.

One of the most versatile types of synthetics is polyalkylene glycol (PAG) lubricants. PAGs are generally known as compressor lubricants, and their use in industry has increased since the 1980s. Increasing performance standards in the automotive and industrial markets peg these sectors as areas that show promise for growth.

$$H—(O—CH_2—CH_2)n—OH$$

PAGs were one of the first synthetic lubricants to be developed and commercialized. They were created under mandate from the U.S. Navy in response to hydraulic fluid fires on ships resulting from ordnance strikes during

World War II. In 1942 and for the next 30 years, the Navy began to use PAG-based water glycol hydraulic fluids exclusively because they were fire-resistant and could operate over a wide temperature range. Later, PAGs began to see extensive use as textile lubricants and as quenchants in metal heat treating.

The terms *polyalkylene glycol* and *polyglycol* are used interchangeably. These fluids can be manufactured to be either water soluble or water insoluble (oil soluble). The most common are the water-soluble fluids, and thus they can have some very different properties. Polyglycols are moderately polar, which gives them moderate film-strength properties. They have a very high VI (180–280) and good low- and high-temperature abilities. They burn off cleanly, leaving no residue, and they have been used as a carrier oil for solid lubricants for high-temperature chain lubrication. Some versions are food grade and biodegradable. They are used as compressor oils in rotary screw and reciprocating units, worm-gear oils, fire-resistant lubricants, metalworking fluids, and brake fluids.

### What are the characteristics, applications, and compatibility of polyalkylene glycol?

Polyalkylene glycols, also known as PAGs, fall under the category of polyglycols. They have low sludge buildup because they decompose completely under high-temperature conditions. They have good temperature/viscosity characteristics with high thermal conductivity.

They are generally not compatible with mineral oils or additives used in mineral oils. While seal swelling is low, there is also a low solubility for hydrocarbon gases and some refrigerants.

The applications are usually divided into water soluble and water insoluble because this property is dictated by the ratio of ethylene oxide to propylene oxide. Some water-soluble applications include hydraulic brake fluid, metalworking lubricants, and fire-resistant hydraulic fluids. Other water-insoluble applications include heat-transfer fluids, high-temperature gear and bearing oils, and screw-type refrigeration compressors (using R12 and hydrocarbon gases).[25]

### Explain what silicone oils are, list their properties, and provide application examples.

A silicone oil is any liquid polymerized siloxane with organic side chains. The most important member is polydimethylsiloxane. This polymer is of commercial interest because of its relatively high thermal stability and lubricating properties. As with all siloxanes (e.g., hexamethyldisiloxane), the polymer backbone consists of alternating silicon and oxygen atoms (. . . Si—O—Si—O—Si . . .). Many groups can be attached to the tetravalent silicon centers, but the dominant substituent is methyl or sometimes phenyl. Many silicone liquids are linear polymers end-capped with trimethylsilyl groups. Other silicone liquids are cyclosiloxanes.

Silicone oils are used primarily as lubricants, thermic fluid oils, or hydraulic fluids. They are excellent electrical insulators and, unlike their carbon analogues, are nonflammable. Their temperature stability and good heat-transfer characteristics make them widely used in laboratories for heating baths (oil baths) placed on top of hot-plate stirrers, as well as in freeze dryers as refrigerants. Silicone oil is also commonly used as the working fluid in dashpots, wet-type transformers, diffusion pumps, and oil-filled heaters. Aerospace use includes the external coolant loop and radiators of the International Space Station *Zvezda* module, which rejects heat into the vacuum of space.

Some silicone oils, such as simethicone, are potent antifoaming agents because of their low surface tension. They are used in industrial applications such as distillation and fermentation, where excessive amounts of foam can be problematic. They are sometimes added to cooking oils to prevent excessive foaming during deep frying. Silicone oils used as lubricants can be inadvertent defoamers (contaminants) in processes where foam is desired, such as in the manufacture of polyurethane foam. Silicone oil is also one of the two main ingredients in Silly Putty, along with boric acid.

*What are the characteristics, applications, and compatibility of silicones?*

Silicones are synthetic lubricants with high VIs, low pour points, and low volatility. They are inert, nontoxic, fire resistant, and water repellent with low seal swelling. They have good thermal and oxidation stability at high temperatures. However, if oxidation does occur, it can lead to silicon oxides, which are very abrasive.

Silicones are used for wide- and high-temperature-range greases, specialty greases (for elastomeric materials), and specialty hydraulic fluids (such as liquid springs and torsional dampers). They are additionally used quite often in hydraulic brake fluids and as antifoaming agents.

*Explain what fluorocarbons are, list their properties, and provide application examples.*

Fluorocarbons, sometimes referred to as perfluorocarbons or PFCs, are, strictly speaking, organofluorine compounds with the formula $C_xF_y$; that is, they contain only carbon and fluorine, though the terminology is not strictly followed. Compounds with the prefix *perfluoro-* are hydrocarbons, including those with heteroatoms, wherein all C—H bonds have been replaced by C—F bonds. Fluorocarbons can be perfluoroalkanes, fluoroalkenes and fluoroalkynes, and perfluoroaromatic compounds. Fluorocarbons and their derivatives are used as fluoropolymers, refrigerants, solvents, and anesthetics.

Perfluoroalkanes are very stable because of the strength of the carbon-fluorine bond, one of the strongest in organic chemistry.

Fluorocarbons are colorless and have high density, up to more than twice that of water. They are not miscible with most organic solvents (e.g., ethanol, acetone, ethyl acetate, and chloroform) but are miscible with some hydrocarbons (e.g., hexane in some cases). They have very low solubility in water, and water has a very low solubility in them (on the order of 10 ppm). They have low refractive indices. Fluoroalkanes are generally inert and nontoxic.

Krytox is a group of colorless synthetic lubricants (oils and greases) with a variety of applications. Invented by researchers at DuPont, Krytox oils are fluorocarbon ether polymers of polyhexafluoropropylene oxide with a chemical formula F—(CF(CF$_3$)—CF$_2$—O)$n$—CF$_2$CF$_3$, where the degree of polymerization $n$ generally lies within the range of 10–60. These compounds are collectively known by many names, including perfluoropolyether (PFPE), perfluoroalkylether (PFAE), and perfluoropolyalkylether (PFPAE).

*Explain what polyphenyl ethers are, list their properties, and provide application examples.*

Phenyl ether polymers are a class of polymers that contain a phenoxy and/or a thiophenoxy group as the repeating group in ether linkages. Commercial phenyl ether polymers belong to two chemical classes: polyphenyl ethers (PPEs) and polyphenylene oxides (PPOs).

The important attributes of PPEs include their thermal and oxidative stability and stability in the presence of ionizing radiation. PPEs have the disadvantage of having somewhat high pour points. For example, PPEs that contain two and three benzene rings are actually solids at room temperatures. The melting points of the ordinarily solid PPEs are lowered if they contain more *m*-phenylene rings and alkyl groups or are mixtures of isomers. PPEs that contain only *o*- and *p*-substituted rings have the highest melting points. PPEs also have high surface tension;

hence these fluids have a lower tendency to wet metal surfaces.

While originally PPEs were developed for use in extreme environments that were experienced in aerospace applications, they are now used in other applications requiring low volatility and excellent thermo-oxidative and ionizing radiation stability. Such applications include use as diffusion pump fluids and high-vacuum fluids and in formulating jet engine/turbine lubricants, high-temperature hydraulic lubricants and greases, and heat-transfer fluids. In addition, because of excellent optical properties, these fluids have found use in optical devices.

PPEs possess good optical clarity, a high refractive index, and other beneficial optical properties. Because of these beneficial properties, PPEs have the ability to meet the rigorous performance demands of signal processing in advanced photonics systems.

### What are the characteristics, applications, and compatibility of polyphenyl ethers?

PPEs are thermally stable above 800°F (450°C) and have great radiation resistance. They also have excellent resistance to oxidation at high temperatures. Generally, small quantities are used as heat-transfer fluids, lubricants for vacuum pumps, and the fluid component of radiation-resistant greases.[26]

### Explain what food-grade lubricant classifications are, list their properties, and provide application examples.

Until 1998, food-grade lubricants were regulated by the U.S. Department of Agriculture (USDA) and the Food Safety and Inspection Services (FSIS). Presently, and globally, risk and liability for food manufacturers are controlled by the Hazard Analysis and Critical Control Point (HACCP) Program.

HACCP process:

- Conduct a hazard analysis.
- Identify and implement controls.
- Establish monitoring requirements with limits, and identify corrective actions.
- Ensure record keeping and program validation.

Food-grade lubricants must perform the same technical functions as any other lubricant: provide protection against wear, friction, corrosion, and oxidation; dissipate heat and transfer power; be compatible with rubber and other sealing materials; and provide a sealing effect in some cases.

The USDA created the original food-grade designations H1, H2, and H3. The approval of a new lubricant and its registration in one of these categories depend on the list of ingredients.

- H1 lubricants are food-grade lubricants used in food-processing environments where there is the possibility of incidental food contact.
- H2 lubricants are food-grade lubricants used on equipment and machine parts in locations where there is no possibility of contact with food.
- H3 lubricants are food-grade lubricants, typically edible oils, used to prevent rust on hooks, trolleys, and similar equipment.

Deciding whether there is a possibility of contact is tough, and many have erred on the side of safety with respect to selecting H1 over H2.

Acceptable H1-approved lubricant base stocks can be either mineral or synthetic. Mineral oils used are either technical white mineral or U.S. Pharmacopeia (USP)–type white mineral oils. Synthetic lubricant base stocks are usually PAOs or PAGs. These base stocks are used primarily in lubricants designed for temperature extremes. Dimethylpolysiloxane (silicones) with a viscosity greater than 300 cSt is also permitted. Acceptable H1-approved grease-thickening agents are aluminum stearate, aluminum complex, organoclay, polyurea, and calcium sulfonate complex.

*Explain the difference between surface-active and fluid-active additive types and their functions, list their properties, and provide application examples.*

The word *surfactant* is a contraction of the three words "surface active agents." What is a surfactant? Surfactants are materials that lower the surface tension (or interfacial tension) between two liquids or between a liquid and a solid.

Most surfactants' tails are fairly similar, consisting of a hydrocarbon chain that can be branched, linear, or aromatic. Fluorosurfactants have fluorocarbon chains. Siloxane surfactants have siloxane chains. Recent advances in surfactant technology have seen the development of mixed chains or/and complex structures. There are four types of surfactants, and this classification is based on the polarity of the head group: nonionic, anionic, cationic, amphosteric.

*What are the functions of additives?*

Additives are incorporated into lubricants to perform the following functions:

1. Impart new properties. These can be extreme-pressure (EP) additives, detergents, metal deactivators, or tackiness agents.
2. Enhance existing properties. These can be antioxidants, corrosion inhibitors, antifoam agents, or demulsifying agents.
3. Suppress undesirable properties. These can be pour-point depressants or VI improvers.[27]

*What are the types and specific functions of each additive?*

Here is a brief list of the various additives and their functions:

- Antioxidants. These are sacrificial in nature and react with free radicals to prevent oxidation of the base oil.
- VI improvers. These help the viscosity of the oil by allowing the oil to retain its viscosity and reduce shearing at higher temperatures.
- Pour-point depressants. These form densely packed crystals that do not affect the flow properties at temperatures below the pour point.
- Detergents. These suspend particulate matter (such as abrasive wear or soot particles). They are usually metal-containing compounds that can neutralize acids because of their alkaline reserves.
- Dispersants. These keep particles and soot suspended so that they do not form large deposits.
- Antifoam agents. These have a low interfacial tension that significantly weakens the walls of bubbles to cause them to burst.
- Demulsifiers. These change the interfacial tension of an oil to allow water to separate more easily from it.
- Emulsifiers. These are used in oil-and-water-based metalworking fluids and fire-resistant fluids to create a stable oil-water emulsion.
- Rust and corrosion inhibitors. These neutralize acids and form a protective layer to repel moisture from metal surfaces, thus reducing rust and corrosion.
- Antiwear agents. These are activated by heat during metal-to-metal contact, and they adhere to the metal by forming

a protective film that minimizes wear. They can also protect the base oil from oxidation and metals from corrosive acids. One example is ZDDP.

■ Extreme-pressure additives. These react with the metal surface to form a film that prevents welding and seizure of the asperities typically caused by adhesive wear. They are usually activated by high loads and high contact temperatures.[28]

### What are the modes of additive depletion?

Additives get consumed and depleted by

- Decomposition or breakdown
- Adsorption onto metal, particle, and water surfaces
- Separation due to settling or filtration

The adsorption and separation mechanisms involve mass transfer or physical movement of the additive. For many additives, the longer the oil remains in service, the less effective is the remaining additive package in protecting the equipment. When the additive package weakens, viscosity increases, sludge begins to form, corrosive acids start to attack bearings and metal surfaces, and/or wear begins to increase. If oils of low quality are used, the point at which these problems begin will occur much sooner.

### Why should solid additives be used?

Solid additives are used where surfaces are prevented from contact with each other mechanically (usually in the boundary- or mixed-lubrication regimes).

There are two main types of additives: graphite and molybdenum disulfide. These additives are usually used in powder form. They have high thermal stability and can be used in extreme-temperature applications. Other types of solid additives include Teflon (modified tetrafluoroethane), polyethylene, and some metal

oxides. Molybdenum disulfide is usually used in greases for heavy loads, low surface speeds, and restricted or oscillating motions.[29]

## IV. Grease Application and Performance

### Can consistency affect the application of grease?

*Consistency* is the degree to which a plastic material resists deformation under the application of force. When applied to greases, it generally translates to the measure of the harness or softness of the grease and can indicate its flow or dispensing properties.

Consistency is measured through penetration (Cone Penetration ASTM D217).

A high consistency relates to a high NLGI grade (4–6). These greases are harder in nature and can be used in applications such as journal bearings and slow-speed, high-speed ball/roller bearings, applications that avoid water washout, excessive leakage, and bleed or are within high ambient or operating temperatures and seal out environment dust.[30]

### Are there applications that require a low consistency?

A low consistency of grease indicates a low NLGI grade (000, 00, 0, or 1). These grades are more fluid and can be used in applications such as low-speed rolling-element bearings, cold temperature operation, and gearboxes lubed for life.[31]

### List the causes of grease separation.

Oil separation, also known as *oil puddling* or *oil bleed*, is a common phenomenon that occurs naturally when storing grease. Oil separation also occurs in reaction to stress, which is a good thing because this is precisely the mechanism that allows oil to be released under mechanical stress, keeping your equipment well lubricated. This process occurs in much the same way as when a sponge releases water after being squeezed. Specific stressors contributing to oil separation during storage include an uneven grease surface or changes in altitude or temperature. Reversibility must take place to ensure grease integrity.

### What can cause grease separation?

Grease separation is also called *oil bleed*. This usually occurs naturally because the thickeners are responsible for holding the oil until it is needed to lubricate the surface. The lower the viscosity of the base oil, the more separation will be present.

Separation can be caused by cyclic temperatures (not necessarily temperatures that are too high). Additionally, if there is pressure on pumping devices when not in use, this also can cause separation.

### What exactly is a multipurpose grease, and which is the best thickener?

Generally, a multipurpose grease is one that can perform well in various characteristics such as high-temperature operating range, good oxidation stability, water resistance, and compatibility with most greases. The most popular thickener is a lithium complex thickener because it is one of the more compatible thickeners and performs satisfactorily in most areas (i.e., mechanical stability, high temperature, dropping point, water resistance, and extreme pressure). Another candidate is calcium sulfonate. While this compound may outperform lithium complex greases, it is not as compatible with other thickeners.

### List the thickeners and functions of multipurpose greases.

- Lithium and calcium. In simple terms, a multipurpose grease can be defined as a grease that combines the properties of two or more specialized greases that can be applied in more than one application. For example, lithium grease can be applied in both chassis and wheel bearing applications of transport vehicles. Traditionally, calcium greases were used for chassis, and sodium-based greases were used for wheel bearings.
- Calcium-based greases. These have been rated high for water resistance but poor for elevated temperatures. In contrast, sodium-based greases cover high temperatures better but are not as good in regard to water resistance. When lithium greases emerged in the marketplace, they were found to be superior to calcium and sodium greases and soon became the most popular multipurpose greases in industry.

### Do specific grease thickeners have particular performances and applications?

Yes, every grease thickener has particular properties that make them special.

Sulfonates, polyureas, and clays are known for high-temperature applications.

Barium and sulfonates operate well under extreme pressure. Aluminum complexes, clays, and lithium or calcium soaps have better pumpability than calcium sulfonate or polyurea.[32]

### What are the characteristics of high-temperature greases?

High-temperature greases are greases that can withstand very high temperatures. Typically, these greases can withstand temperatures between 200 and 300°C (or 392 and 572°F). In these applications, the grease must have good oxidation, thermal, and mechanical stability.

Either mineral or synthetic oils can be used as the base oils, preferably Groups II–V because these have higher oxidative stability.

Additionally, the thickeners that can perform above or around the 180°C threshold are lithium complex, polyurea, modified clay, sodium complex, and calcium sulfonate complex.

### What are the requirements for coupling greases?

Coupling greases, unlike bearing or general-purpose greases, must withstand the centrifugal forces created by a rotating coupling. These greases—KSG and KHP—are specifically formulated to resist the heavy centrifugal forces associated with all applications. These forces can cause the all-important base oil to separate from the soap thickeners and additives. Unlike greases with lithium-based thickeners, KSG and KHP use polyethylene thickeners, with a density closer to that of the base oil, and therefore, they are much less susceptible to separation. KSG grease is intended for speeds up to 3600 rpm, and KHP is best suited for speeds above 3600 rpm.

### What are coupling greases, and what makes them special?

Coupling greases must be able to withstand the centrifugal forces that are applied to them. Additionally, they must be able to perform in very high rotational speeds and carry high-torque loads.

Typically, this type of grease has a high-viscosity base oil to keep the grease in place and prevent separation. The American Gear Manufacturers Association (AGMA) has a standard for coupling lubrication: AGMA 9001-B97: Flexible Couplings Lubrication. The most frequently used thickeners for coupling grease is lithium soap or polymer thickener.

## V. Lubricant Selection

*What are the viscosity selection adjustments based on machinery condition and environmental conditions?*

Viscosity is the most important physical property of a fluid lubricant. Although it is relatively easily measured with the proper equipment, the characteristics and behavior of viscosity can be quite complex. Viscosity is characterized by coefficient of friction, typed as boundary lubrication, full- and mixed-film lubrication, hydrodynamic lubrication, and elastohydrodynamic lubrication. Viscosity properties of fluids can be Newtonian, having low shear rates, or non-Newtonian, with high shear rates. But all this can be broken down into some basic principles when it comes to selecting the proper viscosity for your operating equipment.

Optimal viscosity selection maximizes reduced friction by creating and maintaining a proper fluid film that separates moving parts or minimizes metal contact based on the design and application. Several variables come into play, including startup and operating temperatures, the load on the fluid film, and equipment operating speed. Temperature, load, and speed interplay, and this affects whether a higher- or lower-viscosity lubricant is required.

One basic rule of thumb is to follow the equipment manufacturer's recommendations and warranty requirements. In certain situations, lubrication challenges arise in which your equipment and lubricant supplier may need to make adjustments to viscosity requirements. In these situations, there are some relevant points to consider: generally, a lubricant is used with a viscosity as low as possible that is known to provide ample protection under operating conditions by maintaining a proper fluid-film thickness and keeping moving parts separated.

Lower-viscosity oils also will provide better protection at startup and improved energy efficiency. When there is a level of doubt, a higher-viscosity oil can create a margin of safety, but this is compromised by higher operating temperatures, greater startup and pump wear, and potential for viscosity decreases due to shearing.

The speed of the one moving surface to the other in operating equipment affects viscosity requirements and performance. Higher speeds may require a lower-viscosity oil in order for the fluid to continue to flow sufficiently and disipate heat. Higher speeds allow lower-viscosity oils to carry a higher load. The slower the speed, the higher the viscosity requirement to maintain a fluid film. Higher-viscosity oil at slow speeds generally does not contribute to noticeable heat generation. As the load decreases at a given speed, the viscosity will generally decrease. As loads increase at a given speed, the viscosity will generally increase.

Higher operating temperatures also necessitate higher-viscosity lubrication so that the fluid does not thin down too much. But, as noted earlier, higher-viscosity oils will also create more heat, so this should be taken into consideration.

When testing for viscosity, changes can indicate depletion of the lubricant's service life or contamination. Increases in viscosity can be due to extended drain intervals, contamination, forms of oxidation and oxidation materials, product mixture, and buildup of soot or solids. Viscosity decrease can be attributed to contamination (such as fuel dilution or solvents), shearing, product mixture, and in some particular cases hydrocracking. Checking viscosity is a primary parameter in monitoring a lubricant's service life. Changes in viscosity can be correlated with other test results for contaminants, oxidation, or acid number.

*Is there a specific time to use synthetics?*

The use of synthetics can be attributed to their performance characteristics. Synthetics generally are used in applications that require high-temperature stability, long life, high VI, and low volatility.[33]

*Are biodegradable lubricants suited for all applications? When should they be used?*

Biodegradable lubricants must meet two important environmental characteristics: speed at which the products biodegrade and the toxicity that can affect bacteria or aquatic life. As such, these are the characteristics that are heavily focused on when developing these types of lubricants.

Typically, biodegradable lubricants are more expensive than mineral oils. Therefore, application of biodegradable lubricants will be in areas that are environmentally sensitive. These can include dredging operations for waterways, offshore drilling, marine equipment, agricultural operations, forestry and logging, mining, and construction sites near waterways or groundwater systems.

Economic justifications for biodegradable lubricants can be based on the cost reductions associated with remediation, spills, and fines.[34]

*What are the steps and process for lubricant consolidation?*

Consolidating or optimizing the number of lubricants used is an important part of designing and maintaining an effective lubrication program. The advantages of optimizing the different lubricants are plentiful, including lower required inventory levels, a reduced chance of availability issues, fewer purchase orders, and fewer chances for misapplication.

Follow these steps:

- Group lubricants for basic consolidation based on generic type and performance (e.g., ISO VG 32 turbine oils would be one group and ISO VG 460 EP gear oils would be another group).
- For machines currently using lubricants from the same group (ISO VG 32 turbine oil, for instance), select a single lubricant of suitable robustness based on volume of use and brand performance. Robustness relates to such things as oxidation stability, film strength, VI, demulsibility, dispersancy, and so on.
- De-risk machines selected for switchover based on compatibility between the old lubricant and the switchover lubricant. Compatibility needs to be verified.
  - Rolling change (lowest compatibility risk)
  - Drain and change (moderate compatibility risk)
  - Drain, flush, and change (high compatibility risk)
- After the switchover, monitor machines carefully for any change in operating conditions (e.g., leakage, hot running, noisy, vibration, etc.).

*What are the fundamental requirements for selecting lubricating oils?*

Consider the application as a tribologic system:

- Type of motion
- Speed
- Temperature
- Load
- Operating environment

Many applications have special requirements that go beyond the tribologic system and must be taken into consideration. Some applications are limited to oils, whereas others require grease. Applications that involve the use of sintered bearings or special sealing arrangements will require additional analysis. Material compatibility is another important issue.

Lubricant selection also can be affected by a variety of other specialized requirements:

- Design life
- Lubrication equipment
- Acceptable relubrication intervals
- Cost
- Special certifications such as National Science Foundation registration
- Biodegradability

*What are the fundamental requirements for selecting lubricating oils for fire-resistant applications?*

Considerations include cost, function, and risk. A balance should be reached between the functional sacrifices sometimes made by using a fire-resistant oil instead of a normal oil, the properties of the fire-resistant oil, the need for the oil being fire resistant, and cost and risk.

*What are the characteristics that one should look for when selecting fire-resistant lubricants?*

There are three main types of fire-resistant fluids: synthetics, emulsions, and water glycol. The fire-resistant characteristics depend on the water content of emulsions and water glycols.

Synthetic fire-resistant fluids are generally the most expensive and have compatibility issues with paints, elastomers, and other system materials. They are also incompatible with emulsions and water glycols. These synthetics have high densities but low VIs and are typically phosphate esters.

Emulsions, in contrast, can be classified into two categories: conventional and invert. Conventional emulsions contain approximately 95% water and their viscosity decreases as water content increases. Invert emulsions contain approximately 40%–45% water, and as the water content increases, the viscosity increases. Operating temperatures should fall within the range of 60–49°C to maintain proper viscosity and prevent evaporation of water.

Water glycols are typically 40%–50% water and glycol but contain liquid- and vapor-phase rust inhibitors as well as antiwear additives. Water content will control the viscosity of the fluid and its fire-resistant characteristics. Maximum operating temperatures should not exceed 60°C.[35]

*What are the fundamental requirements for selecting lubricating oils for mobile hydraulic/industrial applications?*

When choosing a hydraulic fluid, consider the following characteristics: viscosity, VI, oxidation stability, and wear resistance. These must match the type of machine or pump and its application.

*What are the characteristics that one should look for when selecting hydraulic applications in mobile industrial equipment?*

When we think about hydraulic applications, one aspect comes to mind, and this is the manner in which the hydraulic applications work. Hydraulic applications work through fluid technology and can be divided into hydrostatics and hydrodynamics.

For hydrostatic systems, static pressure is required for the transfer of energy. In these applications, pressures are high but flow rates are low. These fluids are called *hydraulic oils*.

In contrast, hydrodynamic systems require kinetic energy from the flowing fluid. Thus, in this case, the pressures are low but the flow rates are high. These fluids are called *power transmission fluids*.

A hydraulic fluid should have the following functions and properties:

- Pressure and motion energy transfer
- Force and moment transfer
- Wear and friction minimization
- Protection of components against corrosion
- Heat dissipation
- Suitability for a wide range of temperatures
- Compatibility with elastomers
- Good thermal stability
- Low foaming, good air separation
- Good filterability, good water release

When selecting a hydraulic fluid, the following must be considered:

- Application. Working temperature range, design of hydraulic system, type of pump, working pressure, and environmental considerations
- Required fluid life. Availability and economic and ecological factors

With hydraulic oils, it is always advised to select the lowest viscosity of the fluid because this can guarantee an instant response when the equipment has been set into motion. However, a minimum viscosity is required to reduce leakage and guarantee adequate lubrication of the moving parts. The working temperature of the system should be kept narrow to ensure that there is little room for viscosity change. Typically, hydraulic oils use a VI of 100, but higher-VI oils are used for specialty applications.

Reference can be made to the various International Organization for Standardization (ISO) and German Institute for Standardization (DIN) specifications as they relate to mobile and industrial equipment (Table 2.2).

**Table 2.2** Classification of Mineral Oil–Based Hydraulic Fluids
(categories according to DIN 51 502 and ISO 6743/4)

| Category (Symbol) | | Composition Typical Characteristics | Field of Application Operation Temperatures |
|---|---|---|---|
| DIN | ISO-L | | |
| — | HH | Non-inhibited refined mineral oils | Hydraulic systems without specific requirements (rarely used nowadays) −10 to 90°C |
| HL | HL | Refined mineral oils with improved antirust and anti-oxidation properties | Hydrostatic drive systems with high thermal stress; need for good water separation −10 to 90°C |
| HLP | HM | Oils of HL type with improved anti-wear properties | General hydraulic systems, which include highly loaded components; need for good water separation −20 to 90°C |
| — | HR | Oils of HL type with additives to improve viscosity-temperature behavior | Enlarged range of operating temperatures compare with HL oils −35 to 120°C |
| HVLP | HV | Oils of MH type with additives to improve viscosity-temperature behavior | For example, hydrostatic power units in construction and marine equipment −35 to 120°C |
| — | HS | Synthetic fluids with no specific inflammability characteristics and no specific fire-resistance properties | Special application in hydrostatic systems, special properties −35 to 120°C |
| — | HG | Oils of HM type with additives to improve stick-slip behavior and anti-stick-slip properties | Machines with combined hydraulic and plain bearing way lubrication systems where vibration or intermittent sliding (stick/slip) at low speed must be minimized −30 to 120°C |
| HLPD | — | Oils of HM type with detergent/dispersant (DD) additives; DD additives reduce friction | Hydrostatic drive units with high thermal stress, which require EP/AW additives: DD additives keep contaminants in suspension, e.g., machine tools and mobile hydraulic equipment |

Category L: Lubricants, industrial oils, and related products
Category H: Hydrostatic hydraulic systems
Source: Th. Mang and W. Dresel, *Lubricants and Lubrication*, 2nd ed. Weinheim, Germany: Wiley-VCH, 2007, p. 274.

*What are the fundamental requirements for selecting lubricating oils for turbines?*

Steam, gas, and hydroelectric turbines operate on a family of lubricating oils known as *rust-and oxidation-inhibited oils* (R&O oils). Turbine equipment geometry, operating cycles, loads and speed, maintenance practices, operating temperatures, and potential for system contamination present unique lubricating oil demands versus other lubricating oil applications such as gasoline and diesel engines.

*What characteristics should one look for when selecting turbine oils?*

Turbine oils should meet a number of characteristics such as:

- Ensuring hydrodynamic lubrication of all bearings and the lubrication of gearboxes
- Avoiding friction and wear on gear tooth flanks
- Aging stability for long periods
- Hydrolytic stability
- Corrosion protection
- Demulsibility
- Rapid air release and low foaming
- Good filterability and purity (free of ash, zinc-free)
- Good viscosity/temperature characteristics
- Antioxidants
- Mild extreme-pressure (EP) additives

As per DIN 51 515, turbine oils can be classified into four main groups:

- Normal turbine oils. Steam turbine oils without EP and with EP (DIN 51 515-1) or FZG load stage min 8 (Annex A)
- High-temperature turbine oils. Without EP and with EP (DIN 51 515-2) or FZG load stage min 8 (Annex A)[36]

*What are the fundamental requirements for selecting lubricating oils for compressors?*

Lubrication requirements vary considerably depending not only on the type of compressor but also on the gas that is being compressed. To combat these stressors, a compressor lubricant needs several defenses. Oxidation and thermal stability are very important, along with corrosion inhibitors, detergents, demulsifying agents, and foam suppressors, to increase the life of the machinery.

Special considerations must be made when selecting the proper lubricant:

- Long life without the need for change-out (high oxidation stability)
- Prevention of acidity and sludge and deposit formation
- Excellent protection against rust and corrosion, even during shutdown
- Good demulsibility to shed water that enters the lubrication system
- Easy filterability without additive depletion
- Good foam control

*What characteristics should one look for when selecting compressor oils?*

The correct characteristics for compressor oils depend heavily on the type of compressor and its functions. Particular attention must be paid to the pressure involved, outlet temperatures, and the type of gas being compressed.

Typically, reciprocating-piston compressors require lubricants with higher viscosities (ISO VG 100 or 150), extremely low carbon residues, and no or mild EP and antiwear performance additives. In contrast, screw compressors require lubricants with lower viscosities (ISO VG 46 or 68), excellent oxidation stability, and mild to high EP and antiwear performance additives.[37]

*What are the fundamental requirements for selecting lubricating oils for bearings?*

Design, function, load, speed, environment, viscosity, additives, and if it is a grease, the type of soap, plus seal compatibility and additives for specific applications.

*What characteristics should one look for when selecting lubricants for bearings?*

When selecting lubricants for roller bearings, one should consider the following:

- Oil selection:
  - Excellent resistance to oxidation
  - Proper viscosity at operating speeds and temperatures to protect against friction and wear
  - Antirust properties (to protect against rust)
  - Good antiwear properties (for heavy shock loads)
  - Good demulsibility (separation of water)
- Grease selection:
  - Resistance to oxidation
  - Mechanical stability (resist softening or hardening)
  - Proper consistency
  - Controlled oil bleeding
  - Enhanced antiwear properties

- Protection from rusting
- Good compatibility with system and components
- Ability to resist corrosion

For journal bearings, the following must be considered:

- Oil selection:
  - Good chemical stability to resist oxidation
  - Protection against rust and corrosion
  - Ready separation from water
  - Foaming resistance
  - Viscosity controls (where dilutions may occur, look at three factors: speed, load, and operating temperature)
- Grease selection (not subjected to long-term service use):
  - Good mechanical stability
  - Adequate dispensing and pumpability characteristics
  - Corrosion protection
  - Adequate film strength
  - Adequate high-temperature characteristics[38]

*What are the fundamental requirements for selecting lubricating oils for chains/conveyors?*

Appropriate oil with appropriate character, light or heavy, depending on ambient temperature, lubrication method, chain size, and speed, can be selected from Table 2.3.

**Table 2.3** Chains and Conveyors Lubrication Selection

| Chain Pitch | Operating Temperature | Operating Environment** | Base Oil Type** | Lubricant Vis. at 40°C | Additive Type | Evaporation, % Loss at 8 hours* |
|---|---|---|---|---|---|---|
| <0.75" | −10°C to 20°C | No dust/moisture | Mineral oil | ISO 22 | Solid film AW | N.A. |
|  | −10°C to 20°C | Moisture present | Solvent/oil blend | ISO 22 | Solid film AW | N.A. |
|  | 20°C to 80°C | No dust/moisture | Mineral oil | ISO 68 | Soluble AW | N.A. |
|  | 20°C to 80°C | High dust | Solvent carrier | ISO 68 | Solid film AW | N.A. |
|  | 20°C to 80°C | High moisture | Tackified mineral oil | ISO 68 | Soluble AW | N.A. |
|  | 80°C to 220°C | No dust/moisture | Low residue synthetic | ISO 100 | Soluble AW | ±50% |
|  | 80°C to 220°C | High dust | Low residue synthetic | ISO 100 | Solid film AW | ±50% |
|  | >220°C |  | Low residue synthetic | ISO 150 | Soluble AW | ±50% |
|  | >220°C | High dust | Low residue synthetic | ISO 150 | Solid film AW | ±50% |
| 0.75– 1.25" | −10°C to 20°C | No dust/moisture | Mineral oil | ISO 32 | Solid film AW | N.A. |
|  | −10°C to 20°C | Moisture present | Solvent/oil blend | ISO 32 | Solid film AW | N.A. |
|  | 20°C to 80°C | No dust/moisture | Mineral oil | ISO 100 | Soluble AW | N.A. |
|  | 20°C to 80°C | High dust | Solvent carrier | ISO 68 | Solid film AW | N.A. |
|  | 20°C to 80°C | High moisture | Tackified mineral oil | ISO 100 | Soluble AW | N.A. |
|  | 80°C to 220°C | No dust/moisture | Low residue synthetic | ISO 100 | Soluble AW | ±50% |
|  | 80°C to 220°C | High dust | Low residue synthetic | ISO 100 | Solid film AW | ±50% |
|  | >220°C | No dust/moisture | Low residue synthetic | ISO 150 | Soluble AW | ±50% |
|  | >220°C | High dust | Low residue synthetic | ISO 150 | Solid film AW | ±50% |
| >1.25" | −10°C to 20°C | No dust/moisture | Mineral oil | ISO 32 | Solid film AW | N.A. |
|  | −10°C to 20°C | Moisture present | Solvent/oil blend | ISO 32 | Solid film AW | N.A. |
|  | 20°C to 80°C | No dust/moisture | Mineral oil | ISO 150 | Soluble AW | N.A. |
|  | 20°C to 80°C | High dust | Solvent carrier | ISO 68 | Solid film AW | N.A. |
|  | 20°C to 80°C | High moisture | Tackified mineral oil | ISO 100 | Soluble AW | N.A. |
|  | 80°C to 220°C | No dust/moisture | Low residue synthetic | ISO 220 | Soluble AW | ±50% |
|  | 80°C to 220°C | High dust | Low residue synthetic | ISO 220 | Solid film AW | ±50% |
|  | >220°C | No dust/moisture | Low residue synthetic | ISO 320 | Soluble AW | ±50% |
|  | >220°C | High dust | Low residue synthetic | ISO 320 | Solid film AW | ±50% |

\* Volatility testing: Thin film evaporation loss, 220°C, for 8-, 24-, and 48 -hour intervals.

\*\* High speed chain operations require automatic application, may require lower viscosities and higher frequencies.

## What are the fundamental requirements for selecting lubricating oils for oil mist applications?

Oil mist is an aerosol mixture of very small oil droplets (1–5 μm) suspended in air with the appearance of smoke. This mist is generated by passing compressed air through a venturi or vortex to siphon oil from a small central reservoir. It must be possible to convert the oil from a liquid into a vapor spray and still have the required characteristics, such as adequate viscosity.

## What characteristics should one look for when selecting oil mist applications?

In oil mist systems, oil must be atomized by low-pressure compressed air typically at 10–50 psi or 70–350 kPa to be forced into droplets. These droplets are usually very small (from 1–3 μm) and form mist. This mist is transported very long distances, where it becomes condensed or coalesced at the application point to form larger particles and provide lubrication. Heaters are typically used to lower the viscosity of the oils in the system. Systems without heaters can handle oils up to 800–1000 Saybolt universal seconds (SUS) at 100°F (173–216 cSt at 38°C). Oils with a greater viscosity than 1000 SUS (216 cSt) will require heating to form the mist.[39]

## What are the fundamental requirements for selecting lubricating oils for automotive/ industrial gears?

Gear oil is a lubricant made specifically for transmissions, transfer cases, and differentials in automobiles, trucks, and other machinery. It is of a high viscosity and usually contains organosulfur compounds. Some modern automatic transaxles (integrated transmission and differentials) do not use a heavy oil at all but lubricate with lower-viscosity hydraulic fluid, which is available at pressure within the automatic transmission. Gear oils account for about 20% of the lubricant market.

Most lubricants for manual gearboxes and differentials contain EP and antiwear additives to cope with the sliding action of hypoid bevel gears. Typical additives include dithiocarbonate derivatives and sulfur-treated organic compounds (sulfurized hydrocarbons).

## What characteristics should one look for when selecting oils for automotive/industrial gears?

Gear oils require additives such as antiwear agents, EP agents, oxidation stabilizers, metal deactivators, foam suppressors, corrosion inhibitors, pour-point depressants, dispersants, and VI improvers. Depending on the application and the types of gears being used, gear oil is chosen accordingly.

For automotive gear oils, three main technical societies dictate the type of oil. The Society of Automotive Engineers (SAE) has established the viscosity classification system (SAE J306), whereas the American Society for Testing and Materials (ASTM) has established the test methods and criteria for judging performance levels and defining test limits. In contrast, the American Petroleum Institute (API) defines the performance category language. U.S. military specification MIL-PRF-2105E also exists.

- API GL1. Lubricants for manual transmissions operating under mild service conditions. These oils contain corrosion inhibitors, oxidation inhibitors, pour-point depressants, and antifoam agents. They do not contain antiwear, EP, or friction-modifying additives.
- API GL4. Lubricants designed for differentials containing spiral bevel or hypoid gearing operating under moderate or severe conditions. These are usually used in manual transmissions and transaxles where EP oils are acceptable.
- API GL5. Lubricants designed for differentials containing hypoid gears operat-

ing under severe conditions of torque and occasional shock loading. These contain high levels of antiwear and EP additives.

- API MT-1. Lubricants for manual transmissions that do not contain synchronizers. They are formulated to provide higher levels of oxidation and thermal stability compared with API GL1, GL4, and GL5 products.

For industrial gears, three classes of gears are categorized based on the action between the gear teeth:

1. Spur, bevel, helical, herringbone, and spiral bevel
2. Worm gears
3. Hypoid gears

Both rolling and sliding actions occur in these gears. For bevel and conventional spur gears, sliding occurs at right angles to the lines of contact (across the tooth faces). In contrast, helical, herringbone, and spiral bevel gears do not have sliding at right angles to the lines of contact. For worm gears, sliding and rolling are slow because of the low rotational speed of the worm wheel.

For enclosed gears, we need to consider the following factors: gear type, speed, reduction ratio, operating and startup temperatures, transmitted power, surface finish, load characteristics, drive type, application method, contamination, and lubricant leakage. Enclosed gears must have the correct viscosity at operating temperature, adequate low-temperature fluidity, good chemical stability to minimize oxidation, good demulsibility to allow rapid separation of water, antirust properties, foam resistance, and good compatibility with system components (seals and paints).

Worm gears require lubricants that have mild wear- and friction-reducing properties.[40]

*What are the fundamental requirements for selecting lubricating oils for diesel/gas/gasoline engines?*

Make sure that the oil meets the requirements of the desired standards:

- Consider viscosity.
- Consider base-oil type.
- Consider additives.
- Consider original equipment manufacturer (OEM) recommendations.

*What are the fundamental requirements for selecting lubricating oils for pneumatic tools?*

Modern air tools and various machines operated by compressed air require specialized lubrication to resist the formulation of acids, sludge, varnish, and rust that robs tools of speed and power.

*What are the fundamental requirements for selecting lubricating oils for spindles?*

Spindle bearings could be greased for life, grease maintained, grease injected, oil mist, or under raceway lubrication. Spindle oils are a type of low-viscosity mineral oil marketed for use in lubrication of high-speed machine spindles. Spindle oil is free from gumming properties. Because the viscosity is so low that the oil runs off the surface of the spindle during shutdown periods, the spindle oil may be doped with additives that prevent rusting. Because spindle oil is often used in textile factories, it is also important that it doesn't stain the textiles.

*What are the fundamental requirements for selecting lubricating oils for ways/slides?*

Should have excellent frictional properties, very good slideway adhesion, and excellent antiwear performance.

*What are the fundamental requirements for selecting grease?*

- Composition: base oil, soap, and additives.
- NGLI consistency
- Viscosity
- Drop point
- Maximum usable temperature
- $dN$ Value, where $d$ is diameter and $N$ is speed
- Bleed rate
- Evaporation
- Water washout resistance
- Water spray resistance
- For wheel bearings, bearing leakage
- Mobility at low temperatures
- Four-ball wear
- Four-ball EP
- Timken EP
- Corrosion prevention

*What are the specific requirements for selecting lubricants for a chassis grease?*

The product should contain antioxidants, corrosion inhibitors, and EP and antiwear additives. Because of its special formulation and semifluid consistency, grease should exhibit excellent flow even at very low temperatures and in long pipelines. The oil film adheres well and does not run off, even during stoppages. There is rapid on-flow to the greasing points. The good protection against rust also covers the use of "spreading" salt and the related danger of rust on the lubrication points of the chassis.

*What are the specific requirements for selecting lubricants for coupling grease?*

Flexible couplings. Both lubricating oils and greases can be selected to lubricate flexible couplings. Unless specifically noted by the coupling designer, couplings for most industrial components are grease lubricated. Coupling components are protected primarily by an oil film that bleeds from the grease thickener and seeps into the loading zone.

Lubricated flexible couplings require protection from the low-amplitude relative motion that develops between their components. Other concerns include centrifugal stress on the lubricant (particularly grease), which causes premature separation of the oil from the thickener; poor oil distribution within the housing; and oil leakage from the housing.

Fluid couplings. The dissipation of energy that makes fluid couplings so tolerant of shock loading creates the potential for rapid and extreme increases in fluid temperature. The energy dissipated during stall and slip is converted to heat through the viscous shearing of the fluid (fluid internal friction). In extreme applications, such as a torque converter in an automobile climbing a steep hill or pulling a heavy load, the fluid temperature can rise substantially above the normal 93°C operating temperature in less than a minute.

Oxidation and thermal degradation resistance are important qualities of oils used for fluid couplings because of the potential for drastic temperature increases. Similarly, a high VI is also useful to prevent severe decreases in operating viscosity at temperature spikes and excessively high operating viscosity during low-temperature conditions.

Low-viscosity fluids are ordinarily used in these applications to reduce the power lost to heat due to fluid friction. Fluid coupling viscosities may fall between 2.5 and 72 cSt at 40°C. For fluid couplings designed to operate at high temperatures, such as in automobile automatic transmission fluid (ATF) applications, viscosity limits may be given at 100°C. Typical automotive ATF requirements would be 3–7 cSt at 100°C.

*What are the specific requirements for selecting lubricants for journal bearings (oil or grease)?*

Oil (especially where cooling is required and typically at high speeds):

■ The viscosity grade required depends on bearing revolutions per minute (rpm), oil temperature, and load.
■ ISO grade number
■ Operating temperature
■ VI

Grease (cooling not required, typically lower speeds and where shock loads may be occurring):

■ EP additives
■ NGLI consistency

*What are the specific requirements for selecting lubricants for automotive bearings?*

■ Position and application
■ Corrosion resistance
■ Shock loads
■ Temperature
■ Viscosity and VI

*What are the specific requirements for selecting lubricants for automatic lubrication systems?*

Should be suitable for the method of application while still having the correct lubrication properties.

*Can a lubricant be selected based on one parameter only?*

Lubricant is usually selected based on the tribologic system present. This can be translated into the following factors:

■ Type of motion. If it is sliding, then hydrodynamic lubrication is required. If it is rolling, then elastohydrodynamic lubrication would be required.
■ Speed. The Stribeck curve, which shows the relationship between coefficient of friction and $hn/P$ (where $h$ is dynamic viscosity, $n$ is speed, and $P$ is load per projected area).
■ Temperature. Lubricants can perform within different temperature ranges. These must be known before selecting the viscosity.
■ Load. Depending on the size of the load, this would indicate the types of additives needed (heavy load, EP additives).
■ Operating environment. This indicates whether water or moisture is present (need for anticorrosion and water washout) or other chemicals are present (lubricant must be resistant to these).[41]

*List some of the procedures for testing and quality assurance of incoming lubricants.*

ANSI/AGMA 9005-E02 states the following for new oil cleanliness. For general industrial applications, the following tests are recommended:

■ ISO Particle Count, ISO 4406:99
■ Elemental Analysis, ASTM D5185

- Viscosity at 40°C, ASTM D445
- Viscosity at 100°C, ASTM D445
- Karl Fischer moisture, ASTM D1744 or ASTM D6304
- Fourier Transform infrared (FTIR) spectroscopy
- Acid number, ASTM D664

*List the steps for approval of candidate lubricants.*

- Obtain the data sheets.
- Verify that the candidate lubricant meets the required functional performance specifications.
- Verify that the lubricant manufacturer is managing quality assurance, environmental aspects, and health and safety aspects according to international standards such as ISO9001, 14001, and 18001.

## VI. Lubricant Testing and Performance Analysis

### What is the flash/fire point?

The *flash point* is the lowest temperature to which a lubricant must be heated before its vapor, when mixed with air, will ignite but not continue to burn. The *fire point* is the temperature at which lubricant combustion will be sustained. The flash and fire points are useful in determining a lubricant's volatility and fire resistance. The flash point can be used to determine the transportation and storage temperature requirements for lubricants.

Manufacturers and toll blenders also can use the flash point to detect potential product contamination. A lubricant exhibiting a flash point that is significantly lower than normal should be suspected of contamination with a volatile product. Products with a flash point that is less than 38°C (100°F) usually require special precautions

for safe handling. The fire point for a lubricant is usually 8% to 10% above the flash point.

The flash and fire points should not be confused with the *autoignition temperature* of a lubricant, which is the temperature at which a lubricant will ignite spontaneously without an external ignition source.

### What is the pour point (ASTM D97)/cloud point (ASTM D2500)?

The pour point (ASTM D97) or cloud point (ASTM D2500) of a petroleum specimen, a petroleum product, or biodiesel fuels is an index of the lowest temperature of its utility for certain applications.

### What is the foam tendency test (ASTM D892)?

The foam test measures the foaming tendency of a lubricant. According to this test, also referred to as ASTM D892, the tendency of oils to foam can be a serious problem in systems such as high-speed gearing, high-volume pumping, and splash lubrication.

### What are air release properties according to ASTM D3427?

The ASTM D3427 test of air release properties of hydrocarbon-based oils is an important test for determining the ability of a turbine, a gear, or hydraulic oil to separate entrained air. The air release test forms larger bubbles than the foam test. The main difference is how the air is introduced into the sample.

### What is the neutralization number?

The neutralization number is a measure of the acid or alkaline content of new oils and an indicator of the degree of oxidation degradation of used oils.

### What is the acid number (ASTM D664/D974)?

The *acid number* or *neutralization number* is a measure of the amount of potassium hydroxide required to neutralize the acid contained in a lubricant. Acids are formed as oils oxidize with age and service duration. The acid number for an oil sample is indicative of the age of the oil and can be used to determine when the oil must be changed.

### What is the definition of base number according to ASTM D974/D2896?

Base number or total base number (TBN) is a measurement of basicity that is expressed in terms of number of milligrams of potassium hydroxide per gram of oil sample (mg KOH/g). TBN is an important measurement in petroleum products, and the value varies depending on its application. TBN generally ranges from 6–8 mg KOH/g in modern lubricants, 7–10 mg KOH/g for general internal combustion engine use, and 10–15 mg KOH/g for diesel engine operations. TBN is typically higher for marine grade lubricants, approximately 15–80 mg KOH/g, as the higher TBN values are designed to increase the operating period under harsh operating conditions, before the lubricant requires replacement.

### What is the turbine oil oxidation stability test (ASTM D943)?

The most commonly accepted oxidation test for lubricants is the ASTM D943 oxidation test. This test measures the neutralization number of oil as it is heated in the presence of pure oxygen, a metal catalyst, and water. Once started, the test continues until the neutralization number reaches a value of 2.0.

### What is the rotary pressure vessel oxidation test (ASTM D2272)?

The Rotary Pressure Vessel Oxidation Test is also known as the Rotary Bomb Oxidation Test (ASTM D2272) is a rapid method of comparing the oxidation life of lubricants in similar formulations. It is used to evaluate the oxidation characteristics of turbine, hydraulic, transformer, and gear oils.

The test apparatus consists of a pressurized bomb axially rotating at an angle of 30 degrees from the horizontal in a bath at 150°C (302°F). Fifty grams of test oil and 5 g of water are charged to the bomb containing a copper catalyst coil. The bomb is initially pressurized with oxygen to 90 psi at room temperature. The 150°C bath temperature causes this pressure to increase to approximately 200 psi. As oxidation occurs, the pressure drops, and the usual failure point is taken as a 25-psi drop from the maximum pressure attained at 150°C. The results are reported as the number of minutes to the 25-psi loss. Longer time means better performance.

### What is the turbine oil rust test (ASTM D665)?

Rust inhibitors are tested according to the ASTM D665 rusting test. This test subjects a steel rod to a mixture of oil and salt water that has been heated to 60°C (140°F). If the rod shows no sign of rust after 24 hours, the fluid is considered satisfactory with respect to rust-inhibiting properties.

### What is the Vickers wear pump test (ASTM D2882)?

Quality assurance of antiwear properties is determined through standard laboratory testing. Laboratory tests to evaluate antiwear properties of a hydraulic fluid are performed in accordance with ASTM D2882. This test procedure is generally conducted with a variety of high-speed, high-pressure pump models manufactured by Vickers or Denison.

Throughout the tests, the pumps are operated for a specified period. At the end of each period, the pumps are disassembled and specified components are weighed. The weight of each component is compared with its initial weight; the difference reflects the amount of wear experienced by the pump for the operating period. The components are also inspected for visual signs of wear and stress.

### How do you measure grease consistency (ASTM D217)?

Measurement is in tenths of a millimeter that a standard cone penetrates a grease sample. It may be reported as:

- Unworked
- Worked (60 strokes)
- Prolonged worked (typically 10,000 or 100,000 strokes)

### How do you measure NGLI grade?

- Grease is brought to a standard temperature in a standard container.
- The container with grease is placed below cone tip. The cone is released and drops into the grease. The depth of penetration relates to a scale on the penetrometer.
- The NLGI consistency grade corresponds to a range of depth of penetration measured in tenths of a millimeter: 1/10 mm = 100 μm. For less penetration or a stiffer grease, the NLGI number is bigger.

### What is the grease drop point (ASTM D566, D2265)?

This is the temperature at which a grease starts to melt or first drops when heated. It should not be used as the upper end-temperature limit. A good rule of thumb is 50°C below dropping point as maximum operating temperature.

### What is the maximum usable temperature?

The maximum usable temperature of a grease does not depend on the thickener. It is normally much lower than the drop point. The upper temperature limit of use of grease is therefore not determined by the drop point.

### What is the grease dN value?

The $dN$ value or $DN$ value is a number representing the diameter of the bearing and the shaft speed. It is a *speed factor* based on the bearing size and operational speed. It provides an indication of an upper limit in terms of speed before the grease will start to degrade due to the amount of shearing.

### What is the grease bleed rate (ASTM D1742)?

This test measures resistance to oil separation. Some greases have a tendency to separate while in storage. ASTM D1742 requires measurement as the percent (weight) of bleed in 24 hours at 25°C.

### What is grease evaporation (ASTM D972, D2595)?

This test evaluates the loss of volatile materials from greases (oil) that may lead to thickening or hardening of the grease. ASTM D972 and D2595 require measurement as the percent (weight) loss after 22 hours at a given temperature.

### What is grease water washout resistance (ASTM D1264)?

This test determines the resistance of a grease to water washout from ball bearings. ASTM D1264 requires measurement of the percent of grease removed after spraying for one hour at 38 or 79°C.

*What is grease water spray resistance
(ASTM D4049)?*

This test evaluates the ability of a grease to adhere to a metal surface. ASTM D4049 requires measurement of the percent (weight) of grease remaining after spraying for five minutes at 38°C.

*What is grease wheel bearing leakage
(ASTM D1263, D4290)?*

This test evaluates the high-temperature leakage characteristics of a grease in an automotive wheel bearing. ASTM D1263 and D4290 require reporting of the weight of grease (grams) expelled from the bearing.

*What is grease mobility at low temperature (US
Steel DM43)?*

This test evaluates the pumpability of a grease at low temperatures. US Steel DM43 requires reporting of the volume of grease in grams per minute.

*What is grease four-ball wear (ASTM D2266)?*

This test evaluates the wear-preventive properties of a grease. ASTM D2266 states that normal conditions of 75°C, 1200 rpm, and 40 kgf for 60 minutes must be used. Results report the average scar diameter of the three stationary balls.

*What is grease four-ball EP (ASTM D2596)?*

This test evaluates the EP properties of a grease. ASTM D2596 requires tests at 1770 rpm, 25°C, and 10 s at increasing load until welding occurs. Results report the weld point (load at which welding occurs) and the load wear index (average of the last loads prior to seizure).

*What is Timken EP (ASTM D2509)?*

This test evaluates the load-carrying ability of a grease. ASTM D2509 reports the "OK value," maximum weight added at which no scoring or seizure occurs.

*What is grease corrosion prevention
(ASTM D1743, D5969, D6138)?*

This test evaluates the corrosion-preventive capabilities of a grease. ASTM D1743, D5969, and D6138 require that the bearing is exposed to water for 48 hours at 52°C and 100% humidity. Results report pass or fail (corrosion spot > 1 mm).

# VII. Lubricant Application

*Explain the procedure for bearing cleaning.*

1. Clean work area.
2. Open the housing.
3. Remove old grease from the housing using a palette knife, and clean the housing with solvent.
4. Clean the bearing with solvent and allow it to dry.
5. Fill the free space between the rolling elements with a grease packer.
6. Fill 30%–50% of the housing with grease.
7. Flush the reservoir/system.
8. Reassemble the housing cap.
9. Disassemble/clean reservoirs and sumps.

*Explain how to select and manage optimal
equipment systems for lubricant application
based on machinery requirements.*

Document temperature, shaft orientation, vibration, outer`ring rotation, contamination, load, bearing size, application, oil specifications, and physical properties.

*What are the safety/health requirements for lubricant application?*

- Use the right PPE.
- Use the right lubrication tools.
- Use the right sampling tools.
- Train the workers who carry out the work.
- Have a working plan.

*What are the elements of proper maintenance of lubrication equipment?*

- Right lubricant selection
- Lubricant storage and handling
- Lubricant dispensers and grease gun storage
- Building the lubrication preventive maintenance program
- Training and education
- Create role of lubrication technician
- Performance measurement
- Continuous improvement

*What are the elements of proper maintenance of automatic lubrication systems?*

Maintenance of automated lubrication systems varies greatly from one system to the next. However, there are some common factors that can impact system performance and reliability:

- Check whether compatible with the intended application.
- Ensure that the new lubricant is chemically compatible with the old lubricant.
- Be sure to follow correct handling and storage procedures.
- Do a routine check of all fittings and piping.
- Keep systems filled with lubricant by topping them off on a routine basis.

*How would you create or update a lube survey?*

Make sure that you have the right sample, right machine, right sampling frequency, right lubricant, right sampling location, proper training, health and safety measures in place, and right tag and labeling, and involve other continuous maintenance techniques.

## VIII. Preventive and Predictive Maintenance

*What are the elements involved in creating and managing lube preventive maintenance and routes?*

Design your oil analysis program by listing the questions you want answered.

**Lube Quality and Oil Type:**

- Synthetic or mineral
- EP, antiwear, or R&O
- Viscosity
- Health/condition
- Contamination
- Particles
- Water
- Glycol
- Heat
- Root cause
  - Defective part
  - Operator abuse
  - Wrong oil
  - Overloaded
  - Maintenance interval
  - Oil changed on time
  - Excessive load
  - Washing down equipment during preventive maintenance
  - Negligence

*What are the key elements to creating and managing a lubrication inspection checklist?*

Right sample, right machine, right sampling frequency, right lubricant, right sampling location, proper training, health and safety provisions, right tag and labeling, and involve other continuous maintenance techniques.

*Explain predictive maintenance strategies?*

The aim of predictive maintenance is first to predict when equipment failure might occur and second to prevent the occurrence of the failure by performing maintenance. When predictive maintenance is working effectively as a maintenance strategy, maintenance is only performed on machines when it is required.

*Explain proactive maintenance strategies?*

Proactive maintenance is a preventive maintenance strategy that works to correct the root causes of failure and avoid breakdowns caused by underlying equipment conditions. Resolve problems before they become failures. Extend the life of machinery and reduce downtime as a result of machine failure.

## IX. Lubricant Condition Control

*What are at least five proper sampling procedures?*

- Sample when machine is hot.
- Sample while operating or after shutdown within a couple of hours max.
- Sample from pressurized stream or midpoint of reservoir, not from the bottom.
- Sample in a region of turbulence.
- Sample point must be free of dirt.
- Sample directly to a clean, dry sample bottle.

- Use a new plastic tube for each sample.
- Flush the sampling tube.
- Flush the oil bottle.

*Where would you find the proper sampling locations?*

After the oil pump and before the filter or after the return line and before the filter.

*What is the filter rating/beta ratio?*

A method of comparing filter performance based on efficiency. This is done using the multipass test, which counts the number of particles of a given size before and after fluid passes through a filter.

$$\beta = \text{number of particles upstream} >$$
$$\text{Xum/number of particles downstream}$$
$$(\text{X size in microns per number of particles}) >$$
$$\text{Xum}$$

$$\eta = (\beta - 1)/\beta \times 100 \text{ filter efficiency}$$

The ratio of the number of particles greater than a given size in the influent fluid to the number of particles greater than the same size in the effluent fluid under specified test conditions.

*What steps would you used for sump/tank management to reduce failures?*

1. Air entrainment/foam: antifoam agent and good breather
2. Particles: filters
3. Water: water separator
4. Sediments: filter and good recirculation
5. Heat/cooling system
6. Silt/sediments: filter and good recirculation
7. Unnecessary lubricant volume: reservoir

## X. Lube Storage and Management

*What are the design elements for an optimal lube storage room?*

- Clean and temperature-controlled environment
- Color-coded lubes
- Spill containment
- Breathers
- Dedicated transfer pumps
- Safety gear
- Storage for grease guns and other equipment
- Filtration equipment
- Material Safety Data Sheets
- Self-closing cabinet, if applicable

*How do you define the maximum storage time according to environmental conditions/lubricant type?*

- 6 months outdoors, 1 year indoors
- Oil turnover at 12 months
- Grease turnover at 2–3 years

*What are safety and health requirements for lube storage?*

- Do not dispose of on the ground or in sewers.
- Do not contaminate with volatile materials.
- Collect for reprocessing.
- Avoid contact with skin.
- Wash hands before handling food.
- Understand that food grade does not mean edible.

*Explain each of the following lubricant failure mechanisms:*

### Oxidative Degradation

The oxidation process is an undesirable series of chemical reactions involving oxygen that degrade the quality of oil. Oxidation eventually produces rancidity in oil with accompanying off flavors and smells. All oil is in a state of oxidation—you cannot stop it completely—but there are ways to reduce it. Oxidation is the reaction of oil molecules with oxygen molecules. It can lead to an increase in viscosity and the formation of varnish, sludge, and sediment. In addition, rust and corrosion can form on the equipment as a result of oxidation.

Causes of oxidation include:

- High temperatures
- Water contamination
- Oil aeration
- Wear metal

Oxidative degradation increases wear and corrosion rate and causes sludge, increased acids, lower viscosity, filter plaque, and oil flow problems.

### Thermal Degradation

The thermal failure process is caused by:

- Very high temperatures
- Electrostatic spark discharges
- High temperatures associated with compression of entrained air

The effects of thermal degradation include:

- Oil oxidation
- Sludge formation
- Deterioration of water separation
- Air entrainment

*What would you look for when testing for wrong or mixed lubricants?*

- Viscosity change
- Additive elements
- Color change
- Different properties from the reference oil
- Increased viscosity and decreased limits
- Different properties from the oil on site

*Name the techniques for controlling particle contamination.*

- Filtration
- Additives
- High-quality breathers
- Oil change

*Explain how moisture contamination can affect machine rust, wear, erosion, and corrosion.*

It can affect the lubricant-fuel dilution, air entrainment, oxidation rate, and emulsification.

States of coexistence include suspended, dissolved, and free. Methods and units for measuring moisture contamination: Infrared Spectroscopy (IR), water by Karl Fischer, crackle test.

*What techniques are used for controlling moisture contamination?*

Breathers, corrosion and rust inhibitors, maintenance.

*What happens when you have glycol coolant contamination?*

- Effects on the machine: sludge formation, wear increase.
- Effects on the lubricant: viscosity increases, increased insoluble material.

*Name techniques for controlling glycol contamination.*

- Maintenance
- Corrosion inhibitors
- Drain oil

*What are the effects of soot contamination on machines and lubricants?*

Effects on the machine:

- Sludge
- Wear
- Filter blockage

Effects on the lubricant:

- Varnish
- Increased oxidation
- Increased viscosity
- Insoluble materials

*Name techniques for controlling soot contamination.*

- Lubricant dispersancy
- Filters
- Maintenance of injectors

*What are the effects of fuel contamination (fuel dilution in oil)?*

- Effects on the machine: excessive wear
- Effects on the lubricant: lower viscosity and higher flash and fire points

*What are at least three techniques for controlling fuel contamination?*

Maintenance of injectors, fuel filter changes, increase the lubricity of the fuel.

*What are the different types of air contaminations (air in oil) and their effects?*

Effects on the machine:

- Cavitation
- Vibration
- Low pressure
- Shutdown

Effects on the lubricant:

- Foam formation
- Increased oxidation

## Notes

1. W. Dreseland and R.-P. Heckler, Lubricating greases, Chap. 16 in Th. Mang and W. Dresel (eds.), *Lubricants and Lubrication*, 2nd ed. Weinheim, Germany: Wiley-VCH, 2007, p. 648.

2. W. Dreseland and R.-P. Heckler, Lubricating greases, Chap. 16 in Th. Mang and W. Dresel (eds.), *Lubricants and Lubrication*, 2nd ed. Weinheim, Germany: Wiley-VCH, 2007, p. 648.

3. D. M. Pirro and A. A. Wessol, *Lubrication Fundamentals*, 2nd ed., New York: Marcel Dekker, 2001, p. 44.

4. Pradeep L. Menezes, Carlton J. Reeves, and Michael R. Lovell, *Tribology for Scientists and Engineers: From Basics to Advanced Concepts*, New York: Springer, 2013.

5. *Friction, Lubrication and Wear Technology*, ASTM *Handbook*, Vol. 18: p. 12.

6. *Friction, Lubrication and Wear Technology*, ASTM *Handbook*, Vol. 18: p. 12.

7. Bharat Bhushan, *Introduction to Tribology*, 2nd ed. Hoboken, NJ: John Wiley & Sons, 2013, pp. 316, 329, 342, 349, 357, 359, 361, 363.

8. Bharat Bhushan, *Introduction to Tribology*, 2nd ed. Hoboken, NJ: John Wiley & Sons, 2013, p. 403.

9. Th. Mang and W. Dresel (eds.), *Lubricants and Lubrication*, 2nd ed. Weinheim, Germany: Wiley-VCH, 2007, p. 56.

10. Bharat Bhushan, *Introduction to Tribology*, 2nd ed. Hoboken, NJ: John Wiley & Sons, 2013, pp. 401–402.

11. Bharat Bhushan, *Introduction to Tribology*, 2nd ed. Hoboken, NJ: John Wiley & Sons, 2013, pp. 402–403.

12. D. M. Pirro and A. A. Wessol, *Lubrication Fundamentals*, 2nd ed. New York: Marcel Dekker, 2001.

13. John Rosenbaum, "Base Oil Groups I–V: Manufacturing, Properties and Performance," webinar originally presented at STLE University, October 8, 2014; Stuart F. Brown, "Base Oil Groups: Manufacture, Properties and Performance," *Tribology & Lubrication Technology*, April 2015, available at https://www.stle.org/images/pdf/STLE_ORG/BOK/OM_OA/Base%20Oils/Base%20Oil%20Groups_Manufacture_Prop_Perform_April15%20TLT.pdf.

14. R. J. Prince, Base oils from petroleum, Chap. 1 in Roy M. Mortier, Malcom F. Fox, and Stefan T. Orszuilk (eds.), *Chemistry and the Technology of Lubricants*, 3rd ed. The Netherlands: Springer, Dordrecht, 2010, p. 27.

15. R. J. Prince, Base oils from petroleum, Chap. 1 in Roy M. Mortier, Malcom F. Fox, and Stefan T. Orszuilk (eds.), *Chemistry and the Technology of Lubricants*, 3rd ed. The Netherlands: Springer, Dordrecht, 2010, p. 28.

16. D. M. Pirro and A. A. Wessol, *Lubrication Fundamentals*, 2nd ed. New York: Marcel Dekker, 2001, p. 17.

17. D. M. Pirro and A. A. Wessol, *Lubrication Fundamentals*, 2nd ed. New York: Marcel Dekker, 2001, p. 11.

18. D. M. Pirro and A. A. Wessol, *Lubrication Fundamentals*, 2nd ed. New York: Marcel Dekker, 2001, p. 11.

19. R. J. Prince, Base oils from petroleum, Chap. 1 in Roy M. Mortier, Malcom F. Fox, and Stefan T. Orszuilk (eds.), *Chemistry and the Technology of Lubricants*, 3rd ed. The Netherlands: Springer, Dordrecht, 2010, p. 7.

20. J. Crawford, A. Psaila, and S. T. Orszulik, Miscellaneous additives and vegetable oils, Chap. 6 in Roy M. Mortier, Malcom F. Fox, and Stefan T. Orszuilk (eds.), *Chemistry and the Technology of Lubricants*, 3rd ed. The Netherlands: Springer, Dordrecht, 2010, pp. 209–211.

21. D. M. Pirro and A. A. Wessol, *Lubrication Fundamentals*, 2nd ed. New York: Marcel Dekker, 2001, pp. 100, 268.

22. D. M. Pirro and A. A. Wessol, *Lubrication Fundamentals*, 2nd ed. New York: Marcel Dekker, 2001, p. 104.

23. D. M. Pirro and A. A. Wessol, *Lubrication Fundamentals*, 2nd ed. New York: Marcel Dekker, 2001, p. 105.

24. D. M. Pirro and A. A. Wessol, *Lubrication Fundamentals*, 2nd ed. New York: Marcel Dekker, 2001, p. 107.

25. D. M. Pirro and A. A. Wessol, *Lubrication Fundamentals*, 2nd ed. New York: Marcel Dekker, 2001, p. 106.

26. D. M. Pirro and A. A. Wessol, *Lubrication Fundamentals*, 2nd ed. New York: Marcel Dekker, 2001, p. 109.

27. "Lubricant Additives: A Practical Guide," *Machinery Lubrication*, n.d., available at https://www.machinerylubrication.com/Read/31107/oil-lubricant-additives.

28. "Lubricant Additives: A Practical Guide," *Machinery Lubrication*, n.d., available at https://www.machinerylubrication.com/Read/31107/oil-lubricant-additives.

29. D. M. Pirro and A. A. Wessol, *Lubrication Fundamentals*, 2nd ed. New York: Marcel Dekker, 2001, p. 81.

30. D. M. Pirro and A. A. Wessol, *Lubrication Fundamentals*, 2nd ed. New York: Marcel Dekker, 2001, p. 81.

31. https://www.machinerylubrication.com/Read/29223/selecting-grease-consistency.

32. Anoop Kumar and Nancy McGuire, "Selecting Lubricating Greases: What You Should Know," *Plant Engineering*, October 4, 2019.

33. D. M. Pirro and A. A. Wessol, *Lubrication Fundamentals*, 2nd ed. New York: Marcel Dekker, 2001, p. 98.

34. D. M. Pirro and A. A. Wessol, *Lubrication Fundamentals*, 2nd ed. New York: Marcel Dekker, 2001, p. 111.

35. D. M. Pirro and A. A. Wessol, *Lubrication Fundamentals*, 2nd ed. New York: Marcel Dekker, 2001, p. 156.

36. Th. Mang and W. Dresel, *Lubricants and Lubrication*, 2nd ed. Weinheim, Germany: Wiley-VCH, 2007, p. 367.

37. Th. Mang and W. Dresel, *Lubricants and Lubrication*, 2nd ed. Weinheim, Germany: Wiley-VCH, 2007, p. 338.

38. D. M. Pirro and A. A. Wessol, *Lubrication Fundamentals*, 2nd ed. New York: Marcel Dekker, 2001, p. 186.

39. D. M. Pirro and A. A. Wessol, *Lubrication Fundamentals*, 2nd ed. New York: Marcel Dekker, 2001, p. 259.

40. D. M. Pirro and A. A. Wessol, *Lubrication Fundamentals*, 2nd ed. New York: Marcel Dekker, 2001, pp. 75, 221.

41. Dennis Lauer, "Tribology: The Key to Proper Lubricant Selection," *Machinery Lubrication*, March 2008.

# 3 | Statements of Truth and Exam Development

In this section, you will be taught how to think like an examiner. This section has been an influencing reason why so many applicants have obtained their certifications the first time. It is a fantastic way to truly understand the content as well as the logic behind the questions asked. Take your time with this section.

The content is from the recommended MLT® Body of Knowledge with space to develop your multiple-choice questions. This section is vital to complete because it helps you develop your critical thinking process and understand how the exam questions are structured. You are about to learn how to think like an examiner!

The following are the rules that examiners use to develop questions for exams. You will be given a "Statement of Truth" and asked to develop a question from that content. There are specific rules to follow, but by understanding how examiners think, you will have better insight into how to take the exam.

You will be asked to develop multiple-choice questions consisting of four possible answers. The questions will consist of:

- A question, based on a specific, approved, and documented source statement (the Statement of Truth), and the answer selection should consist of a correct answer and three incorrect answers. The incorrect answers are often considered to be *distractors*.
- A plausible distractor that is incorrect in only *one concept* or technical aspect.
- A plausible distractor that is incorrect in *more than one concept* or technical aspect.
- A technically *implausible* distractor.

When developing a question, it may help to imagine that you are taking the exam and having to answer the question. Ask yourself, "Is there any possibility of two or more of these answers being correct?"

In writing exam questions, do *not* use:

- The "all of the above" option
- The "none of the above" option
- Only X and Y are correct
- Double negatives
- Confusing or ambiguous words or phrases
- Deceptive or tricky wording—keep it simple and straightforward
- Humor

# I. Maintenance Strategies (MLT® I & II)

Industrial and process plants typically use two types of maintenance management: (1) run to failure or (2) preventive maintenance.

## Run-to-Failure Management

The logic of run-to-failure management is simple and straightforward. When a machine breaks, fix it. This "if it ain't broke, don't fix it" method of maintaining plant machinery has been a major part of plant maintenance operations since the first manufacturing plant was built, and on the surface, it sounds reasonable. A plant using run-to-failure management does not spend any money on maintenance until a machine or system fails to operate. Run to failure is a reactive management technique that waits for machine or equipment failure before any maintenance action is taken. It is in truth a no-maintenance approach of management. It is also the most expensive method of maintenance management.

*Question:*

_____

_____

_____

A. _____

   _____

B. _____

   _____

C. _____

   _____

D. _____

   _____

*Correct Answer:* _____

The major expenses associated with this type of maintenance management are:

- High spare parts inventory cost
- High overtime labor costs
- High machine downtimes
- Low production availability

Because there is no attempt to anticipate maintenance requirements, a plant that uses true run-to-failure management must be able to react to all possible failures within the plant. This reactive method of management forces the maintenance department to maintain extensive spare parts inventories that include spare machines or at least all major components for all critical equipment in the plant.

*Question:*

_____

_____

_____

A. _____

_____

B. _____

_____

C. _____

_____

D. _____

_____

*Correct Answer:* _____

# II. Lubricants and Lubricant Formulation (MLT® I & II)

## Crude Distillation

Crude oil is sometimes used in its unprocessed form as fuel in power plants and in some internal combustion engines, but in most cases it is separated or converted into different fractions, which, in turn, require further processing to supply the large number of petroleum products needed. In many cases, the first step is to remove from the crude certain inorganic salts suspended as minute crystals or dissolved in entrained water. These salts break down during processing to form acids that severely corrode refinery equipment, plug heat exchangers and other equipment, and poison catalysts used in subsequent processes. Therefore, the crude oil is mixed with additional water to dissolve the salts, and the resulting brine is removed by settling.

*Question:*

_____

_____

_____

A. _____

_____

B. _____

_____

C. _____

_____

D. _____

_____

*Correct Answer:* _____

Lube base stocks make up a significant portion of finished lubricants, ranging from 70% of automotive engine oils to 99% of some industrial oils. The base stocks contribute significant performance characteristics to finished lubricants in areas such as thermal stability, viscosity, volatility, ability to dissolve additives and contaminants (oil degradation materials, combustion by-products, etc.), low-temperature properties, demulsibility, air release/foam resistance, and oxidation stability. This list indicates the importance of base-stock processing and selection, along with the use of proper additives and blending procedures, in achieving balanced performance in the finished lubricant.

*Question:*

_____

_____

_____

_____

A.  _____

_____

B.  _____

_____

C.  _____

_____

D.  _____

_____

*Correct Answer:* _____

The American Petroleum Institute (API) has defined five categories of lubricant base stocks to try to separate conventional, unconventional, synthetic, and other classifications of base stocks. Of these five categories, Groups I–III are mineral oils and are classified by the amounts of saturates and sulfur and by the viscosity index (VI) of each. Group IV is reserved for polyalphaolefins, and Group V is for esters and other base stocks not included in Groups I–IV. The API classification system is based on base-stock characteristics, as just mentioned, not on the refining process used. Group III base stocks are very high VI products that are typically achieved through a hydrocracking process.

*Question:*

_____

_____

_____

A.  _____

_____

B.  _____

_____

C.  _____

_____

D.  _____

_____

*Correct Answer:* _____

If a given base stock falls under a Group I classification, it does not necessarily mean that it is better or worse than a base stock that falls under a Group II classification. Although the Group II base stock would have lower levels of sulfur and aromatics, increased potential for improved oxidation stability, and a higher VI, it may provide poorer solubility of additives and contaminants than a conventionally refined base stock that falls under Group I. The real measurement of base-stock suitability for formulating finished lubricants is in the performance of the finished lubricants.

*Question:*

_____

_____

_____

A. _____

_____

B. _____

_____

C. _____

_____

D. _____

_____

*Correct Answer:* _____

The most common processes used to produce lube base stocks in refineries worldwide involve separation, that is, processes that operate by dividing feedstock, which is a complex mixture of chemical compounds, into products. Usually this results in two sets of products: the desired lube product and by-products. Thus, although the products themselves are complex mixtures, the compounds in each of the products are similar in either physical or chemical properties. In contrast, the fastest-growing method for lube manufacture is by the alternate conversion process, which involves converting undesirable structures to desirable lube molecules under the influence of hydrogen pressure and selected catalysts.

*Question:*

_____

_____

_____

A. _____

_____

B. _____

_____

C. _____

_____

D. _____

_____

*Correct Answer:* _____

Five processes are involved in lube separation (extraction) in conventional lube oil refining:

1. Vacuum distillation
2. Propane deasphalting
3. Furfural extraction (solvent extraction)
4. Methyl ethyl ketone (MEK) dewaxing/hydrodewaxing
5. Hydrofinishing

*Question:*

_____

_____

_____

A.  _____

_____

B.  _____

_____

C.  _____

_____

D.  _____

_____

*Correct Answer:* _____

## Vacuum Distillation

Assuming that an acceptable crude oil has been processed properly in the atmospheric distillation column for recovery of light products, the residuum from the atmospheric distillation column is the feedstock for the vacuum distillation column. Vacuum distillation is the first step in refining lubricating base stocks. This is a separation process that segregates crude oil into products that are similar in boiling-point range.

*Question:*

_____

_____

_____

A.  _____

_____

B.  _____

_____

C.  _____

_____

D.  _____

_____

*Correct Answer:* _____

## Propane Deasphalting

Propane deasphalting (PD) operates on the very bottom of the barrel—the residuum. This is the product of the highest boiling portion of the crude. The residuum has now been divided into two products, one containing almost all the resins and asphaltenes, called *PD tar*, and the other containing compounds that are similar chemically to those in the lube distillates but of a higher boiling point. This material is called *deasphalted oil* and is refined in the same way as lube distillates.

*Question:*

_____

_____

_____

A.  _____

   _____

B.  _____

   _____

C.  _____

   _____

D.  _____

   _____

*Correct Answer:* _____

## MEK Dewaxing

MEK dewaxing of the raffinate from the furfural extraction is another horizontal cut that produces a by-product wax that is almost completely paraffins and a dewaxed oil that contains paraffins, naphthenes, and some aromatics. This dewaxed oil is the base stock for many fluid lubricants. For certain premium applications, however, a finishing step is needed.

*Question:*

_____

_____

_____

A.  _____

   _____

B.  _____

   _____

C.  _____

   _____

D.  _____

   _____

*Correct Answer:* _____

## Hydrofinishing

Hydrofinishing by chemical reaction of the oil with hydrogen changes the polar compounds slightly but retains them in the oil. In this process, the most obvious result is oil of a lighter color. From the simplified description of crude oil, the product is indistinguishable from the dewaxed oil feedstock. Hydrofinishing has superseded the older clay processing of conventional stocks because of simpler, lower-cost operation.

*Question:*

_____

_____

_____

A.  _____

    _____

B.  _____

    _____

C.  _____

    _____

D.  _____

    _____

*Correct Answer:* _____

## Lube Conversion Process

Hydroprocessing offers a number of methods to further aid in the removal of aromatics, sulfur, and nitrogen from lube base stocks. Two of the hydroprocessing methods use the feedstock from the solvent refining process (hydrofinishing and hydrotreating), and a third, hydrocracking, uses the vacuum gas oil (VGO) from the crude distillation unit.

*Question:*

_____

_____

_____

A.  _____

    _____

B.  _____

    _____

C.  _____

    _____

D.  _____

    _____

*Correct Answer:* _____

The lube base stocks produced by hydrofinishing and hydrotreating would be classified as API Group I and II base stocks; the base stocks produced using the hydrocracking process generally would be classified in Groups II and III. As mentioned earlier in this section, classification into an API group is based on the base-stock specifications, not the process for refining.

There are four primary processes:

- Vacuum distillation
- Hydrocracking
- Hydrodewaxing
- Hydrotreating

*Question:*

_____

_____

_____

A. _____

_____

B. _____

_____

C. _____

_____

D. _____

_____

*Correct Answer:* _____

## Vacuum Distillation

While the vacuum distillation process is the same one used in extractive processing, in this case, the propane deasphalter is omitted because the vacuum tower residuum is normally not hydroprocessed; rather, it is processed conventionally by solvent treatment.

*Question:*

_____

_____

_____

A. _____

_____

B. _____

_____

C. _____

_____

D. _____

_____

*Correct Answer:* _____

## Hydrocracking

The hydrocracking unit is a catalytic processing unit that converts less desirable hydrocarbon species to more desirable species in the presence of hydrogen at pressures up to 3000 psi. Typically, aromatic and naphthene rings are opened to produce a higher portion of saturated paraffinic molecules. The hydrogen also removes heteroatoms as gases—hydrogen sulfide, ammonia, and water.

*Question:*

_____

_____

_____

A. _____

_____

B. _____

_____

C. _____

_____

D. _____

_____

*Correct Answer:* _____

## Hydrodewaxing

The hydrodewaxing unit, like the hydrocracker, is also a catalytic hydrogenation unit, but in this case the catalyst employed is specific to converting waxy normal paraffins to more desirable iso-paraffin structures.

*Question:*

_____

_____

_____

A. _____

_____

B. _____

_____

C. _____

_____

D. _____

_____

*Correct Answer:* _____

## Hydrotreating

Because the hydrocracking and hydrodewaxing processes involve breaking of carbon–carbon bonds, it is necessary to have a final hydrotreating stage to induce saturation of any remaining unsaturated molecules, which could cause thermal or oxidative instability in the base oil and finished products blended from it.

*Question:*

_____

_____

_____


A. _____

_____

B. _____

_____

C. _____

_____

D. _____

_____

*Correct Answer:* _____

## Lubricating Oils

Additives are chemical compounds added to lubricating oils to impart specific properties to the finished oils. Some additives impart new and useful properties to the lubricant, some enhance properties already present, and some act to reduce the rate at which undesirable changes take place in the product during its service life.

*Question:*

_____

_____

_____


A. _____

_____

B. _____

_____

C. _____

_____

D. _____

_____

*Correct Answer:* _____

## Pour-Point Depressants

Certain high-molecular-weight polymers function by inhibiting the formation of a wax crystal structure that would prevent oil flow at low temperatures.

*Question:*

_____

_____

_____

A. _____

_____

B. _____

_____

C. _____

_____

D. _____

_____

*Correct Answer:* _____

## VI Improvers

VI improvers are long-chain, high-molecular-weight polymers that function by causing the relative viscosity of an oil to increase more at high temperatures than at low temperatures.

*Question:*

_____

_____

_____

A. _____

_____

B. _____

_____

C. _____

_____

D. _____

_____

*Correct Answer:* _____

## Defoamants (Antifoaming Agents)

Silicone polymers used at a few parts per million are the most widely used defoamants. These materials are essentially insoluble in oil, and the correct choice of polymer size and blending procedures is critical if settling during long-term storage is to be avoided. Also, these additives may increase air entrainment in the oil. Organic polymers are sometimes used to overcome these difficulties with the silicones, although much higher concentrations are generally required.

*Question:*

_____

_____

_____

A. _____

_____

B. _____

_____

C. _____

_____

D. _____

_____

*Correct Answer:* _____

## Oxidation Inhibitors

When oil is heated in the presence of air, oxidation occurs. As a result of this oxidation, both the oil's viscosity and the concentration of organic acids in the oil increase, and varnish and lacquer deposits may form on hot metal surfaces exposed to the oil. In extreme cases, these deposits may be further oxidized to form hard, carbonaceous materials.

*Question:*

_____

_____

_____

A. _____

_____

B. _____

_____

C. _____

_____

D. _____

_____

*Correct Answer:* _____

## Rust and Corrosion Inhibitors

A number of kinds of corrosion can occur in systems served by lubricating oils. Probably the two most important types are corrosion by organic acids that develop in the oil itself and corrosion by contaminants that are picked up and carried by the oil.

*Question:*

_____

_____

_____

A.  _____

   _____

B.  _____

   _____

C.  _____

   _____

D.  _____

   _____

*Correct Answer:* _____

## Detergents and Dispersants

In internal combustion engine service, a number of effects tend to cause oil deterioration and the formation of harmful deposits. These deposits can interfere with oil circulation, build up behind piston rings to cause ring sticking and rapid ring wear, and affect clearances and proper functioning of critical components such as hydraulic valve lifters. Once formed, such deposits are generally hard to remove except by mechanical cleaning. The use of detergents and dispersants in the oil can delay the formation of deposits and reduce the rate at which they accumulate on metal surfaces.

*Question:*

_____

_____

_____

A.  _____

   _____

B.  _____

   _____

C.  _____

   _____

D.  _____

   _____

*Correct Answer:* _____

## Antiwear Additives

Antiwear additives are used in many lubricating oils to reduce friction, wear, and scuffing and scoring under boundary lubrication conditions, that is, when full lubricating films cannot be maintained. As the oil film becomes progressively thinner as a result of increasing loads or temperatures, contact through the oil film is first made by minute surface irregularities or asperities.

*Question:*

_____

_____

_____

A.   _____

    _____

B.   _____

    _____

C.   _____

    _____

D.   _____

    _____

*Correct Answer:* _____

Mild antiwear and friction-reducing additives, sometimes called *boundary lubrication additives*, are polar materials such as fatty oils, acids, and esters. They are long-chain materials that form an adsorbed film on metal surfaces with the polar ends of the molecules attached to the metal and the molecules projecting more or less normally to the surface.

*Question:*

_____

_____

_____

A.   _____

    _____

B.   _____

    _____

C.   _____

    _____

D.   _____

    _____

*Correct Answer:* _____

Wear is reduced under mild sliding conditions, but under severe sliding conditions, the layers of molecules can be rubbed off, with the result that their wear-reducing effect is lost.

*Question:*

_____

_____

_____

A.  _____

_____

B.  _____

_____

C.  _____

_____

D.  _____

_____

*Correct Answer:* _____

## Extreme-Pressure (EP) Additives

At high temperatures or under heavy loads where more severe sliding conditions exist, compounds called *EP additives* are required to reduce friction, control wear, and prevent severe surface damage. These materials function by chemically reacting with the sliding metal surfaces to form relatively oil-insoluble surface films. The kinetics of the reaction are a function of the surface temperatures generated by the localized high temperatures that result from rubbing between opposing surface asperities and the breaking of junctions between these asperities.

*Question:*

_____

_____

_____

A.  _____

_____

B.  _____

_____

C.  _____

_____

D.  _____

_____

*Correct Answer:* _____

EP agents are usually compounds containing sulfur, chlorine, or phosphorus either alone or in combination. The compounds used depend on the end use of the lubricant and the chemical activity required in it. Sulfur compounds, sometimes with chlorine or phosphorus compounds, are used in many metal-cutting fluids. Sulfur–phosphorus combinations are used in most industrial and automotive gear lubricants. These materials provide excellent protection against gear tooth scuffing and have the advantages of better oxidation stability, lower corrosivity, and often lower friction than other combinations that have been used in the past.

*Question:*

_____

_____

_____

A.  _____

_____

B.  _____

_____

C.  _____

_____

D.  _____

_____

*Correct Answer:* _____

## Synthetic Lubricants

The terms *synthetic* and *synthesized* are both used to describe the base fluids used in these lubricants. A *synthesized* material is one that is produced by combining or building individual units into a unified entity. The production of synthetic lubricants starts with synthetic base stocks that are often manufactured from petroleum. The base fluids are made by chemically combining (synthesizing) low-molecular-weight compounds that have adequate viscosity for use as lubricants. Unlike mineral oils, which are a complex mixture of naturally occurring hydrocarbons, synthetic base fluids are human made and tailored to have a controlled molecular structure with predictable properties.

*Question:*

_____

_____

_____

A.  _____

_____

B.  _____

_____

C.  _____

_____

D.  _____

_____

*Correct Answer:* _____

The primary performance advantage of synthetic lubricants is the extended-service-life capability and the ability to handle a wider range of application temperatures. Their outstanding flow characteristics at low temperatures and their stability at high temperatures mark the preferred use of these lubricants.

*Question:*

_____

_____

_____

    A.  _____

        _____

    B.  _____

        _____

    C.  _____

        _____

    D.  _____

        _____

    *Correct Answer:* _____

# III. Lubricants and IV. Grease Application and Performance (MLT® I & II)

## Lubricating Greases

The reasons for the use of greases in preference to fluid lubricants are well stated by the Society of Automotive Engineers in SAE Information Report J310, Automotive Lubricating Grease. This report states:

> Greases are most often used instead of fluids where a lubricant is required to maintain its original position in a mechanism, especially where opportunities for frequent relubrication may be limited or economically unjustifiable. This requirement may be due to the physical configuration of the mechanism, the type of motion, the type of sealing, or the need for the lubricant to perform all or part of any sealing function in the prevention of lubricant loss or the entrance of contaminants. Because of their essentially solid nature, greases do not perform the cooling and cleaning functions associated with the use of a fluid lubricant. With these exceptions, greases are expected to accomplish all other functions of fluid lubricants.

*Question:*

_____

_____

_____


A. _____

    _____

B. _____

    _____

C. _____

    _____

D. _____

    _____


*Correct Answer:* _____


A satisfactory grease for a given application is expected to:

1. Provide adequate lubrication to reduce friction and prevent harmful wear of components
2. Protect against rust and corrosion
3. Act as a seal to prevent entry of dirt and water
4. Resist leakage, dripping, or undesirable throw-off from the lubricated surfaces
5. Retain apparent viscosity or relationship between viscosity, shear, and temperature over useful life of the grease in a mechanical component that subjects the grease to shear forces
6. Not stiffen excessively to cause undue resistance to motion in cold environments

7. Have suitable physical characteristics for the method of application
8. Be compatible with elastomer seals and other materials of construction in the lubricated portion of the mechanism
9. Tolerate some degree of contamination, such as moisture, without loss of significant characteristics

*Question:*

_____

_____

_____


A. _____

    _____

B. _____

    _____

C. _____

    _____

D. _____

    _____


*Correct Answer:* _____

## Composition of Grease

In the definition of a lubricating grease given here, the liquid portion of the grease may be a mineral or synthetic oil or any fluid that has lubricating properties. The thickener may be any material that in combination with the selected fluid will produce the solid to semifluid structure. The other ingredients are additives or modifiers that are used to impart special properties or modify existing ones.

*Question:*

_____

_____

_____

A.  _____

_____

B.  _____

_____

C.  _____

_____

D.  _____

_____

*Correct Answer:* _____

Most of the greases produced today have mineral oils as their fluid components. These oils may range in viscosity from as light as mineral seal oil up to the heaviest cylinder stocks. In the case of some specialty greases, products such as waxes, petrolatum, and asphalts may be used. Although perhaps these latter materials are not precisely describable as *liquid lubricants*, they perform the same function as the fluid components in conventional greases.

*Question:*

_____

_____

_____

A.  _____

_____

B.  _____

_____

C.  _____

_____

D.  _____

_____

*Correct Answer:* _____

Greases made with mineral oils generally provide satisfactory performance in most automotive and industrial applications. In very low- or high-temperature applications or in applications where temperature may vary over a wide range, greases made with synthetic fluids generally are now used.

*Question:*

_____

_____

_____

A. _____

     _____

B. _____

     _____

C. _____

     _____

D. _____

     _____

*Correct Answer:* _____

## Grease Thickeners

The principal thickeners used in greases are metallic soaps. The earliest greases were made with calcium soaps; then greases made with sodium soaps were introduced. Later, soaps such as aluminum, lithium, clay, and polyurea came into use. Some greases made with mixtures of soaps, such as sodium and calcium, are usually referred to as *mixed-base greases*. Soaps made with other metals have been used but have not received commercial acceptance either because of cost, health, and safety issues, or environmental concerns or performance problems.

*Question:*

_____

_____

_____

A. _____

     _____

B. _____

     _____

C. _____

     _____

D. _____

     _____

*Correct Answer:* _____

The earlier forms of greases were hydrated metallic soaps, which were made by combining steric acid with a soap. These low-cost greases provided good water resistance, fair low-temperature properties, and fair shear stability but limited temperature performance. Improvements to hydrated greases were necessary to provide higher-temperature capability. These improvements were made by use of 12-hydroxysteric acid with the metallic soaps to produce the next class of greases, anhydrous metallic soaps. This change increased dropping points above 290°F, but the products were also more costly to make than the earlier hydrated metallic soap greases.

*Question:*

_____

_____

_____

A. _____

_____

B. _____

_____

C. _____

_____

D. _____

_____

*Correct Answer:* _____

Modifications of metallic soap greases, called *complex greases*, are continuing to gain popularity. These complex greases are made by using a combination of a conventional metallic soap-forming material with a complexing agent. The complexing agent may be either organic or inorganic and may or may not involve another metallic constituent. Among the most successful of the complex greases are the lithium complex greases. These are made with a combination of conventional lithium soap-forming materials and a low-molecular-weight organic acid as the complexing agent. Greases of this type are characterized by very high dropping points, usually above 500°F (250°C), and also may have excellent load-carrying properties. Other complex greases—aluminum and calcium—are also manufactured for certain applications.

*Question:*

_____

_____

_____

A. _____

_____

B. _____

_____

C. _____

_____

D. _____

_____

*Correct Answer:* _____

## Additives and Modifiers (Greases)

Additives and modifiers commonly used in lubricating greases are oxidation or rust inhibitors, pour-point depressants, EP additives, antiwear agents, lubricity or friction-reducing agents, and dyes or pigments. Most of these materials have much the same function as similar materials added to lubricating oils.

*Question:*

_____

_____

_____

A. _____

_____

B. _____

_____

C. _____

_____

D. _____

_____

*Correct Answer:* _____

Boundary-film lubricants such as molybdenum disulfide or graphite may be added to greases to enhance specific performance characteristics such as load-carrying ability. An EP agent reacts with the lubricated surface to form a chemical film. Molybdenum disulfide is used in many greases for applications in which loads are heavy, surface speeds are low, and restricted or oscillating motion is involved. In these applications, the use of molysulfide (or *moly* as it is sometimes called) reduces friction and wear without adverse chemical reactions with the metal surfaces. Polyethylene and modified tetrafluoroethane (Teflon) also may be used for applications of this type.

*Question:*

_____

_____

_____

A. _____

_____

B. _____

_____

C. _____

_____

D. _____

_____

*Correct Answer:* _____

The manufacture of a grease, whether by a batch or continuous process, involves the dispersion of the thickener in the fluid and the incorporation of additives or modifiers. This is accomplished in a number of ways. In some cases, the thickener is purchased by the grease manufacturer in a finished state and then mixed with oil until the desired grease structure is obtained. In most cases with metallic soap thickeners, the thickener is produced, through reaction, during the manufacture of the grease from the following five steps:

1. Saponification
2. Dehydration
3. Cutback
4. Milling
5. Deaeration

Question:

_____

_____

_____

A. _____

_____

B. _____

_____

C. _____

_____

D. _____

_____

Correct Answer: _____

The general description of a grease is in terms of the materials used in its formulation and physical properties, some of which are visual observations. The type and amount of thickener and the viscosity of the fluid lubricant are formulation properties. Color and texture, or structure, are observed visually. There is some correlation between these descriptive items and performance. For example:

- Certain types of thickeners usually impart specific properties to a finished grease.
- The viscosity of the fluid lubricant is very important in selecting greases for some applications.
- Light-colored or white greases may be desirable in certain applications (e.g., in the textile and paper industries, where staining is a consideration).

Question:

_____

_____

_____

A. _____

_____

B. _____

_____

C. _____

_____

D. _____

_____

Correct Answer: _____

## Grease Consistency and Physical Properties

### Penetration Point (Greases)

*Consistency* is defined as the degree to which a plastic material resists deformation under the application of a force. In the case of lubricating greases, it is a measure of the relative hardness or softness and may indicate something of flow and dispensing properties. Consistency is reported in terms of ASTM D217, Cone Penetration of Lubricating Grease, or National Lubricating Grease Institute (NLGI) grade. Consistency is measured at a specific temperature, 77°F (25°C), and degree of shear (working).

*Question:*

_____

_____

_____

A. _____

_____

B. _____

_____

C. _____

_____

D. _____

_____

*Correct Answer:* _____

*NLGI Grease Classification*

| NLGI Grade | ASTM Worked Penetration |
|---|---|
| 000 | 445–475 |
| 00 | 400–430 |
| 0 | 355–385 |
| 1 | 310–340 |
| 2 | 265–295 |
| 3 | 220–250 |
| 4 | 175–205 |
| 5 | 130–160 |
| 6 | 85–115 |

Ranges are the penetration in tenths of a millimeter after 5 s at 77°F (25°C).

*Question:*

_____

_____

_____

A. _____

_____

B. _____

_____

C. _____

_____

D. _____

_____

*Correct Answer:* _____

## Dropping Point (Greases)

The *dropping point* of a grease is the temperature at which a drop of material falls from the orifice of a test cup under prescribed test conditions. Two procedures are used (ASTM D566 and ASTM D2265) that differ in the type of heating units and therefore the upper temperature limits. An oil bath is used for ASTM D566 with a measurable dropping-point limit of 500°F (260°C); ASTM D2265 uses an aluminum block oven with a dropping-point limit of 625°F (330°C). Greases thickened with organoclay soaps do not have a true melting point; instead, they have a melting range during which they become progressively softer. Some other types of greases may, without change in state, separate oil. In either case, only an arbitrary, controlled test procedure can provide a temperature that can be established as a characteristic of the grease.

Question:

_____

_____

_____

A. _____

_____

B. _____

_____

C. _____

_____

D. _____

_____

*Correct Answer:* _____

The dropping point of a grease is only loosely related to the upper operating temperature to which a grease can successfully provide adequate lubrication. Additional factors must be taken into account in high-temperature lubrication with grease. It is useful for characterization and also as a quality control during grease manufacture.

Question:

_____

_____

_____

A. _____

_____

B. _____

_____

C. _____

_____

D. _____

_____

*Correct Answer:* _____

## Grease Compatibility

Greases are available with many thickener types, additives, and base oils. As a result, mixing of different greases could result in altering the performance or physical properties (incompatibility), which could lead to a grease (mixture) that exhibits characteristics inferior to those of either grease before mixing. The mixing of incompatible greases will alter properties such as consistency, pumpability, shear stability, oil separation, and oxidation stability. Generally, when two incompatible greases are mixed, the result is a softening, which can lead to increased leakage as well as loss of other performance features.

*Question:*

_____

_____

_____

A. _____

_____

B. _____

_____

C. _____

_____

D. _____

_____

*Correct Answer:* _____

Equipment performance problems as a result of mixing incompatible greases could manifest after a relatively short period of operation but usually occur over longer time periods, sometimes making it difficult to trace the source of the problem back to the mixing of incompatible greases. When it becomes necessary to use different greases, it is best to mix greases of the same thickener type, but in all cases, compatibility charts and the grease suppliers should be consulted. However, the safest practice is to avoid mixing of greases.

*Question:*

_____

_____

_____

A. _____

_____

B. _____

_____

C. _____

_____

D. _____

_____

*Correct Answer:* _____

# IV. & V. Lubricant Selection (MLT® I & II)

## Viscosity (Hydraulics)

The single most important physical characteristic of a hydraulic fluid is its viscosity. Viscosity is a measure of the oil's resistance to flow, so in hydraulic systems that depend on flow, viscosity is important with respect to both lubrication and energy transmission. Although viscosity requirements are to some extent dictated by the components (i.e., pumps, valves, motors, etc.) and by system manufacturers, certain effects of improper viscosity selection need to be recognized. Too low a viscosity can lead to excessive metal-to-metal contact of moving parts as well as to wear and leakage. Too high a viscosity can result in excessive heating, sluggish operation (particularly at startup), higher energy consumption, lower mechanical efficiencies, and increased pressure drops in transmission lines and across filters.

*Question:*

_____

_____

_____

A. _____

_____

B. _____

_____

C. _____

_____

D. _____

_____

*Correct Answer:* _____

## Hydraulic Oil Additives and Properties

*Antiwear/Wear Protection (Hydraulics)*

To ensure satisfactory hydraulic component life, the hydraulic fluid must minimize wear. Wear results in loss of mechanical efficiency as well as higher costs due to shorter component life. In some hydraulic systems, such as low-pressure, low-temperature systems with gear pumps, antiwear additives are not necessary.

*Question:*

_____

_____

_____

A. _____

_____

B. _____

_____

C. _____

_____

D. _____

_____

*Correct Answer:* _____

Other high-pressure, high-temperature systems using vane pumps do require antiwear additives in the hydraulic fluid.

*Question:*

_____

_____

_____

A. _____

_____

B. _____

_____

C. _____

_____

D. _____

_____

*Correct Answer:* _____

Piston pumps may or may not require antiwear additives depending on the metallurgy and design. A properly refined petroleum oil has naturally good wear protection (without the use of antiwear additives).

*Question:*

_____

_____

_____

A. _____

_____

B. _____

_____

C. _____

_____

D. _____

_____

*Correct Answer:* _____

Antiwear fluids generally are required in gear and vane pumps operating at pressures above 1000 psi and more than 1200 rpm. Piston pumps may or may not require antiwear additives depending on the specific manufacturer and the metallurgy used.

*Question:*

_____

_____

_____

A. _____

_____

B. _____

_____

C. _____

_____

D. _____

_____

*Correct Answer:* _____

*Oxidation Stability (Hydraulics)*

Oil is circulated over and over during long periods in hydraulic systems. It is heated by the churning and shearing action in pumps, valves, tubing, and actuators. Also, the energy released as the oil goes from high pressure to low pressure in a relief valve is converted to heat, which raises its temperature. Oil temperature can be further increased by convection or conduction heating while performing its work in applications such as the hot molds in plastic injection-molding operations and continuous caster hydraulics in steel mills.

*Question:*

_____

_____

_____

A. _____

_____

B. _____

_____

C. _____

_____

D. _____

_____

*Correct Answer:* _____

The oil is in contact with warm air in the reservoir. Air is also dissolved or entrained in the oil. Because of this contact with air, oxygen is intimately mixed with the oil. Under these conditions (exposure to temperature and oxygen), the oil tends to chemically combine with the oxygen, creating oxidation products. The tendency to oxidize is greatly increased as temperatures increase, as agitation or splashing becomes excessive, and by exposure to certain materials that catalyze the oxidation reactions. Catalysts such as iron, copper, rust, and other metallic materials are commonly present in hydraulic systems.

*Question:*

_____

_____

_____

A. _____

_____

B. _____

_____

C. _____

_____

D. _____

_____

*Correct Answer:* _____

## Antifoam/Air-Separation Characteristics (Hydraulics)

Air causes spongy or erratic motion, which will result in poor system performance, particularly during the production of close-tolerance parts. Antifoaming and air-separation characteristics are two different concepts, although somewhat connected. *Air separation* means that the entrained air is released from the oil, whereas *antifoam* means that the air bubbles getting to the surface of the oil are readily dissipated. Both aspects are important to the performance of a hydraulic oil. Contamination can alter both these characteristics, so it is not only important to select an oil that will provide good antifoam and air-separation performance, but it is also necessary to minimize contamination in order to maintain this good performance.

*Question:*

_____

_____

_____

A. _____

_____

B. _____

_____

C. _____

_____

D. _____

_____

*Correct Answer:* _____

## Demulsibility/Water-Separating Ability (Hydraulics)

Water contamination is sometimes a problem in hydraulic systems. It may be present as a result of water leaks in heat exchangers or wash-down procedures, but more commonly, it accumulates because of condensation of atmospheric moisture. Most condensation occurs above the oil level in reservoirs as machines cool during idle or shutdown periods. A clean hydraulic oil of suitable type will have little tendency to mix with water; and in a still reservoir, the water will tend to settle at a low point. During operation, the water may be picked up by oil circulation, broken up into droplets, and mixed with the oil, forming an emulsion. The water and oil in such an emulsion should separate quickly in the reservoir, but when solid contaminants or oil oxidation products are present, emulsions tend to persist and to join with other deposit-forming materials present to form sludge. The emulsion may be drawn into the pump and made more permanent by the churning action of the pump and the mixing effect of flow at high velocities through control devices.

Question:

_____

_____

_____

A. _____

   _____

B. _____

   _____

C. _____

   _____

D. _____

   _____

*Correct Answer:* _____

## Rust Protection (Hydraulics)

Water and oxygen can cause rusting of ferrous surfaces in hydraulic systems. In as much as air is always present (except in specialized nitrogen blanket systems), oxygen is available, and some water is often present. The possibility of rusting is greatest during shutdown, when surfaces that are normally covered with oil may be unprotected and subject to gathering condensation as they cool. This is particularly important for operations that experience high-humidity conditions and temperature changes within the reservoirs.

*Question:*

_____

_____

_____

A. _____

B. _____

C. _____

D. _____

*Correct Answer:* _____

Rusting results in surface destruction, and rust particles may be carried into the system, where they will contribute to wear and the formation of sludge-like deposits certain to interfere with the operation of pumps, actuators, and control mechanisms. Rusting of piston rods or rams causes rapid seal or packing wear, resulting in increased leakage and system contamination; moreover, external contaminants can enter the hydraulic fluid through worn seals or packing.

*Question:*

_____

_____

_____

A. _____

B. _____

C. _____

D. _____

*Correct Answer:* _____

## Hydraulic Maintenance

The degree of system maintenance is based on specific performance expectations, the fluid used, and the system operating parameters. The various hydraulic fluids, ranging from mineral oil–based to synthetic- to water-containing fire-resistant fluids, demand various levels of maintenance to ensure performance. Water-containing fluids require higher levels of maintenance to ensure not only that the fire-protection properties are retained but that the fluid will provide proper lubrication characteristics while in service. This topic was discussed earlier in this chapter, but selecting the proper fluid matched to system needs and understanding the limitations of that fluid comprise the basic starting point for an effective maintenance program.

*Question:*

_____

_____

_____

A.  _____

_____

B.  _____

_____

C.  _____

_____

D.  _____

_____

*Correct Answer:* _____

Once the proper fluid has been selected, the equipment and operating conditions will dictate the degree of maintenance required to keep that fluid in service for long periods of time while retaining its lubrication characteristics. These maintenance procedure objectives should include the following:

- Keeping fluid clean/controlling contamination
- Maintaining proper temperatures
- Maintaining proper oil levels
- Periodic oil analysis
- Routine inspections
- Noise levels
- Vibration pressures
- Shock loads
- Leakage
- Fluid odor and color
- Filtration
- Temperatures
- Foaming

*Question:*

_____

_____

_____

A.  _____

_____

B.  _____

_____

C.  _____

_____

D.  _____

_____

*Correct Answer:* _____

## Oil Selection: Rolling-Element Bearings

Oils for the lubrication of rolling-element bearings should have the following characteristics:

- Excellent resistance to oxidation at operating temperatures
- Proper viscosity at operating speeds and temperatures to protect against friction and wear
- Antirust properties to protect against rust when moisture is present
- Good antiwear properties where required because of heavy or shock loads
- Good demulsibility to allow separation of water in circulation systems

*Question:*

_____

_____

_____

A.  _____

_____

B.  _____

_____

C.  _____

_____

D.  _____

_____

*Correct Answer:* _____

## Grease Selection: Rolling-Element Bearings

Grease for the lubrication of rolling-element bearings should have the following characteristics:

- Excellent resistance to oxidation to resist the formation of deposits or hardening that might shorten bearing life
- Mechanical stability to resist excessive softening or hardening as a result of shearing in service
- Proper consistency for the method of application
- Controlled oil bleeding
- Enhanced antiwear properties
- Ability to protect surfaces against rusting
- Good compatibility with system and system components
- Depending on application, the ability to resist corrosion or deterioration where acids or caustic materials in small quantities can get into the grease

*Question:*

_____

_____

_____

A.  _____

_____

B.  _____

_____

C.  _____

_____

D.  _____

_____

*Correct Answer:* _____

## Gears: Factors Affecting Lubrication of Enclosed Gears

In selecting the lubricant for enclosed gear sets, in addition to the requirement for adequate oxidation resistance, the following factors of design and operation require consideration:

- Gear type
- Gear speed
- Reduction ratio
- Operating and startup temperatures
- Transmitted power
- Surface finish
- Load characteristics
- Drive type
- Application method
- Contamination (water, metalworking fluids, dirt, etc.)
- Lubricant leakage

*Question:*

_____

_____

_____

A. _____

_____

B. _____

_____

C. _____

_____

D. _____

_____

*Correct Answer:* _____

## Factors Affecting Lubrication of Enclosed Gears

The necessary characteristics of lubricants for enclosed gears may be summarized as follows:

- Correct viscosity at operating temperature to ensure distribution of oil to all rubbing surfaces and formation of protective oil films at prevailing speeds and pressures
- Adequate low-temperature fluidity to permit circulation at the lowest expected start temperature
- Good chemical stability to minimize oxidation under conditions of high temperature and agitation in the presence of air and to provide long service life for the oil
- Good demulsibility to permit rapid separation of water and protect against the formation of harmful emulsions
- Antirust properties to protect gear and bearing surfaces from rusting in the presence of water, entrained moisture, or humid atmospheres
- A noncorrosive nature to prevent gears and bearings from being subjected to chemical attack by the lubricant
- Foam resistance to prevent the formation of excessive amounts of foam in reservoirs and gear cases

*Question:*

_____

_____

_____

A. _____

_____

B. _____

_____

C. _____

_____

D. _____

_____

*Correct Answer:* _____

## Factors Affecting Lubrication of Open Gears

In contrast to enclosed gears that are flood lubricated by splash or circulation systems, there are many gears for which it is not practical or economical to provide oil-tight housings. These so-called open gears can only be sparingly lubricated and perhaps only at infrequent intervals. Gears of this type are lubricated by either a continuous or an intermittent method. Some of the more sophisticated open gearing systems may be lubricated with a full circulation system that captures and filters the oil for reuse.

*Question:*

_____

_____

_____

A. _____

_____

B. _____

_____

C. _____

_____

D. _____

_____

*Correct Answer:* _____

The three most common continuous methods are splash, idler gear immersion, and pressure. In the first two, the lubricant is lifted from a reservoir or sump (sometimes referred to as a *slush pan*) by the partially submerged gear or an idler. Pressure systems require a shaft or independently driven pump to draw oil from a sump and spray it over the gear teeth. With the continuous methods of application, lubrication of open gears is similar to that with enclosed gears. Because the gears are usually large and relatively slow moving, very high-viscosity lubricants are required. Most gears lubricated in any of these ways are equipped with relatively oil-tight enclosures.

*Question:*

_____

_____

_____

A. _____

_____

B. _____

_____

C. _____

_____

D. _____

_____

*Correct Answer:* _____

# V. Lubricant Application and Lubricant Selection (MLT® I & II)

## Factors Affecting Lubrication

Whether oil or grease is used for rolling-element bearings depends on a number of factors. For bearings installed in machines that require oil for other elements, the same oil is often supplied to the bearings. In other cases, it may be more convenient to lubricate the bearings separately, and grease is often used. Groups of similar bearings may be lubricated with oil by a circulation system, an oil mist system, or with grease in a centralized lubrication system. Many bearings are now *packed for life* with grease by the bearing manufacturer and need no further lubrication in service. The characteristics of these various methods of application have some influence on the oil or grease selected.

*Question:*

_____

_____

_____

A. _____

_____

B. _____

_____

C. _____

_____

D. _____

_____

*Correct Answer:* _____

The thickness of the elastohydrodynamic films formed in the contact areas between rolling elements and the raceways is a function of the speed at which the surfaces roll together, the load, the oil viscosity, and the operating temperature. The film thickness increases with increases in speed or oil viscosity and decreases with increases in load or operating temperature (because higher temperatures reduce oil viscosity). For any given set of operating conditions, the oil viscosity (or the viscosity of the oil in a grease) should be selected to provide a safe minimum film thickness.

*Question:*

_____

_____

_____

A.  _____

_____

B.  _____

_____

C.  _____

_____

D.  _____

_____

*Correct Answer:* _____

## Effect of Speed

The speed at which the surfaces of a rolling-element bearing roll together is a fairly complex calculation, so an approximation called the *bearing speed factor nDm* is usually used. The bearing speed factor is determined by multiplying the rotational speed in revolutions per minute $n$ by the pitch diameter in millimeters $Dm$.

*Question:*

_____

_____

_____

A.  _____

_____

B.  _____

_____

C.  _____

_____

D.  _____

_____

*Correct Answer:* _____

## Effect of Load

Under extra-heavy-load (EHL) conditions, the effect of load on film thickness is not as great as the effect of speed or oil viscosity. For example, while doubling the speed or oil viscosity might result in a film-thickness increase exceeding 50%, doubling the load might result in only about a 10% decrease in film thickness. As a result, under steady-load conditions, the viscosity of oil generally can be selected on the basis of the bearing speed factor and operating temperature without regard to the load.

*Question:*

_____

_____

_____

A.  _____

_____

B.  _____

_____

C.  _____

_____

D.  _____

_____

*Correct Answer:* _____

## Effect of Temperature

Because both viscosity and grease consistency are functions of temperature, the operating temperature of bearings must always be considered during selection of lubricants. Bearing operating temperatures may be increased above normal by heat that is conducted to the bearing from a hot shaft or spindle or by heat radiated to the housing from a hot surrounding atmosphere. Excessive churning of grease resulting from overfilling can also raise bearing temperatures.

*Question:*

_____

_____

_____

A.  _____

_____

B.  _____

_____

C.  _____

_____

D.  _____

_____

*Correct Answer:* _____

## Contamination

Solid particles of any kind that are trapped between the rolling elements and raceways are the most frequent cause of shortened bearing life. Consequently, dirt should be kept out of bearings as much as possible (including during periods of storage prior to installation), and lubricants should be changed before oxidation has progressed to a point at which deposits begin to form. The use of premium-quality, oxidation-inhibited lubricants can greatly extend the period of time that lubricants may be left in service without excessive oxidation.

*Question:*

_____

_____

_____

A. _____

_____

B. _____

_____

C. _____

_____

D. _____

_____

*Correct Answer:* _____

Water that gets into a bearing tends to reduce the fatigue life and cause rusting that can quickly ruin a bearing. Water, when mixed with some greases, may cause them to soften and perhaps leak from the bearings. Large quantities of water can wash the lubricant out. Sometimes fluids such as acids get into the bearings and cause corrosion. Any of these conditions usually involves special precautions and possibly specialty lubricants.

*Question:*

_____

_____

_____

A. _____

_____

B. _____

_____

C. _____

_____

D. _____

_____

*Correct Answer:* _____

# V. & VI. Lubricant Application (MLT® I & II)

## Lubricant Application Methods

### All-Loss Method

Most open gears and wire ropes, many drive chains and rolling-element bearings, and some cylinders, bearings, and enclosed gears are lubricated by all-loss methods. Nearly all grease lubrication (except sealed, packed rolling-element bearings) represents the all-loss approach. Only relatively small amounts of grease are applied, mainly to replenish the lubricating films, but in some cases to flush away some or all of the old lubricant and contaminants.

*Question:*

_____

_____

_____

A. _____

_____

B. _____

_____

C. _____

_____

D. _____

_____

*Correct Answer:* _____

## Oiling Devices: Drop Feed and Wick Feed Cups

Drop feed and wick feed cups are often used to supply the small amount of oil required by high-speed rolling-element bearings, thin film-plain bearings and slides, and some open gears. The rate of oil feed from the drop feed cup can be adjusted with a needle valve, whereas the wick feed can be adjusted by changing the number of strands in the wick. Both devices have the disadvantage of requiring to be started and stopped by hand when the machine is started or stopped.

*Question:*

_____

_____

_____

A. _____

_____

B. _____

_____

C. _____

_____

D. _____

_____

*Correct Answer:* _____

## Oiling Devices: Bottle Oilers

In a typical bottle oiler, the spindle of the oiler rests on the journal and is vibrated slightly as the journal rotates. This motion results in a pumping action, which forces air into the bottle, causing minute amounts of oil to feed downward along the spindle to the bearing. The oil feed is more or less continuous but stops and starts when the machine is stopped or started.

*Question:*

_____

_____

_____

A. _____

_____

B. _____

_____

C. _____

_____

D. _____

_____

*Correct Answer:* _____

## Oiling Devices: Wick and Pad Oilers

In one type of wick oiler, a felt wick is held against the journal by a spring. The wick draws oil up from the reservoir by capillary action, and the turning journal wipes oil from the wick. No or little oil is fed when the journal is not turning. In another variation, the wick carries oil up to the slinger, which throws it into the bearing in the form of a fine spray. In both cases, oil leaking out along the shaft drains back to the reservoir, so the devices possess some elements of a reuse system.

*Question:*

_____

_____

_____

A. _____

_____

B. _____

_____

C. _____

_____

D. _____

_____

*Correct Answer:* _____

## Oiling Devices: Mechanical Force-Feed Lubricators

Force-feed lubricators are used in applications requiring positive feed of lubricants under pressure. A number of force-feed lubricators are in use. The oil is drawn from the reservoir in the base on the down stroke of the single-plunger pump and forced under pressure on the upstroke through the liquid-filled sight glass to the delivery line. The pump is operated by an eccentric cam and lever, which can be driven from a shaft on the machine or operated by a hand crank. The stroke of the pump can be regulated to adjust the oil feed rate, which can be estimated by counting the drops as they pass through the liquid-filled sight glass.

*Question:*

_____

_____

_____

A. _____

_____

B. _____

_____

C. _____

_____

D. _____

_____

*Correct Answer:* _____

## Oiling Devices: Air-Line Oilers

Air-powered cylinders and tools are often lubricated by "lubricating" the compressed air supply. Pneumatic cylinders used for actuating parts of machines may be lubricated by means of an oil fog lubricator. The flow of air through the lubricator creates an air–oil fog that carries sufficient oil to lubricate the cylinders.

*Question:*

_____

_____

_____

A. _____

_____

B. _____

_____

C. _____

D. _____

_____

*Correct Answer:* _____

## Oiling Devices: Air-Spray Application

Some open-gear and wire-rope lubricants, including some grease-type materials, are applied by means of hand-operated air-spray equipment that uses either external mixing nozzles or airless atomizing equipment. This approach, however, has the same disadvantages as hand oiling. A number of automatic or semiautomatic units have been developed to overcome some of these problems.

*Question:*

_____

_____

_____

A. _____

_____

B. _____

_____

C. _____

_____

D. _____

_____

*Correct Answer:* _____

## Circulation Systems

Although the many types of application for circulation systems require considerable variations in size, arrangement, and complexity, in a general way, circulation systems can be considered in one of three groups:

1. Systems comprising a compact arrangement of pump, reservoir, and oil passages built into the housing of the lubricated parts
2. Systems employing a multicompartment tank combining reservoir and purification facilities
3. Systems comprising an assembly of individual units (reservoir, oil cooler, oil heater, oil pumps, purification equipment, etc.)

*Question:*

_____

_____

_____

A. _____

_____

B. _____

_____

C. _____

_____

D. _____

_____

*Correct Answer:* _____

## Splash Oiling

Splash oiling is encountered mainly in gear sets or in compressor or steam engine crankcases. Gear teeth, or projections on connecting rods, dip into the reservoir and splash oil to the parts to be lubricated or to the casing walls, where pockets and channels are provided to catch the oil and lead it to the bearings.

*Question:*

_____

_____

_____

A.  _____

_____

B.  _____

_____

C.  _____

_____

D.  _____

_____

*Correct Answer:* _____

## Bath Oiling

The bath system is used for the lubrication of vertical-shaft hydrodynamic thrust bearings and for some vertical-shaft journal bearings. The lubricated surfaces are submerged in a bath of oil, which is maintained at a constant level. When necessary, cooling coils are placed directly in the bath. The bath system for a thrust bearing may be a separate system or may be connected into a circulation system.

*Question:*

_____

_____

_____

A.  _____

_____

B.  _____

_____

C.  _____

_____

D.  _____

_____

*Correct Answer:* _____

## Ring, Chain, and Collar Oiling

In a ring-oiled bearing, oil is raised from a reservoir by means of a ring that rides on and turns with the journal. Some of the oil is removed from the ring at the point of contact with the journal and is distributed by suitable grooves in the bearing. The oil flows through the bearing and drains back to the reservoir for reuse.

*Question:*

_____

_____

_____

A. _____

B. _____

C. _____

D. _____

*Correct Answer:* _____

Ring oiling is applied to a wide variety of medium-speed bearings in stationary service. At high surface speeds, too much slip occurs between the ring and journal, and not enough oil is delivered. Also, at high speeds, in large, heavily loaded bearings, not enough cooling may be provided.

*Question:*

_____

_____

_____

A. _____

B. _____

C. _____

D. _____

*Correct Answer:* _____

Oil rings are usually made about 1.5–2 times the journal diameter. Bearings more than about 8 in. (200 mm) long usually require two or more rings. The oil level in reservoirs is usually maintained so that the rings dip less than one-quarter their diameter. The oil level within a given range is not usually critical; too low a level may result in inadequate oil supply, however, and too high a level, because of excessive viscous drag, may cause ring slip or stalling. As a result, too little oil may reach the bearing, and "flats" may wear on the rings to such an extent that satisfactory performance is no longer possible.

*Question:*

_____

_____

_____

A.  _____

_____

B.  _____

_____

C.  _____

_____

D.  _____

_____

*Correct Answer:* _____

## Central Lubrication Systems

A number of types of central lubrication systems have been developed. Most can apply either oil or grease depending on the type of reservoir and pump used. Greases generally require higher pump pressures because greater pressure losses occur in lines, metering valves, and fittings. Pump and reservoir capacities vary depending on the number of application points to be served, ranging from small capacities to units that install on standard drums and systems that operate directly from bulk tanks or bins requiring large volumes of lubricant.

*Question:*

_____

_____

_____

A.  _____

_____

B.  _____

_____

C.  _____

_____

D.  _____

_____

*Correct Answer:* _____

In some systems, called *direct systems*, the pump serves to pressurize the lubricant and also to meter it to the application points. In *indirect systems*, the pump pressurizes the lubricant, but valves in the distribution lines meter it to the application points.

*Question:*

_____

_____

_____

A. _____

_____

B. _____

_____

C. _____

_____

D. _____

_____

*Correct Answer:* _____

Two basic types of indirect systems are in common use, and each type in turn has two variations. In parallel systems, also called *header* or *nonprogressive systems*, the metering valves or feeders are actuated by bringing the main distribution line up to operating pressure. All the metering valves operate more or less simultaneously. This type of system has the disadvantage that if one valve fails, no indication of failure is given at the pumping station. However, all the other application points will continue to receive lubricant. In series, or progressive, systems the valves are "in" the main distribution line. When the main distribution line is brought up to pressure, the first valve operates. After it has cycled, flow passes through it to the second valve and then to each succeeding valve in turn. Thus, if one valve fails, all fail, and the pressure rise at the pump or the distribution block can be used to signal that a failure has occurred.

*Question:*

_____

_____

_____

A. _____

_____

B. _____

_____

C. _____

_____

D. _____

_____

*Correct Answer:* _____

## Two-Line System

One variation of the parallel system uses two supply lines. A four-way valve directs pressure alternately to the two lines, at the same time relieving pressure in the line that is not receiving flow from the pump. The valve can be operated manually from the machine, cycled by a timer, or controlled by a counter that measures the volume delivered by the pump. The valves are designed to deliver a charge of lubricant to the application point each time the flow in the lines is reversed.

*Question:*

_____

_____

_____

A. _____

_____

B. _____

_____

C. _____

_____

D. _____

_____

*Correct Answer:* _____

## Single-Line Spring Return

In the second variation of the parallel system, only a single distribution line is used. As with the four-way valve of the preceding system, the three-way valve may be operated manually from the machine, by a timer, or by a counter measuring pump output. The valves deliver a charge of lubricant when system pressure is applied to them and reset themselves when the system pressure is relieved.

*Question:*

_____

_____

_____

A. _____

_____

B. _____

_____

C. _____

_____

D. _____

_____

*Correct Answer:* _____

## Series Manifolded System

In a series manifolded type of system, a single supply line is used. No relief valve is required because the manifold-type valves automatically reset themselves and continue cycling as long as pressure is applied to them through the supply line. The system can be cycled by starting and stopping the pump.

*Question:*

_____

_____

_____

A. _____

_____

B. _____

_____

C. _____

_____

D. _____

_____

*Correct Answer:* _____

## Series System, Reversing Flow

The second series system uses a single supply line with a four-way valve to reverse the flow in it. The valves are designed to deliver a charge of lubricant and then permit lubricant flow to pass through to the next valve. When the flow in the supply line is reversed, the valves cycle again in sequence in the reverse order.

*Question:*

_____

_____

_____

A. _____

_____

B. _____

_____

C. _____

_____

D. _____

_____

*Correct Answer:* _____

## Mist Oiling Systems

In oil mist lubricators, oil is atomized by low-pressure compressed air (10–50 psi [70–350 kPa]) into droplets so small that they float in the air, forming practically dry mist, or fog, that can be transported relatively long distances in small tubing. When the mist reaches the application point, it is condensed, or coalesced, into larger particles that wet the surfaces and provide lubrication.

*Question:*

_____

_____

_____

A. _____

_____

B. _____

_____

C. _____

_____

D. _____

_____

*Correct Answer:* _____

# V. Lubricant Selection (MLT® II)

## Internal Combustion Engines: Viscosity, Viscosity Index

In reciprocating engines, viscosity of the engine oil is extremely important. It has a bearing on wear prevention, sealing, oil economy, frictional power losses (fuel economy), and deposit formation. For some engines, particularly vehicle engines, it also is a factor in cranking speed and starting ease. Too high a viscosity may cause excessive viscous drag, reduction in cranking speed, and increased fuel consumption after the engine is started.

*Question:*

_____

_____

_____

A. _____

_____

B. _____

_____

C. _____

_____

D. _____

_____

*Correct Answer:* _____

Viscosity index (VI) is important in engines that must be started and operated over a wide temperature range. In these cases, all other factors being equal, oils with higher VIs give less viscous drag during starting and provide thicker oil films for better sealing and wear prevention; moreover, oil consumption at operating temperatures is lower.

*Question:*

_____

_____

_____

A. _____

_____

B. _____

_____

C. _____

_____

D. _____

_____

*Correct Answer:* _____

## Internal Combustion Engines: Low-Temperature Fluidity

When an oil is to be used in engines operating at low ambient temperatures, the oil must have low temperature fluidity that is adequate to permit immediate flow to the oil pump suction when the engine is started. The pour point of an oil is an adequate indication of whether it will flow to the pump suction. Most conventional oils will flow to the pump suction at temperatures below their pour points because the pump suction creates a considerably greater pressure head than is present in the pour-point test. However, many multigrade oils will not circulate adequately at temperatures considerably above their pour points. The correlation between pour point and flow in instrumented engines is poor, and at best, pour point is only a rough guide to the minimum temperature at which an oil may be used safely.

*Question:*

_____

_____

_____

A. _____

_____

B. _____

_____

C. _____

_____

D. _____

_____

*Correct Answer:* _____

The ASTM has introduced two tests that measure the low-temperature performance of an oil. The test that simulates cold cranking of an engine is the *cold cranking simulator* (CCS), and the test for measuring low-temperature pumping is the *minirotary viscometer* (MRV). Both the CCS and MRV show good correlation with low-temperature performance. The maximum values for these tests are listed for the W grades in the SAE engine oil viscosity classification. The CCS is designed to reproduce the elements of viscous drag that affect cranking speed. The MRV is designed to predict the low-temperature pumpability and therefore the ability of the oil to reach critical components under low-temperature starting conditions.

*Question:*

_____

_____

_____

A. _____

_____

B. _____

_____

C. _____

_____

D. _____

_____

*Correct Answer:* _____

## Internal Combustion Engines: Oxidation Stability (Chemical Stability)

High resistance to oxidation is an important requirement of a good engine oil in view of the high temperatures the oil is exposed to and, in the crankcase, the agitation of the oil in the presence of air. Deterioration of an engine oil by oxidation tends to increase viscosity, create deposit-forming materials, and promote corrosive attack on some hard alloy bearings. Where the engine oil capacity is relatively small, the rate of deterioration tends to be higher, other factors being equal.

*Question:*

_____

_____

_____

A. _____

_____

B. _____

_____

C. _____

_____

D. _____

_____

*Correct Answer:* _____

An oil's natural oxidation stability is determined in part by the crude oil from which it is made and the refining processes to which it is subjected. Where engine design or operating conditions require a high degree of oxidation stability, oxidation inhibitors are used. As a general rule, the need for greater oxidation stability increases as oil service temperatures and drain intervals increase.

*Question:*

_____

_____

_____

A. _____

   _____

B. _____

   _____

C. _____

   _____

D. _____

   _____

*Correct Answer:* _____

Among the principal factors that make enhanced oxidation stability necessary are high engine-specific power output (high horsepower per unit of displacement), small crankcase charge volume, long oil drain intervals, and modifications and devices to control emissions that result in high operating temperatures. For example, a heavily loaded truck engine requires an oil with excellent oxidation stability because of the high operating temperatures involved, whereas a large, low-speed diesel engine in central station (stationary) service with a large crankcase oil supply at a moderate temperature requires an oil with good stability because the oil is expected to remain in service for thousands of hours.

*Question:*

_____

_____

_____

A. _____

   _____

B. _____

   _____

C. _____

   _____

D. _____

   _____

*Correct Answer:* _____

## Internal Combustion Engines: Thermal Stability

Thermal stability, or resistance to cracking and decomposition under high-temperature conditions, is a fundamental characteristic of the lubricating base oil that cannot be substantially improved by means of additives. However, careful selection of additives is important in formulating thermally stable oils because decomposition of the additives can contribute to the formation of deposits under operating conditions that promote thermal cracking of the oil.

*Question:*

_____

_____

_____

A. _____

_____

B. _____

_____

C. _____

_____

D. _____

_____

*Correct Answer:* _____

## Internal Combustion Engines: Detergency and Dispersancy

In nearly all current internal combustion engine applications, oils with enhanced detergency and dispersancy are necessary to control engine deposits and maintain engine performance. The levels of detergency and dispersancy required depend on a number of factors such as engine design, operating temperatures, type of fuel, continuity of operation, and exposure to low ambient temperatures. In general, conditions that tend to promote oil oxidation, such as supercharging or the use of high-sulfur fuels, dictate the use of oils with higher levels of detergency. Conditions that promote condensation of water and unburned or partially burned fuel in the engine require the use of oils with higher dispersancy.

*Question:*

_____

_____

_____

A. _____

_____

B. _____

_____

C. _____

_____

D. _____

_____

*Correct Answer:* _____

## Internal Combustion Engines: Alkalinity

Most detergents and, to a lesser extent, many dispersants have some ability to neutralize the acidic end products of fuel combustion and oil oxidation. When a considerable ability to neutralize acids is required, as in oils for diesel engines burning high-sulfur fuels, however, highly alkaline (overbased) detergent-type materials are used. The concentration of these materials in an oil, and an indication of the oil's ability to neutralize acids, is given by the *total base number* (TBN), also called the *alkalinity value*. There is only a general relationship between TBN and the ability of an oil to control wear and corrosion caused by strong acids because it has been found that some newer additive systems are more effective in this respect than would be predicted by consideration of the TBN level alone.

*Question:*

_____

_____

_____

A. _____

B. _____

C. _____

D. _____

*Correct Answer:* _____

## Internal Combustion Engines: Antiwear

In addition to the corrosive wear caused by acidic products of combustion, metallic wear may occur in areas where loads or operating conditions prevent the maintenance of effective lubricating films. The main areas of concern in this respect are cylinder walls and rings, particularly of large, high-output trunk piston engines, and the valve-train mechanisms of small, high-speed engines.

*Question:*

_____

_____

_____

A. _____

B. _____

C. _____

D. _____

*Correct Answer:* _____

## Internal Combustion Engines: Rust and Corrosion Protection

All petroleum oils have some ability to prevent rusting and corrosion of engine metals. However, in most cases, this natural ability is not sufficient to do the following:

1. Protect hard alloy bearings from corrosion caused by oil oxyacids
2. Prevent rusting and corrosion due to condensation of water and combustion products in low-temperature or stop-and-go service
3. Control the corrosive wear caused by acidic end products of combustion

Because one or more of these conditions that can cause troublesome corrosion are encountered to some extent in nearly all internal combustion engine service, most oils for internal combustion engines are formulated to provide additional protection against corrosion.

*Question:*

_____

_____

_____

    A.  _____

       _____

    B.  _____

       _____

    C.  _____

       _____

    D.  _____

       _____

*Correct Answer:* _____

## Internal Combustion Engines: Foam Resistance

All oils will foam to some extent when agitated. If excessive foaming occurs in an internal combustion engine, several problems may result. Overflow and spillage of oil are, of course, two of the most obvious, but foaming also can result in starvation at the oil pump inlet or slugs of foam being drawn into the pump with the oil. Foam entrained in the oil can cause failure of lubricating films and noisy, erratic operation of hydraulic valve lifters.

*Question:*

_____

_____

_____

    A.  _____

       _____

    B.  _____

       _____

    C.  _____

       _____

    D.  _____

       _____

*Correct Answer:* _____

## Automotive Gear Oils: Load-Carrying Capacity

One of the most important performance characteristics of a gear lubricant is its load-carrying capacity, that is, its ability to prevent or minimize wear, scuffing, or scoring of gear tooth surfaces. The load-carrying capacity of straight mineral oil is adequate for the conditions under which some gears are operated, but most gears require lubricants with higher load-carrying capacities. This higher capacity is provided via the use of additives. Lubricants of this type are generally referred to as *EP lubricants*.

*Question:*

_____

_____

_____

A. _____

_____

B. _____

_____

C. _____

_____

D. _____

_____

*Correct Answer:* _____

## API Lubricant Service Designations

The gear lubricant service designations are as follows:

- API GL-1. Designates service characteristics for automotive spiral bevel and some truck manually operated transmissions that have components sensitive to additive materials such as EP additives. These transmissions are designed for operation under conditions of low unit pressures and sliding velocities so mild that straight mineral oil can be used satisfactorily. Oxidation and corrosion inhibitors, defoamants, and pour depressants may be used to improve the characteristics of lubricants for this service. Frictional modifiers and EP agents should not be used.

- API GL-2. Designates service characteristics for automotive-type worm-gear axles operating under conditions of load, temperature, and sliding velocities that cannot be accommodated by lubricants satisfactory for API GL-1 service. The API GL-2 gear oils generally contain fatty-type additives, making them satisfactory for worm gears and other types of industrial gearing.

- API GL-3. Designates service characteristics for manual transmissions and spiral bevel axles operating under moderately severe conditions of speed and load. These service conditions require a lubricant having a load-carrying capacity greater than those that will satisfy API GL-1 service but below the requirements of lubricants satisfying API GL-4 service.

- API GL-4. A classification still used commercially to describe lubricants for differentials and transmissions operating under moderate to severe conditions such as hypoid gears. These oils may be used in manual transmissions and transaxles where EP oils are acceptable. Limited-slip differentials generally have special lubrication requirements. The supplier should be consulted regarding the suitability of a given lubricant for such differentials. Information helpful in evaluating lubricants for this type of service may be found in ASTM STP-512.

- API GL-5. Designates service characteristics for gears, particularly hypoid, in passenger cars and other automotive equipment operated under conditions of high speed and shock load, high speed and low torque, and low speed and high torque.

- API MT-1. Designates lubricants for manual transmissions that do not contain synchronizers. These oils are formulated to provide levels of oxidation and thermal stability higher than those obtainable from API GL-1, GL-4, and GL-5 category oils.

*Question:*

_____

_____

_____

A. _____

_____

B. _____

_____

C. _____

_____

D. _____

_____

*Correct Answer:* _____

## Oxidation Resistance (Axles, Manual Transmissions)

Operating temperatures of drive axles and manual transmissions in normal passenger car service are usually moderate, and thus the oxidation resistance of the lubricant may not be a major consideration. In heavy-duty service, such as trailer towing and the use of many commercial vehicles, high operating temperatures may be encountered. In these applications, to avoid the development of sludge that could restrict oil flow, the lubricant must have adequate oxidation resistance to prevent excessive thickening. Oxidation also may result in the development of materials that are corrosive to some of the metals used.

*Question:*

_____

_____

_____

A. _____

_____

B. _____

_____

C. _____

_____

D. _____

_____

*Correct Answer:* _____

## Chemical Activity or Corrosion (Axle and Manual Transmissions)

EP agents function by reacting chemically with metal surfaces. This reaction normally is initiated when local overheating results from the rubbing of surface asperities through the oil film. If the reactivity of the EP agents is too high, some chemical reaction may occur at normal temperatures, resulting in corrosion and metal loss. Copper and its alloys are particularly susceptible to this type of corrosion, and although copper alloys are not normally used in drive axles, they are frequently used in manual transmission components.

*Question:*

_____

_____

_____

A. _____

_____

B. _____

_____

C. _____

_____

D. _____

_____

*Correct Answer:* _____

## Viscosity (Gear Lubricants)

The viscosity of a gear lubricant has some effect on load-carrying capacity, leakage, and gear noise. At low temperatures, it also determines ease of gear shifting and has considerable influence on flow to gear tooth surfaces and bearings.

*Question:*

_____

_____

_____

A. _____

_____

B. _____

_____

C. _____

_____

D. _____

_____

*Correct Answer:* _____

## Channeling Characteristics

Under low-temperature conditions, a lubricant may *channel*; that is, the lubricant may become so solid that when gear teeth cut a channel through it, it does not flow back rapidly enough to provide fresh material for the gear teeth to pick up. The channeling temperature of a lubricant is somewhat related to its pour point and also to its low-temperature viscosity.

*Question:*

_____

_____

_____

A. _____

   _____

B. _____

   _____

C. _____

   _____

D. _____

   _____

   *Correct Answer:* _____

## Storage Stability (Gear Lubricants)

In extended storage, particularly if the storage temperature is excessively high or low or if moisture is present, some of the additive materials may separate from some highly additized gear lubricants, or reactions that can change the properties of these materials may occur. With some of the older additive systems, such instability was occasionally a severe problem, but the newer additive systems generally exhibit improved oil solubility and a reduced tendency for the components to react with each other.

*Question:*

_____

_____

_____

A. _____

   _____

B. _____

   _____

C. _____

   _____

D. _____

   _____

   *Correct Answer:* _____

## Foaming (Gear Lubricants)

The churning in gear sets, combined with contaminants such as moisture that enter the case through breathers, tends to promote foaming of the gear lubricant. Foaming may cause overflow with loss of lubricant and may interfere with lubricant circulation and the load-carrying ability of lubricant films. Defoamants are used in gear lubricants to reduce the foaming tendency. Some of the commonly used defoamants are not soluble in the oil. As a result, defoamants are carried in suspension. If the density of the defoamer is significantly greater than that of the oil, the defoamer can settle out during extended storage.

*Question:*

_____

_____

_____

A. _____

_____

B. _____

_____

C. _____

_____

D. _____

_____

*Correct Answer:* _____

## Rust Protection (Gear Lubricants)

Some moisture enters gear sets through normal breathing, and heavy water contamination may occur in certain types of off-highway equipment. The gear lubricant must provide rust protection adequate to prevent rusting that might result in gear or bearing damage, particularly during shutdown periods.

*Question:*

_____

_____

_____

A. _____

_____

B. _____

_____

C. _____

_____

D. _____

_____

*Correct Answer:* _____

## Seal Compatibility (Gear Lubricants)

To maintain control of leakage, gear lubricants must be formulated to be compatible with the elastomeric materials used as seals. This means that the lubricant must not cause excessive swelling or shrinkage and hardening. A slight amount of swelling is usually considered desirable to help keep the seals tight.

*Question:*

_____

_____

_____

A. _____

_____

B. _____

_____

C. _____

_____

D. _____

_____

*Correct Answer:* _____

# VI. & VIII. Preventive Maintenance Management (MLT® I & II)

## Preventive Maintenance

There are many definitions of preventive maintenance, but all preventive maintenance management programs are time driven. In other words, maintenance tasks are based on elapsed time or hours of operation. The mean time to failure (MTTF) or *bathtub curve* indicates that a new machine has a high probability of failure because of installation problems during the first few weeks of operation. After this initial period, the probability of failure is relatively low for an extended period of time. Following this normal machine life period, the probability of failure increases sharply with elapsed time. In preventive maintenance management, machine repairs or rebuilds are scheduled on the basis of the MTTF statistic.

*Question:*

_____

_____

_____

A. _____

_____

B. _____

_____

C. _____

_____

D. _____

_____

*Correct Answer:* _____

## Predictive Maintenance

Predictive maintenance is a condition-driven preventive maintenance program. Instead of relying on industrial or in-plant average-life statistics (e.g., MTTF) to schedule maintenance activities, predictive maintenance uses direct monitoring of the mechanical condition, system efficiency, and other indicators to determine the actual MTTF or loss of efficiency for each machine train and system in the plant.

*Question:*

_____

_____

_____

A. _____

B. _____

C. _____

D. _____

*Correct Answer:* _____

The addition of a comprehensive predictive maintenance program can and will provide factual data on the actual mechanical condition of each machine train and the operating efficiency of each process system. These data provide the maintenance manager with actual data for scheduling maintenance activities.

*Question:*

_____

_____

_____

A. _____

B. _____

C. _____

D. _____

*Correct Answer:* _____

There are five nondestructive techniques normally used for predictive maintenance management:

- Vibration monitoring
- Process parameter monitoring
- Thermography
- Tribology
- Visual inspection

Each technique has a unique data set that will assist the maintenance manager in determining the actual need for maintenance.

*Question:*

_____

_____

_____

A. _____

_____

B. _____

_____

C. _____

_____

D. _____

_____

*Correct Answer:* _____

Any moving component is likely to fail at a relatively high rate and is a fine candidate for preventive maintenance. The following are familiar causes of failure:

- Abrasion
- Abuse
- Age deterioration
- Bond separation
- Consumable depletion
- Contamination
- Corrosion
- Dirt
- Fatigue
- Friction
- Operator negligence
- Puncture
- Shock
- Stress
- Temperature extremes
- Vibration
- Wear

*Question:*

_____

_____

_____

A. _____

_____

B. _____

_____

C. _____

_____

D. _____

_____

*Correct Answer:* _____

The primary functions and steps involved in developing a lubrication program are to:

- Identify every piece of equipment that requires lubrication.
- Ensure that every major equipment is uniquely identified, preferably with a prominently displayed number.
- Ensure that equipment records are complete for manufacturer and physical location.
- Determine locations on each piece of equipment that needs to be lubricated.
- Identify lubricant to be used.
- Determine the best method of application.
- Establish the frequency or interval of lubrication.
- Determine whether the equipment can be safely lubricated while operating or if it must be shut down.
- Decide who should be responsible for any human involvement.
- Standardize lubrication methods.
- Package the preceding elements into a lubrication program.
- Establish storage and handling procedures.
- Evaluate new lubricants to take advantage of state of the art.
- Analyze any failures involving lubrication and initiate necessary corrective actions.

*Question:*

_____

_____

_____

A.  _____

_____

B.  _____

_____

C.  _____

_____

D.  _____

_____

*Correct Answer:* _____

## Lubrication Program Implementation

An individual supervisor in the maintenance department should be assigned the responsibility for implementation and continued operation of the lubrication program. This person's primary functions are to:

- Establish lubrication service actions and schedules.
- Define the lubrication routes by building, area, and organization.
- Assign responsibilities to specific persons.
- Train lubricators.
- Ensure supplies of proper lubricants through the storeroom.
- Establish feedback that ensures completion of assigned lubrication and follow up on any discrepancies.
- Develop a manual or computerized lubrication scheduling and control system as part of the larger maintenance management program.
- Motivate lubrication personnel to check equipment for other problems and to create work requests where feasible.
- Ensure continued operation of the lubrication system.

*Question:*

_____

_____

_____

A. _____

_____

B. _____

_____

C. _____

_____

D. _____

_____

*Correct Answer:* _____

## Routine Inspections

Routine inspections of operating systems can provide insight into potential problem areas and suggest ways to initiate corrective actions. These corrective actions will result in longer equipment life and lower-cost operation of the hydraulic system. Commonly included in routine inspection programs are the following:

- Noise levels. Increases in noise levels may signal problems with cavitation. Noise levels also may increase when excessive temperatures allow the oil's viscosity to become too low, resulting in metal-to-metal contact.
- Vibration. Loose mounting or misalignment of components will cause vibration, resulting in accelerated wear or failure.
- Shock loads. System components, such as hoses and fittings, subjected to shock loads as a result of abrupt changes in flows and pressures can lead to leakage and failure. Where shock loads are experienced, accumulators should be installed in the system.
- Leakage. Oil leakage can result in low reservoir oil levels, leading to poor system performance. In addition, leakage can be costly and can create safety problems.
- Temperatures. As discussed earlier, controlling temperatures in the appropriate range is important both from the oil life and system performance standpoints. Excessive temperatures may be caused by plugged or dirty heat exchangers, excessive pressures, high rates of internal leakage, or low oil levels. Use of fluids with too high (excessive shearing) or too low (inadequate films) a viscosity also will result in higher temperatures.
- Filtration. The condition of fill screens, breathers, and filters (indicators or differential pressure gauges) should indicate the need for cleaning or replacement.
- Foaming. A little foam on top of the oil in the reservoir is normal. Excessive foam may indicate air leaks in the suction line, unsatisfactory contamination levels, or inadequate antifoam characteristics of the oil.
- Fluid odor and color. Although odor and color are not characteristics used to judge an oil's ability to provide proper lubrication, changes in these physical properties may indicate contamination (e.g., solvents, wrong oils added, degradation, etc.) or that the oil has reached the end of its service life. If there is doubt about the causes of the color and odor changes, a sample should be submitted for laboratory oil analysis. If these changes are accompanied by undesirable machine operating characteristics, change the oil but still submit a sample for laboratory oil analysis.

*Question:*

_____

_____

_____

A. _____

_____

B. _____

_____

C. _____

_____

D. _____

_____

*Correct Answer:* _____

# VIII. & X. Lube Storage and Management (MLT® I & II)

## Handling, Storing, and Dispensing Lubricants

Deterioration of lubricants can result from exposure to heat or cold, intermixing of brands or types, oxidation, prolonged storage, chemical reaction with fumes or vapor, entrance of dust and abrasive particles, and water contamination.

*Question:*

_____

_____

_____

A. _____

_____

B. _____

_____

C. _____

_____

D. _____

_____

*Correct Answer:* _____

Economic benefits that can be obtained are based mainly on the elimination of waste due to the following preventable factors:

- Leakage or spills from damaged or improperly closed containers
- Contamination due to exposure of lubricants to dust, metal particles, fumes, and moisture
- Deterioration caused by storage in excessively hot or cold locations
- Deterioration due to prolonged storage
- Residual oil or grease left in containers at the time of disposal or return
- Mixing of incompatible brands or types of lubricant
- Leaks, spills, and drips when a reservoir is being charged or when machine-lubricated operating benefits, which also are reflected in dollar savings, include the following:
  - Reduction of machine problems attributable to the lubricant, resulting in fewer downtime occurrences.
  - Reduced material-handling time—it has been estimated that labor costs for lubricant application can be as high as eight times the cost of the lubricant applied.
  - Better housekeeping—oil or grease spilled on floors is a major safety and fire hazard.

*Question:*

_____

_____

_____

A. _____

_____

B. _____

_____

C. _____

_____

D. _____

_____

*Correct Answer:* _____

## Bulk Products

The advantages of bulk delivery and storage of lubricants have resulted in the increased use of this method of operation wherever the quantities of lubricants used are sufficient to justify the installation costs. Bulk handling and storage systems offer both economic and operating benefits to the lubricant user. Economic benefits include the following:

- Reduced handling costs. When delivery is made directly into permanent storage tanks, handling costs are reduced markedly over those for drums or smaller packages.
- Reduced floor space requirements. Permanent tanks occupy much less floor space than is required for an equal volume of oil in drums.
- Reduced contamination hazards. Bulk storage tanks usually are filled directly from the tank car or tank truck through tight fill connections. The exposure of the lubricant to contamination is greatly reduced in comparison to handling and dispensing the same volume of packaged product.
- Reduced residual waste. The user is charged only for the amount of product delivered into a bulk tank. Practice varies with bulk bins, but in many cases the user is allowed credit for residual material in the bin or is charged for only the amount added to the bin when it is refilled. Because the amount of residual oil left in drums is usually up to one gallon and the amount of grease left in drums may range up to 20 lb (9 kg), the amount of lost lubricant attributable to residual waste with these packages can be significant.

- Simplification of inventory control. The use of bulk storage facilities sharply reduces the number of individual items to be ordered, tallied, routed, and processed for return or salvage.
- Reduced container deposit losses. Damaged drums or containers that cannot be returned for other reasons such as excessive or unacceptable materials left in them result in lost deposit value and/or additional disposal costs.

*Question:*

_____

_____

_____

A. _____

_____

B. _____

_____

C. _____

_____

D. _____

_____

*Correct Answer:* _____

The operating benefits to be achieved through the use of bulk facilities include the following:

- Simplification of handling. Less handing time means that personnel are freed for other tasks.
- Constant availability of needed lubricants. Time spent waiting for placement and opening of new drums and installation of pumps is minimized.
- Reduced contamination hazards. Minimizing contamination means fewer lubrication problems on the manufacturing floor.
- Lubricants can be made available readily at strategic points. Pumps and pipelines can be used to move lubricants from permanent tanks to locations at which repetitive lubrication operations are performed.
- Improved personnel and facility safety. Any reduction in handling operations reduces exposure to conditions that could result in injuries to personnel, plant, or environment.

*Question:*

_____

_____

_____

A. _____

_____

B. _____

_____

C. _____

_____

D. _____

_____

*Correct Answer:* _____

## Storage

Proper storage of lubricants and associated products requires that they be protected from sources of contamination and from degradation due to excessive heat or cold and that product identification is maintained. Also important are the ease with which the products can be moved in and out of storage and the ability to operate on a first-in, first-out (FIFO) basis. Increasingly important in the selection, location, and operation of petroleum product storage facilities are applicable fire, safety, and insurance requirements.

*Question:*

_____

_____

_____

A. _____

_____

B. _____

_____

C. _____

_____

D. _____

_____

*Correct Answer:* _____

## Packaged Products: Outdoor Storage

As a general rule, lubricants in drum containers should never be stored outdoors. When drums must be stored outside, a temporary shelter, lean-to, or waterproof tarpaulin will protect them from rain or snow.

*Question:*

_____

_____

_____

A.  _____

    _____

B.  _____

    _____

C.  _____

    _____

D.  _____

    _____

*Correct Answer:* _____

Drums should be laid on their sides with the bungs approximately horizontal. In this position, the bungs are below the level of the contents, thus greatly reducing breathing of water or moisture and preventing water from collecting inside the chime.

*Question:*

_____

_____

_____

A.  _____

    _____

B.  _____

    _____

C.  _____

    _____

D.  _____

    _____

*Correct Answer:* _____

For maximum protection, the drums should be stood on end (as long as product identification is visible) with the bung ends down on a well-drained surface. Where all these approaches are impractical, drum covers can be used. These are available in both metal and plastic. Drums that have a bung on the side should be stored either on end or on the side with the bung down. Regardless of the position of storage, drums should always be placed on blocks or racks several inches above the ground to avoid moisture damage.

*Question:*

_____

_____

_____

A. _____

_____

B. _____

_____

C. _____

_____

D. _____

_____

*Correct Answer:* _____

Drums that must be stored outdoors with the bung end up prior to use should be cleaned carefully to eliminate the hazard of collected rust, scale, or dirt falling into the drum contents.

*Question:*

_____

_____

_____

A. _____

_____

B. _____

_____

C. _____

_____

D. _____

_____

*Correct Answer:* _____

The drum storage area should be kept clean and free of debris that might present a fire hazard.

*Question:*

_____

_____

_____

A. _____

_____

B. _____

_____

C. _____

_____

D. _____

_____

*Correct Answer:* _____

## Warehouse Storage

Racks and shelving should provide adequate protection for all containers. Aisle space should be adequate for maneuvering whatever type of mechanical handling equipment is used. To maintain a FIFO policy, no lubricant should be stocked in a way that blocks access to the older stacks. Care in this regard will minimize the hazard of deterioration due to excessive storage time.

*Question:*

_____

_____

_____

A. _____

_____

B. _____

_____

C. _____

_____

D. _____

_____

*Correct Answer:* _____

## Facilities

The simplicity or complexity of oil house facilities depends largely on the size of the plant and the comprehensiveness of its lubrication program. Speaking generally, a well-equipped oil house will contain adequate stocks of the following items:

- Drum racks, either of the rocker type or tiered
- Oil and grease transfer pumps
- Drum faucets
- Grease gun fillers
- Grease guns
- Oil cans (nongalvanized)
- Portable equipment such as oil wagons, lubrication carts, sump drainers, catch pans, and power grease guns
- Maintenance supplies such as wiping rags, cleaning materials, wicks, replacement screens and filters, and spare grease fittings
- Containers of absorbent materials for cleaning up oil spills
- Grounding straps for use with combustible materials

*Question:*

_____

_____

_____

A. _____

_____

B. _____

_____

C. _____

_____

D. _____

_____

*Correct Answer:* _____

## Lubricant Deterioration in Storage

Lubricants can deteriorate in storage, usually as a result of one of the following causes:

- Contamination, most frequently with water
- Exposure to excessively high temperatures
- Exposure to low temperatures
- Long-term storage

*Question:*

_____

_____

_____

A.  _____

   _____

B.  _____

   _____

C.  _____

   _____

D.  _____

   _____

*Correct Answer:* _____

# 4

# Body of Knowledge Outline

This section is provided for you to work with as a reference to help answer questions that may arise as well as a resource to help perfect your notes.

## Friction

*Friction* is a force that resists relative motion between two surfaces in contact. Depending on the application, friction may be desirable or undesirable. Certain applications, such as tire traction on pavement and braking or when feet are firmly planted to move a heavy object, rely on the beneficial effects of friction for their effectiveness. In other applications, such as operation of engines or equipment with bearings and gears, friction is undesirable because it causes wear and generates heat, which frequently lead to premature failure.

For purposes of this manual, the *energy expended in overcoming friction* is dispersed as heat and is considered to be wasted because useful work is not accomplished. This *waste heat* is a major cause of excessive wear and premature failure of equipment. Two general cases of friction occur: *sliding friction* and *rolling friction*.

### Sliding Friction

To visualize sliding friction, imagine a steel block lying on a steel table. Initially, a force $F$ (action) is applied horizontally in an attempt to move the block. If the applied force $F$ is not high enough, the block will not move because the friction between the block and table resists movement. If the applied force is increased, eventually it will be sufficient to overcome the frictional resistance force *f*, and the block will begin to move. At this precise instant, the applied force *F* is equal to the resisting friction force *f* and is referred to as the *force of friction.*

In mathematical terms, the relation between the normal load *L* (weight of the block) and the friction force *f* is given by the *coefficient of friction*, denoted by the Greek symbol μ. Note that in the present context, *normal* has a different meaning than that commonly used. When discussing friction problems, the *normal load* refers to a load that is perpendicular to the contacting surfaces. For the example used here, the normal load is equal to the weight of the block because the block is resting on a horizontal table. However, if the block were resting on an inclined plane or ramp, the normal load would not equal the weight of the block but would depend on the angle of the ramp. Because the intent here is to provide a means of visualizing friction, the example has been simplified to avoid confusing readers not familiar with statics.

## Laws of Sliding Friction

The following friction laws are extracted from the *Machinery Handbook*, 23rd Revised Edition.

### (1) Dry or Unlubricated Surfaces

Three laws govern the relationship between the frictional force *f* and the load or weight *L* of the sliding object for unlubricated or dry surfaces:

1. "For low pressures (normal force per unit area), the friction force is directly proportional to the normal load between the two surfaces. As the pressure increases, the friction does not rise proportionally; but when the pressure becomes abnormally high, the friction increases at a rapid rate until seizing takes place."

2. The value of *f/L* is defined as the coefficient of friction μ. "The friction both in its total amount and its coefficient is independent of the area of contact so long as the normal force remains the same. This is true for moderate pressures only. For high pressures, this law is modified in the same way as the first case."

3. "At very low velocities, the friction force is independent of the velocity of rubbing. As the velocities increase, the friction decreases." The third law (c) implies that the force required to set a body in motion is the same as the force required to keep it in motion, but this is not true. Once a body is in motion, the force required to maintain motion is less than the force required to initiate motion, and there is some dependency on velocity.

These facts reveal two categories of friction: *static* and *kinetic*. Static friction is the force required to *initiate* motion $F_s$. Kinetic or dynamic friction is the force required to *maintain* motion $F_k$.

### (2) Lubricated Surfaces

The friction laws for well-lubricated surfaces are considerably different from those for dry surfaces as follows:

1. "The frictional resistance is almost independent of the pressure (normal force per unit area) if the surfaces are flooded with oil."

2. "The friction varies directly as the speed, at low pressures; but for high pressures, the friction is very great at low velocities, approaching a minimum at about 2 ft/sec linear velocity, and

afterwards increasing approximately as the square root of the speed."

3. "For well-lubricated surfaces, the frictional resistance depends, to a very great extent, on the temperature, partly because of the change in viscosity of the oil and partly because, for journal bearings, the diameter of the bearing increases with the rise in temperature more rapidly than the diameter of the shaft, thus relieving the bearing of side pressure."

4. "If the bearing surfaces are flooded with oil, the friction is almost independent of the nature of the material of the surfaces in contact. As the lubrication becomes less ample, the coefficient of friction becomes more dependent upon the material of the surfaces."

### (3) Coefficient of Friction

The coefficient of friction depends on the type of material. Tables showing the coefficient of friction of various materials and combinations of materials are available. Common sources for these tables are *Marks Mechanical Engineering Handbooks* and *Machinery's Handbook* (Industrial Press). The tables show the coefficient of friction for clean, dry surfaces and lubricated surfaces. It is important to note that the coefficients shown in these tables can vary.

### (4) Asperities

Regardless of how smooth a surface may appear, it has many small irregularities called *asperities*. In cases where a surface is extremely rough, the contacting points are significant, but when the surface is fairly smooth, the contacting points have a very modest effect. The *real* or *true surface area* refers to the area of the points in direct contact. This area is considerably smaller than the apparent geometric area.

### (5) Adhesion

*Adhesion* occurs at the points of contact and refers to the welding effect that arises when two bodies are compressed against each other. This effect is more commonly referred to as *cold welding* and is attributed to pressure rather than heat, which is associated with welding in the more familiar sense. A shearing force is required to separate cold-welded surfaces.

### (6) Shear Strength and Pressure

As noted previously, the primary objective of lubrication is to reduce friction and wear of sliding surfaces. This objective is achieved by introducing a material with a low shear strength or coefficient of friction between the wearing surfaces.

Although nature provides such materials in the form of oxides and other contaminants, the reduction in friction due to their presence is insufficient for machinery operation. For these conditions, a second relationship is used to define the coefficient of friction:

$$\mu = S/P$$

where $S$ is the shear strength of the material and $P$ is the pressure (or force) contributing to compression. This relationship shows that the coefficient of friction is a function of the force required to shear a material.

### (7) Stick-Slip

To the unaided eye, the motion of sliding objects appears steady. In reality, this motion is jerky or intermittent because the objects slow during shear periods and accelerate following the shear. This process is continuously repeated while the objects are sliding.

During shear periods, the static friction force $F_s$ controls the speed. Once shearing is completed, the kinetic friction force $F_k$ controls the speed, and the object accelerates. This effect is known as *stick-slip*. In well-lubricated machinery operated at the proper speed, stick-slip is insignificant, but it is responsible for the squeaking or chatter sometimes heard in machine operation.

Machines that operate over long sliding surfaces, such as the ways of a lathe, are subject to stick-slip. To prevent stick-slip, lubricants are provided with additives to make $F_s$ less than $F_k$.

## Rolling Friction

When a body rolls on a surface, the force resisting the motion is termed *rolling friction* or *rolling resistance*. Experience shows that much less force is required to roll an object than to slide or drag it. Because force is required to initiate and maintain rolling motion, there must be a definite but small amount of friction involved. Unlike the coefficient of sliding friction, the coefficient of rolling friction varies with conditions and has a dimension expressed in units of length.

Ideally, a rolling sphere or cylinder will make contact with a flat surface at a *single point* or *along a line* (in the case of a cylinder). In reality, the area of contact is slightly larger than a point or line because of *elastic deformation* of either the rolling object or the flat surface or both. Much of the friction is attributed to *elastic hysteresis*. A perfectly elastic object will spring back immediately after relaxation of the deformation. In reality, a small but definite amount of time is required to restore the object to its original shape. As a result, energy is not entirely returned to the object or surface but is retained and converted to heat. The source of this energy is, in part, the rolling frictional force.

A certain amount of *slippage* (which is the equivalent of sliding friction) occurs in rolling friction. If the friction of an unhoused rolling object is measured, slippage effects are minimal. However, in practical applications such as a housed ball or roller bearing, slippage occurs and contributes to rolling friction. Neglecting slippage, rolling friction is very small compared with sliding friction.

## Laws of Rolling Friction

The laws for sliding friction cannot be applied to rolling bodies in equally quantitative terms, but the following generalities can be given:

1. The rolling friction force $F$ is proportional to the load $L$ and inversely proportional to the radius of curvature $r$, or $F = \mu rL/r$, where $\mu r$ is the coefficient of rolling resistance (in meters or inches). As the radius increases, the frictional force decreases.

2. The rolling friction force $F$ can be expressed as a fractional power of the load $L$ times a constant $k$, or $F = kLn$, where the constant $k$ and the power $n$ must be determined experimentally.

3. The friction force $F$ decreases as the smoothness of the rolling element improves.

# Wear

*Wear* is defined as the progressive damage resulting in material loss due to relative contact between adjacent working parts. Although some wear is to be expected during normal operation of equipment, excessive friction causes premature wear, and this creates significant economic costs owing to equipment failure, cost for replacement parts, and downtime. Friction and wear also generate heat, which represents wasted energy that is not recoverable. In other words, wear is also responsible for overall loss in system efficiency.

## Wear and Surface Damage

The *wear rate* of a sliding or rolling contact is defined as the volume of material lost from the wearing surface per unit of sliding length and is expressed in units of length squared. For any specific sliding application, the wear rate depends on the normal load, the relative sliding speed, the initial temperature, and the mechanical, thermal, and chemical properties of the materials in contact.

The effects of wear are commonly detected by visual inspection of surfaces. Surface damage can be classified as follows:

1. Surface damage without exchange of material:
   - Structural changes. Aging, tempering, phase transformations, and recrystallization
   - Plastic deformation. Residual deformation of the surface layer
   - Surface cracking. Fractures caused by excessive contact strains or cyclic variations of thermally or mechanically induced strains
2. Surface damage with loss of material (wear):
   - Characterized by wear scars of various shapes and sizes
   - Can be shear fracture, extrusion, chip formation, tearing, brittle fracture, fatigue fracture, chemical dissolution, and diffusion
3. Surface damage with gain of material:
   - Corrosion. Material degradation by chemical reactions with ambient elements or elements from the opposing surface; can include pickup of loose particles and transfer of material from the opposing surface.

## Mild or Severe Wear

Wear also may be classified as mild or severe. The distinguishing characteristics between mild and severe wear are as follows:

1. Mild:
   - Produces extremely smooth surfaces, sometimes smoother than the original.
   - High electrical contact resistance but very little true metallic contact.
   - Debris is extremely small, typically in the range of 100 nm ($10^{-9}$ meters) in diameter.
2. Severe:
   - Rough, deeply torn surfaces, much rougher than the original.
   - Large metallic wear debris, typically up to 0.01 mm ($3.28 \times 105$ ft) in diameter.
   - Low contact resistance, but true metallic junctions are formed.

## Types of Wear

Ordinarily, wear is thought of only in terms of abrasive wear occurring in connection with sliding motion and friction. However, wear also can result from adhesion, fatigue, or corrosion.

### (1) Abrasive Wear

Abrasive wear occurs when a hard surface slides against and cuts grooves into a surface. This condition is frequently referred to as *two-body abrasion*. Particles cut from the softer surface or dust and dirt introduced between the wearing surfaces also contribute to abrasive wear. This condition is referred to as *three-body abrasion*.

## (2) Adhesive Wear

Adhesive wear frequently occurs because of shearing at points of contact or asperities that undergo adhesion or cold welding, as described previously. Shearing occurs through the weakest section, which is not necessarily at the adhesion plane. In many cases, shearing occurs in the softer material, but such a comparison is based on shear tests of relatively large pure samples. The adhesion junctions, in contrast, are very small spots of weakness or impurity that would be insignificant in a large specimen but in practice may be sufficient to permit shearing through the harder material. In some instances, the wearing surfaces of materials with different hardness can contain traces of material from the other face. Theoretically, this type of wear does not remove material but merely transfers it between wearing surfaces. However, the transferred material is often loosely deposited and eventually flakes away in microscopic particles; these, in turn, cause wear.

## (3) Pitting Wear

1. Pitting wear is due to surface failure of a material as a result of stresses that exceed the endurance (fatigue) limit of the material. Metal fatigue is demonstrated by bending a piece of metal wire, such as a paper clip, back and forth until it breaks. Whenever a metal shape is deformed repeatedly, it eventually fails. A different type of deformation occurs when a ball bearing under a load rolls along its race. The bearing is flattened somewhat, and the edges of contact are extended outward. This repeated flexing eventually results in microscopic flakes being removed from the bearing. Fatigue wear also occurs during sliding motion. Gear teeth frequently fail due to pitting.

2. While pitting is generally viewed as a mode of failure, some pitting wear is not detrimental. During the break-in period of new machinery, friction wears down working-surface irregularities. This condition is considered to be nonprogressive and usually improves after the break-in period. However, parts that are continuously subjected to repeated stress will experience destructive pitting as the material's endurance limit is reached.

## (4) Corrosive Wear

1. Corrosive wear occurs as a result of a chemical reaction on a wearing surface. The most common form of corrosion is due to a reaction between the metal and oxygen (oxidation); however, other chemicals also may contribute. Corrosion products, usually oxides, have shear strengths that are different from those of the wearing-surface metals from which they were formed. The oxides tend to flake away, resulting in pitting of the wearing surfaces. Ball and roller bearings depend on extremely smooth surfaces to reduce frictional effects. Corrosive pitting is especially detrimental to these bearings.

2. American National Standards Institute (ANSI) Standard ANSI/AGMA 1010-E95 provides numerous illustrations of wear in gears and includes detailed discussions of the types of wear mentioned here and more. Electric Power Research Institute (EPRI) Report EPRI GS-7352 provides illustrations of bearing failures.

3. Normal wear is inevitable whenever there is relative motion between surfaces. However, wear can be reduced by appropriate machinery design, precision machining, material selection, and

proper maintenance, including lubrication. The remainder of this section is devoted to discussions on the fundamental principles of lubrication that are necessary to reduce wear.

# Lubrication and Lubricants and the Purpose of Lubrication

The primary purpose of lubrication is to reduce wear and heat between contacting surfaces in relative motion. Although wear and heat cannot be completely eliminated, they can be reduced to negligible or acceptable levels. Because heat and wear are associated with friction, both effects can be minimized by reducing the coefficient of friction between the contacting surfaces.

Lubrication is also used to:

- Reduce oxidation and prevent rust
- Provide insulation in transformer applications
- Transmit mechanical power in hydraulic-fluid power applications
- Seal against dust, dirt, and water

## Lubricants

Reduced wear and heat are achieved by inserting a lower-viscosity (shear-strength) material between wearing surfaces that have a relatively high coefficient of friction. In effect, the wearing surfaces are replaced by a material with a more desirable coefficient of friction. Any material used to reduce friction in this way is a *lubricant*.

Lubricants are available in liquid, solid, and gaseous forms. Industrial machinery ordinarily uses oil or grease. Solid lubricants such as molybdenum disulfide and graphite are used when the loading at contact points is heavy. In some applications, the wearing surfaces of a material are plated with a different metal to reduce friction.

## Hydrodynamic or Fluid-Film Lubrication: General

In heavily loaded bearings such as thrust bearings and horizontal journal bearings, the fluid's viscosity alone is not sufficient to maintain a film between the moving surfaces. In these bearings, higher fluid pressures are required to support the load until the fluid film is established. If this pressure is supplied by an outside source, it is called *hydrostatic lubrication*. If the pressure is generated internally, that is, within the bearing by dynamic action, it is referred to as *hydrodynamic lubrication*.

In hydrodynamic lubrication, a fluid wedge is formed by the relative surface motion of the journals or the thrust runners over their respective bearing surfaces. The guide bearings of a vertical hydroelectric generator, if properly aligned, have little or no loading and will tend to operate in the center of the bearing because of the viscosity of the oil.

### (1) Thrust Bearings

In hydrodynamic lubrication, sometimes referred to as *fluid-film lubrication*, the wearing surfaces are completely separated by a film of oil. This type of lubricating action is similar to a speedboat operating on water. When the boat is not moving, it rests on the supporting water surface. As the boat begins to move, it meets a certain amount of resistance or opposing force due to the viscosity of the water. This causes the leading edge of the boat to lift slightly and allows a small amount of water to come between it and the supporting water surface. As the boat's velocity increases, the wedge-shaped water film increases in thickness until a constant velocity is attained. When the velocity is constant, water entering under the leading edge equals the amount passing outward from the trailing edge. For the boat to remain above the supporting surface, there must be an upward pressure that equals the load.

The same principle can be applied to a sliding surface. Fluid-film lubrication reduces friction between moving surfaces by substituting fluid friction for mechanical friction. To visualize the shearing effect taking place in the fluid film, imagine that the film is composed of many layers similar to a deck of cards. The fluid layer in contact with the moving surface clings to that surface, and both move at the same velocity. Similarly, the fluid layer in contact with the other surface is stationary. The layers in between move at velocities directly proportional to their distance from the moving surface. For example, at a distance of ½h from surface 1, the velocity would be ½V. The force F required to move surface 1 across surface 2 is simply the force required to overcome the friction between the layers of fluid. This internal friction, or resistance to flow, is defined as the viscosity of the fluid. Viscosity will be discussed in more detail later.

The principle of hydrodynamic lubrication also can be applied to a more practical example related to thrust bearings used in the hydropower industry. Thrust-bearing assemblies are also known as *tilting-pad bearings*. These bearings are designed to allow the pads to lift and tilt properly and provide sufficient area to lift the load of the generator. As the thrust runner moves over the thrust shoe, fluid adhering to the runner is drawn between the runner and the shoe, causing the shoe to pivot and forming a wedge of oil.

As the speed of the runner increases, the pressure of the oil wedge increases, and the runner is lifted as full fluid-film lubrication takes place. In applications where the loads are very high, some thrust bearings have high-pressure pumps to provide the initial oil film. Once the unit reaches 100% speed, the pump is switched off.

*(2) Journal Bearings*

Although not as obvious as the above-mentioned plate and thrust-bearing examples, the operation of journal or sleeve bearings is also an example of hydrodynamic lubrication. When the journal is at rest, the weight of the journal squeezes out the oil film so that the journal rests on the bearing surface. As rotation starts, the journal has a tendency to roll up the side of the bearing. At the same time, fluid adhering to the journal is drawn into the contact area. As the journal speed increases, an oil wedge is formed. The pressure of the oil wedge increases until the journal is lifted off the bearing. The journal is not only lifted vertically but also is pushed to the side by the pressure of the oil wedge. The minimum fluid-film thickness at full speed will occur at a point just to the left of center and not at the bottom of the bearing. In both the pivoting-shoe thrust bearing and the horizontal journal bearing, the minimum thickness of the fluid film increases with an increase in fluid viscosity and surface speed and decreases with an increase in load.

*(3) Film Thickness*

The preceding discussion is a very simplified attempt to provide a basic description of the principles involved in hydrodynamic lubrication. For a more precise, rigorous interpretation, refer to *American Society for Metals Handbook*, Volume 11A.[1] Simplified equations have been developed to provide approximations of film thickness with a considerable degree of precision. Regardless of how film thickness is calculated, it is a function of viscosity, velocity, and load. As viscosity or velocity increases, the film thickness increases. When these two variables decrease, the film thickness also decreases. Film thickness varies inversely with the load: as the load increases, film thickness decreases. Viscosity, velocity, and operating temperature are also interrelated. If the oil's viscosity is increased, the operating temperature will increase, and this, in turn, has a tendency to reduce viscosity. Thus an increase in viscosity tends to neutralize itself somewhat. Velocity increases also cause temperature increases that subsequently result in viscosity reduction.

### (4) Factors Influencing Film Formation

The following factors are essential to achieve and maintain the fluid film required for hydrodynamic lubrication:

- The contact surfaces must meet at a slight angle to allow formation of the lubricant wedge.
- The fluid's viscosity must be high enough to support the load and maintain adequate film thickness to separate the contacting surfaces at operating speeds.
- The fluid must adhere to the contact surfaces for conveyance into the pressure area to support the load.
- The fluid must distribute itself completely within the bearing clearance area.
- The operating speed must be sufficient to allow formation and maintenance of the fluid film.
- The contact surfaces of bearings and journals must be smooth and free of sharp surfaces that would disrupt the fluid film.

Theoretically, hydrodynamic lubrication reduces wear to zero. In reality, the journal tends to move vertically and horizontally owing to load changes or other disturbances, and some wear does occur. However, hydrodynamic lubrication reduces sliding friction and wear to acceptable levels.

## Definition of Boundary Lubrication

When a complete fluid film does not develop between potentially rubbing surfaces, the film thickness may be reduced to permit momentary dry contact between wear surface high points or asperities. This condition is characteristic of *boundary lubrication*.

Boundary lubrication occurs whenever any of the essential factors that influence formation of a full fluid film are missing. The most common example of boundary lubrication includes bearings, which normally operate with fluid-film lubrication but experience boundary lubricating conditions during routine starting and stopping of equipment. Other examples include gear-tooth contacts and reciprocating equipment.

### (1) Oiliness

Lubricants required to operate under boundary lubrication conditions must possess an added quality referred to as *oiliness* or *lubricity* to lower the coefficient of friction of the oil between the rubbing surfaces. Oiliness is an oil enhancement property provided through the use of chemical additives known as *antiwear* (AW) *agents*. AW agents have a polarizing property that enables them to behave in a manner similar to a magnet. Like a magnet, the opposite sides of the oil film have different polarities. When an AW oil adheres to metal wear surfaces, the sides of the oil film not in contact with the metal surface have identical polarities and tend to repel each other and form a plane of slippage. Most oils intended for use in heavier machine applications contain AW agents.

Examples of equipment that relies exclusively on boundary lubrication include reciprocating equipment such as engine and compressor pistons and slow-moving equipment such as turbine wicket gates. Gear teeth also rely on boundary lubrication to a great extent.

### (2) EP Gear Oil (ISO 32)

Greases containing additives to protect against extreme pressure (EP) are classified as *EP greases* (e.g., EP grease NGLI 2). EP lubrication is provided by a number of chemical compounds. The most common are compounds of boron, phosphorus, sulfur, chlorine, and combinations thereof. The compounds are activated by the higher temperature resulting from extreme pressure, not by the pressure itself. As the temperature rises, EP molecules become reactive and release derivatives of phosphorus, chlorine, or sulfur (depending on which compound is used) to react with only the exposed metal surfaces to form a new compound such as iron chloride or iron sulfide.

The new compound forms a solid protective coating that fills the asperities on the exposed metal. Thus the protection is deposited at exactly the sites where it is needed. AW agents in the EP oil continue to provide antiwear protection at sites where wear and temperature are not high enough to activate the EP agents.

## Definition of Elastohydrodynamic (EHD) Lubrication

The lubrication principles applied to rolling bodies, such as ball or roller bearings, is known as *elastohydrodynamic* (*EHD*) *lubrication*.

### (1) Rolling-Body Lubrication

Although lubrication of rolling objects operates on a considerably different principle than sliding objects, the principles of hydrodynamic lubrication can be applied, within limits, to explain the lubrication of rolling elements. An oil wedge, similar to that which occurs in hydrodynamic lubrication, exists at the lower leading edge of the bearing. Adhesion of oil to the sliding element and the supporting surface increases pressure and creates a film between the two bodies. Because the area of contact is extremely small in a roller or ball bearing, the force per unit area, or load pressure, is extremely high.

Roller-bearing load pressures may reach 34,450 kPa (5000 psi), and ball-bearing load pressures may reach 689,000 kPa (1 million psi). Under these pressures, it would appear that the oil would be entirely squeezed from between the wearing surfaces.

However, viscosity increases that occur under extremely high pressure prevent the oil from being entirely squeezed out. Consequently, a thin film of oil is maintained.

### (2) Effect of Film Roughness

The roughness of the wearing surfaces is an important consideration in EHD lubrication. *Roughness* is defined as the arithmetic average of the distance between the high and low points of a surface and is sometimes called the *centerline average* (CLA).

### (3) Effect of Film Thickness

As film thickness increases in relation to roughness, fewer asperities make contact. Engineers use the ratio of film thickness to surface roughness to estimate the life expectancy of a bearing system. The relation of bearing life to this ratio is very complex and not always predictable. In general, life expectancy is extended as the ratio increases. Full film thickness is considered to exist when the value of this ratio is between 2 and 4. When this condition prevails, fatigue failure is due entirely to subsurface stress. However, in most industrial applications, a ratio between 1 and 2 is achieved. At these values, surface stresses occur, and asperities undergo stress and contribute to fatigue as a major source of failure in antifriction bearings.

## Viscosity

Technically, the *viscosity* of an oil is a measure of the oil's resistance to shear. Viscosity is more commonly known as *resistance to flow*. If a lubricating oil is considered as a series of fluid layers superimposed on each other, the viscosity of the oil is a measure of the resistance to flow between the individual layers. A high viscosity implies a high resistance to flow, whereas a low viscosity indicates a low resistance to flow.

*Viscosity varies inversely with temperature.* Viscosity is also affected by pressure; higher pressure causes the viscosity to increase, and subsequently, the load-carrying capacity of the oil also increases. This property enables use of thin oils to lubricate heavy machinery. Load-carrying capacity also increases as operating speed of the lubricated machinery is increased.

## Methods of Measuring Viscosity

Two methods for measuring viscosity are commonly employed: *shear* and *time*.

### (1) Shear

When viscosity is determined by directly measuring shear stress and shear rate, it is expressed in centipoise (cP) and is referred to as the *absolute* or *dynamic viscosity*. In the oil industry, it is more common to use *kinematic viscosity*, which is the absolute viscosity divided by the density of the oil being tested. Kinematic viscosity is expressed in centistokes (cSt). Viscosity in centistokes is conventionally given at two standard temperatures: 40°C and 100°C (104°F and 212°F).

### (2) Time

Another method used to determine oil viscosity measures the time required for an oil sample to flow through a standard orifice at a standard tem-perature. Viscosity is then expressed in Saybolt Universal Seconds (SUS). SUS viscosities are also conventionally given at two standard temperatures: 37°C and 98°C (100°F and 210°F). As noted previously, the units of viscosity can be expressed as centipoise (cP), centistokes (cSt), or Saybolt Universal Seconds (SUS) depending on the actual test method used to measure the viscosity.

## Viscosity Index

The *viscosity index* (*VI*) is an arbitrary numbering scale that indicates the changes in oil viscosity with changes in temperature. VI can be classified as follows:

- Low VI. Below 35
- Medium VI. 35–80
- High VI. 80–110
- Very high VI. 110–125
- Super VI. 125–160
- Super high VI. Above 160–200

A high VI indicates small oil viscosity changes with temperature. A low VI indicates high viscosity changes with temperature. Therefore, a fluid that has a high VI can be expected to undergo very little change in viscosity with temperature extremes and is considered to have a stable viscosity. A fluid with a low VI can be expected to undergo a significant change in viscosity as the temperature fluctuates.

For a given temperature range, say −18 to 370°C (0–100°F), the viscosity of one oil may change considerably more than another. An oil with a VI of 95–100 would change less than one with a VI of 80. Knowing the VI of an oil is crucial when selecting a lubricant for an application and is especially critical in extremely hot or cold climates. Failure to use an oil with the proper VI when temperature extremes are expected may result in poor lubrication and equipment failure.

## Pour Point

The *pour point* is the lowest temperature at which an oil will flow. This property is crucial for oils that must flow at low temperatures. A commonly used rule of thumb when selecting oils is to ensure that the pour point is at least 10°C (20°F) lower than the lowest anticipated ambient temperature.

## Cloud Point

The *cloud point* is the temperature at which dissolved solids in an oil, such as paraffin wax, begin to form and separate from the oil. As the temperature drops, wax crystallizes and becomes visible. Certain oils must be maintained at temperatures above the cloud point to prevent clogging of filters.

## Flash Point and Fire Point

The *flash point* is the lowest temperature to which a lubricant must be heated before its vapor, when mixed with air, will ignite but not continue to burn. The *fire point* is the temperature at which lubricant combustion will be sustained. The flash and fire points are useful in determining a lubricant's *volatility* and *fire resistance*. The flash point can be used to determine the transportation and storage temperature requirements for lubricants.

Manufacturers and toll blenders also can use the flash point to detect potential product contamination. A lubricant exhibiting a flash point that is significantly lower than normal will be suspected of contamination with a volatile product. Products with a flash point less than 38°C (100°F) usually require special precautions for safe handling. The fire point for a lubricant is usually 8%–10% above the flash point.

The flash point and fire point should not be confused with the *autoignition temperature* of a lubricant, which is the temperature at which a lubricant will ignite spontaneously without an external ignition source.

## Acid Number or Neutralization Number

The *acid number* or *neutralization number* is a measure of the amount of potassium hydroxide required to neutralize the acid contained in a lubricant. Acids are formed as oils oxidize with age and service. The acid number for an oil sample is indicative of the age of the oil and can be used to determine when the oil must be changed.

## Quality

The *quality* of a lubricant is an indication of the length of time that the lubricant's essential properties will continue to perform as expected. The quality of a hydraulic fluid is the length of time that the fluid's essential properties will continue to perform as expected (i.e., the fluid's resistance to change with time). The following discussion uses lubricants to convey the characteristics of quality.

The primary properties affecting quality are oxidation stability, rust prevention, foam resistance, water separation, and antiwear properties. Many of these properties are achieved through use of chemical additives. However, these additives can enhance one property while adversely affecting another. The selection and compatibility of additives are very important to minimize adverse chemical reactions that may destroy essential properties.

## Oxidation Stability

Oxidation, or the chemical union of oil and oxygen, is one of the primary causes for decreasing the stability of lubricants. Once the reactions begin, a catalytic effect takes place. The chemical reactions result in the formation of acids that can increase the fluid viscosity and cause corrosion.

Polymerization and condensation produce insoluble gum, sludge, and varnish that cause sluggish operation, increase wear, reduce clearances, and plug lines and valves. The most significant contributors to oxidation include temperature, pressure, contaminants, water, metal surfaces, and agitation.

### (1) Temperature

The rate of chemical reactions, including oxidation, approximately doubles for every 10°C (18°F) increase in temperature. The reaction may start at a local area where the temperature is high. However, once started, the oxidation reaction has a catalytic effect that causes the rate of oxidation to increase.

### (2) Pressure

As the pressure increases, the fluid viscosity also increases, causing an increase in friction and heat generation. As the operating temperature increases, the rate of oxidation increases. Furthermore, as the pressure increases, the amount of entrained air and associated oxygen also increases. This condition provides additional oxygen to accelerate the oxidation reaction.

### (3) Contaminants

Contaminants that accelerate the rate of oxidation may be dirt, moisture, joint compounds, insoluble oxidation products, or paints. A 1% sludge concentration in a hydraulic fluid is sufficient to cause the fluid to oxidize in half the time it would take if no sludge were present. Therefore, the contaminated fluid's useful life is reduced by 50%.

### (4) Water and Metal

Certain metals, such as copper, are known to be catalysts for oxidation reactions, especially in the presence of water. Because of the production of acids during the initial stages of oxidation, the viscosity and neutralization numbers increase. The neutralization number for a fluid provides a measure of the amount of acid contained in that fluid.

The most commonly accepted oxidation test for lubricants is the ASTM Method D943 Oxidation Test. This test measures the neutralization number of oil as it is heated in the presence of pure oxygen, a metal catalyst, and water. Once started, the test continues until the neutralization number reaches a value of 2.0.

### (5) Agitation

To reduce the potential for oxidation, oxidation inhibitors are added to the base hydraulic fluid. Two types of inhibitors are generally used: *chain breakers* and *metal deactivators*. Chain breakers interrupt the oxidation reaction immediately after the reaction is initiated. Metal deactivators reduce the effects of metal catalysts.

## Rust and Corrosion Prevention

*Rust* is a chemical reaction between water and ferrous metals. *Corrosion* is a chemical reaction between chemicals (usually acids) and metals.

Water condensed from entrained air in a hydraulic system causes rust if the metal surfaces are not properly protected. In some cases, water reacts with chemicals in a hydraulic fluid to produce acids that cause corrosion. The acids attack and remove particles from metal surfaces, allowing the affected surfaces to leak and in some cases to seize.

To prevent rust, lubricants use *corrosion inhibitors* that deposit a protective film on metal surfaces. The film is virtually impervious to water and completely prevents rust once the film is established throughout the hydraulic system.

Rust inhibitors are tested according to the ASTM D665 Rusting Test. This test subjects a steel rod to a mixture of oil and salt water that has been heated to 60°C (140°F). If the rod shows no sign of rust after 24 hours, the fluid is considered satisfactory with respect to rust-inhibiting properties.

## Air Containment and Foaming

Air enters a hydraulic system through the reservoir or through air leaks within the hydraulic system. Air entering through the reservoir contributes to surface foaming on the oil. Good reservoir design and use of foam inhibitors usually eliminate surface foaming.

*Air entrainment* is a dispersion of very small air bubbles in a hydraulic fluid. Oil under low pressure absorbs approximately 10% air by volume. Under high pressure, the percentage is even greater. When the fluid is depressurized, the air produces foam as it is released from the solution.

Foam and high air entrainment in a hydraulic fluid cause erratic operation of servos and contribute to *pump cavitation*. Oil oxidation is another problem caused by air entrainment. As a fluid is pressurized, the entrained air is compressed and increases in temperature. This increased air temperature can be high enough to scorch the surrounding oil and cause oxidation. The amount of foaming in a fluid depends on the viscosity of the fluid, the source of the base stock, the refinement process, and usage.

*Foam depressants* are added to lubricants to expedite foam breakup and release of dissolved air. However, it is important to note that foam depressants do not prevent foaming or inhibit air from dissolving in the fluid. In fact, some antifoamants, when used in high concentrations to break up foam, actually retard the release of dissolved air from the fluid.

## Demulsibility or Water Separation

Water that enters a hydraulic system can emulsify and promote the collection of dust, grit, and dirt, and this can adversely affect the operation of valves, servos, and pumps; increase wear and corrosion; promote fluid oxidation; deplete additives; and plug filters.

## Antiwear Properties

Conventional lubricants are satisfactory for low-pressure and low-speed applications. However, lubricants for high-pressure (>6900 kPa or 1000.5 psi) and high-speed (>1200 rpm) applications that use vane or gear pumps must contain *AW additives*. These applications do not permit the formation of full fluid-film lubrication to protect contacting surfaces—a condition known as *boundary lubrication*.

Boundary lubrication occurs when the fluid viscosity is insufficient to prevent surface contact. AW additives provide a protective film at the contact surfaces to minimize wear. At best, use of a hydraulic fluid without the proper AW additives will cause premature wear of the pumps and result in inadequate system pressure. Eventually, the pumps will be destroyed.

Quality assurance of AW properties is determined via standard laboratory testing. Laboratory tests to evaluate AW properties of a hydraulic fluid are performed in accordance with ASTM D2882. This test procedure is generally conducted with a variety of high-speed, high-pressure pump models manufactured by Vickers or Denison.

Throughout the tests, the pumps are operated for a specified period. At the end of each period, the pumps are disassembled, and specified components are weighed. The weight of each component is compared with its initial weight; the difference reflects the amount of wear experienced by the pumps for the operating period. The components are also inspected for visual signs of wear and stress.

## Additives

Although the overall performance of an oil can be improved by introducing additives, a poor-quality oil cannot be converted into a premium-quality oil by introducing additives.

*Note:* There are limits to the amount of additives that can be introduced to improve performance. Beyond these limits, the benefits are minimal or may provide no gains in performance. They also may increase the cost of lubricants and, in some cases, are harmful to metals and seals.

## Functions of Additives

An additive may function in any of the following three ways:

- Protecting lubricated surfaces. EP additives, corrosion inhibitors, and rust inhibitors are included in this category. These additives coat the lubricated surfaces and prevent wear, corrosion, and rust.
- Improving performance. VI improvers and antifoaming agents are examples. They make the oil perform in a desired manner for specific applications.
- Protecting base lubricants. Antioxidants reduce the tendency of high-oleic base stocks to oxidize and form sludge and acids.

## Most Common Additives: Surface Additives

The primary purpose of a *surface additive* is to protect lubricated surfaces. EP agents, rust inhibitors, corrosion inhibitors, tackiness agents, AW additives, and oiliness additives are included in this category. These additives coat the lubricated surfaces to prevent wear or rust.

### (1) Rust Inhibitors

Rust inhibitors are added to most industrial lubricants to minimize rusting of metal parts, especially during shipment, storage, and equipment shutdown. Although oil and water do not mix very well, water will emulsify. In some instances, the water will remain either suspended by agitation or will rest beneath the oil on machine surfaces when agitation is absent. Rust inhibitors form a surface film that prevents water from making contact with metal parts. This is accomplished by making the oil adhere better or by emulsifying the water if it is in a low concentration.

### (2) Corrosion Inhibitors

Corrosion inhibitors suppress oxidation and prevent the formation of acids. These inhibitors form a protective film on metal surfaces and are used primarily in internal combustion engines to protect alloy bearings and other metals from corrosion.

### (3) EP Agents

EP agents react with the metal surfaces to form compounds that have a lower shear strength than the metal. The EP reaction is initiated by an increased temperature caused by the pressure between asperities on wearing surfaces. The EP reaction creates a protective coating at the specific points where protection is required. This coating reduces friction, wear, scoring, seizure, and galling of wear surfaces. EP additives are used in heavy-loading and shock-loading applications.

### (4) Tackiness Agents

In some cases, oils must adhere to surfaces extremely well. Adding sticky polymers composed of long-chain molecules increases the tackiness or adhesiveness of base lubricants.

### (5) AW Agents

Additives that cause an oil to resist wear by coating the metal surfaces are called *AW agents*. Molecules of the AW compound are polar and attach (adsorb) themselves to metal surfaces. When boundary-lubrication conditions (direct contact between metal asperities) occur, such as in starting and stopping of machinery, these molecules resist removal more than ordinary oil molecules. This reduces friction and wear.

### (6) Detergents, Dispersants

*Detergents* and *dispersants* are used primarily in hydraulic systems and internal combustion engines to keep metal surfaces clean by preventing deposition of oxidation products.

### (7) Compounded Oils (Blends)

The ability of an oil to provide a lower coefficient of friction at a given viscosity is often called *oiliness* or *lubricity*. When fatty vegetable oils are added to obtain this quality of oiliness, the lubricant is called a *compounded oil*.

Fatty base oils from high-oleic base stocks are proven to adhere to metal more strongly than mineral oils and provide more protective thin, boundary, and fluid-type films. A small amount of animal fat or vegetable oil added to a mineral oil will reduce the coefficient of friction without affecting the viscosity. Compounded oils are generally used in worm gears.

## Performance-Enhancing Additives

VI improvers, antifoaming agents, emulsifiers, demulsifiers, and pour-point depressants are examples of additives used to improve the quality performance of base lubricants.

### (1) Pour-Point Depressants

An oil's *pour point* is the temperature at which the oil ceases to flow under the influence of gravity. In cold weather, oil with a high pour point makes startup of machinery difficult or impossible. Pour-point depressants reduce the size and cohesiveness of the crystal structures, resulting in reduced pour point and increased cold flow at reduced temperatures.

*Note:* The stiffness of cold petroleum-based mineral oil is due to paraffin waxes that tend to form crystal structures.

### (2) VI Improvers

The VI is an indicator of the change in viscosity as the temperature is changed. The higher the VI, the less the viscosity of an oil changes for a given temperature change. *VI improvers* are used to limit the rate of change in viscosity with temperature. These improvers have little effect on oil viscosity at low temperatures. However, when heated, the improvers enable the oil's viscosity to increase within the limited range permitted by the type and concentration of the additive. This quality is most apparent in the application of multigrade motor oils.

### (3) Emulsifiers

Soluble oils are used as lubricants and coolants for cutting, grinding, and drilling applications in machine shops. Soluble oils require *emulsifiers* to promote rapid mixing of oil and water and to form stable emulsions. In most industrial applications, it is undesirable to have emulsified water in the oil.

### (4) Demulsifiers

*Demulsifiers* promote separation of oil and water in lubricants exposed to water.

## Fluid-Protective Additives

*Fluid-protective additives* are employed to protect the base fluid instead of the equipment. Oxidation inhibitors and foam inhibitors are examples.

### (1) Oxidation Inhibitors

Oxidation of lubricants is a major issue in industry because oils tend to oxidize. *Oxidation inhibitors* are used in most industrial lubricant applications where oil is continuously circulated or contained in a housing.

### (2) Foam Inhibitors

*Foam inhibitors* are added to reduce foaming. In many applications, air or other gases may become entrained in oil. Unless these gases are released, foam is produced.

Foaming can result in insufficient oil delivery to bearings, causing premature failure. Foam also may interfere with proper operation of equipment, such as lubricating pumps, and may result in false oil level readings. In some circumstances, foam may overflow from oil reservoirs.

## Precautions When Choosing Petroleum-Based Oils

Additives alone do not establish oil quality with respect to oxidation resistance, emulsification, pour point, and VI. Traditionally, lubricant producers do not usually state which compounds are used to enhance lubricant quality but only specify the generic function, such as AW agents, EP agents, and oxidation inhibitors.

Furthermore, producers of petroleum-based oils do not always use the same additive to accomplish the same goal. Consequently, any two brands selected for the same application may not be chemically identical. Users must be aware of these differences and that these differences may be significant when mixing different products.

### (1) Additive Depletion

Certain precautions must be taken with regard to lubricant additives in petroleum-based oils. Some additives are consumed during use. As these additives are consumed, lubricant performance for the specific application is reduced, and equipment failure may result under continued use. Oil-monitoring programs should be implemented to periodically test oils and verify that the essential additives in petroleum-based oils have not been depleted to unacceptable levels.

### (2) Product Incompatibility

Another important consideration is *incompatibility* of petroleum-based lubricants. Some petroleum-based oils, such as those used in turbine, hydraulic, motor, and gear applications, are naturally acidic. Other oils, such as motor oils and transmission fluids, are alkaline. Acidic and alkaline lubricants are incompatible.

*Application Note:* When servicing any oil lubricating system, the existing and new oils must be compatible. Oils for similar applications but produced by different manufacturers may be incompatible because of the additives used.

When incompatible fluids are mixed, the additives may be consumed because of chemical reactions with one another. The resulting oil mixture may be deficient of essential additives and therefore unsuitable for the intended application.

## Biobased Lubricants

In the past, there have been many published patents and publications using the word *degradable*. *Biobased* and *degradable* are different words with different meanings.

The term *biobased* has derived from the necessity to provide products from vegetable-, plant-, or animal-based materials. However, just because the term *biobased* is used to denote a particular lubricant, it may not mean that the composition is *ultimately degradable* or even *readily degradable* depending on the other base oils and additives that may be included in the formulation.

*Note:* The industry has not identified the word *biobased* to mean a 100% pure vegetable-based oil formula.

Vegetable oils are known as *ultimately degradable*, but many formulations are now being commercialized that are part based that include base materials with other petroleum mineral oils, polyalphaolefins (PAOs), synthetic ester base oils, polymers, and other necessary additives.

## Renewable Lubricants

The natural oily fluid film of these high-viscosity, multigrade products has proven in laboratory and field tests to outperform petroleum-based fluids and commercial vegetable oil formulas in terms of lubricity, oxidation stablility, thermal shear stability, and load-carrying capacity. Renewable lubricants make obsolete the use of harmful petroleum-based fluids.

Testing data include comparative results with all different types of lubricants, including:

- Based on ultimately degradable
- Based on readily degradable
- Mineral oil
- Synthetic esters
- Synthetic ester and mineral oil blends

## Lab Tests

### Oxidation Stability Test

The Oxidation Stability Test is also know as the Rotary Bomb Oxidation Test (ASTM D2272) is a rapid method of comparing the oxidation life of lubricants in similar formulations. It is used to evaluate the oxidation characteristics of turbine, hydraulic, transformer, and gear oils.

The test apparatus consists of a pressurized vessel axially rotating at an angle of 30 degrees from the horizontal in a bath at 150°C (302°F). Fifty grams of test oil and 5 g of water are charged to the vessel containing a copper catalyst coil. The vessel is initially pressurized with oxygen to 90 psi at room temperature. The 150°C bath temperature causes this pressure to increase to approximately 200 psi. As oxidation occurs, the pressure drops, and the usual failure point is taken as a 25 psi drop from the maximum pressure attained at 150°C. The results are reported as the number of minutes to the 25 psi loss. Longer time means better performance.

### Cold Temperature Stability Test

The mini-rotary viscometer (MRV; ASTM D4684) provides a low-shear-rate measurement. Slow sample cooling rate is the key feature of this instrument. A sample is preheated so as to a specified thermal history, which includes warming, slow cooling, and soaking cycles. The MRV measures an apparent yield stress, which is the minimum stress needed to cause oil to flow. It also measures an apparent viscosity under shear rates of 1–50 per second. This procedure was developed to predict low-temperature pumpability of motor and hydraulic oils in field service.

The test shows how long a 15-mL formulation can be stored and how it compares with other formulas at –20°C. This test also can determine the field performance of a hydraulic fluid, gear oil, or motor oil that has been static in the equipment for a period.

An example of a problem area could be in a hydraulic shifting valve that may contain less than 15 mL of fluid that has solidified and prevented the valve from working. This hydraulic valve could be on equipment that must work in critical areas such as in military tactical equipment. Another problem could be solidified oil in the pump suction or return lines, filters, or any other tight-tolerance areas.

## Viscosity

The viscosity of a fluid has a significant impact on the operational performance of a hydraulic system and is the single most important factor. If the viscosity is too high, then friction, pressure drop, power consumption, and heat generation will increase, creating sluggish operation of valves and servos. If the viscosity is too low, increased internal leakage may result under higher operating temperatures. In this case, the fluid film may be insufficient to prevent excessive wear or possible seizure of moving parts. Ultimately, pump efficiency may decrease, and sluggish operation may result.

### Mono-Grade (ISO Viscosity Grades)

If the system has a narrow operating-temperature range, then it is possible to maintain optimal fluid viscosity using a mono-grade fluid. Fluid formulations allow for a range of ISO viscosity grades (e.g., ISO 32, ISO 64, ISO 68, etc.).

### Multigrade Performance

Conversely, multigrade oils are used for systems operating in broad temperature ranges. Multigrade oil is required to maintain viscosity within permissible limits across a wide operating-temperature range.

## Oxidation Stability

The oxidation stability of an oil is perhaps the most important quality affecting fluid function and service life. Physical properties of an oil can be expected to change with time. Factors that influence these changes include:

- Mechanical stress and cavitation, which can break down and reduce fluid viscosity
- Oxidation and hydrolysis, which cause
  - Accelerated chemical decomposition
  - Formation of volatile components
  - Precipitation of insoluble materials
  - Corrosive by-products

## Compressibility

Compressibility is a measure of the amount of volume reduction due to pressure. Compressibility increases with pressure and temperature and has significant effects on high-pressure fluid systems. Compressibility is sometimes expressed by the *bulk modulus*, which is the reciprocal of compressibility.

Problems directly caused by compressibility include the following:

- Servos fail to maintain static rigidity.
- Losses occur in power or energy efficiency.
- Metal fracture and corrosive fatigue occur because of cavitation.

## Oxidation Stability

Oxidation, or the chemical union of oil and oxygen, is one of the primary causes for decreased stability of oil. Once the reactions begin, a catalytic effect takes place. The chemical reactions result in the formation of acids that can increase fluid viscosity and cause corrosion. Polymerization and condensation produce insoluble gum, sludge, and varnish that cause sluggish operation, increased wear, reduced clearances, plugged lines, and clogged valves.

## Factors Contributing to Oxidation

The most significant contributors to oxidation include temperature, pressure, contaminants, water, metal surfaces, and agitation.

### (1) Temperature

The rate of chemical reactions, including oxidation, approximately doubles for every 10°C (18°F) increase in temperature. The reaction may start at a local area where the temperature is high. However, once started, the oxidation reaction has a catalytic effect that causes the rate of oxidation to increase.

### (2) Pressure

As the pressure increases, the fluid viscosity also increases, causing an increase in friction and heat generation. As the operating temperature increases, the rate of oxidation increases. Furthermore, as the pressure increases, the amount of entrained air and associated oxygen also increases. This condition provides additional oxygen to accelerate the oxidation reaction.

### (3) Contaminants

Contaminants that accelerate the rate of oxidation may be dirt, moisture, joint compounds, insoluble oxidation products, and paints.

*Note:* Only a 1% sludge concentration in an oil is sufficient to cause the fluid to oxidize in half the time it would take if no sludge were present. Therefore, the contaminated fluid's useful life is reduced by 50%.

### (4) Water and Metal

Certain metals, such as copper, are known to be catalysts for oxidation reactions, especially in the presence of water. Because of the production of acids during the initial stages of oxidation, the viscosity and neutralization numbers increase. The total acid number (TAN) for a fluid provides a measure of the amount of acid contained in that fluid.

### (5) Agitation

To reduce the potential for oxidation, oxidation inhibitors, namely, patented stabilized technology, are compounded to the base hydraulic fluid. Two types of inhibitors are generally used: *chain breakers* and *metal deactivators*. Chain breakers interrupt the oxidation reaction immediately after the reaction is initiated. Metal deactivators reduce the effects of metal catalysts.

## Rust and Corrosion Prevention

*Rust* is a chemical reaction between water and ferrous metals. *Corrosion* is a chemical reaction between chemicals (usually acids) and metals. Water condensed from entrained air in a hydraulic system causes rust if the metal surfaces are not properly protected. In some cases, water reacts with chemicals in a hydraulic fluid to produce acids that cause corrosion. The acids attack and remove particles from metal surfaces, allowing the affected surfaces to leak and, in some cases, to seize.

## Air Entrainment and Foaming

*Air entrainment* is a dispersion of very small air bubbles in an oil. When the fluid is depressurized, the air produces foam as it is released from a solution. Foam and high air entrainment in a hydraulic fluid cause erratic operation of servos and contribute to pump cavitation. Fluid oxidation is another problem caused by air entrainment. As a fluid is pressurized, the entrained air is compressed and increases in temperature. This increased air temperature can be high enough to scorch the surrounding oil and cause oxidation.

## Demulsibility or Water Separation

Water that enters a system can emulsify and promote the collection of dust, grit, and dirt, and this can adversely affect the operation of valves, servos, and pumps; increase wear and corrosion; promote fluid oxidation; deplete additives; and plug filters.

## AW Properties

AW oils are engineered for high pressure (>6900 kPa or 1000.5 psi) and high speed (>1200 rpm) applications that use vane or gear pumps. These applications do not permit the formation of full fluid-film lubrication to protect contacting surfaces, a condition known as *boundary lubrication.*

Boundary lubrication occurs when the fluid viscosity is insufficient to prevent surface contact. AW additives provide a protective film at the contact surfaces to minimize wear.

At best, use of a hydraulic fluid without the proper AW additives will cause premature wear of the pumps and cause inadequate system pressure. Eventually, the pumps will be destroyed.

## Antifoaming Agents

Antifoaming and air release properties make lubricants highly resistant to foaming and air entrainment. Antifoaming additives prevent reservoir overflow, overheating, and contact between adjacent metal surfaces.

Antifoaming additives also eliminate sponginess in hydraulic systems and prevent damage caused by pump cavitation. Cavitation causes metal erosion, which damages hydraulic components and contaminates the fluid. In extreme

cases, cavitation can cause mechanical failure of system components. Abnormal noise in hydraulic systems is often caused by aeration or cavitation.

### Aeration

Aeration occurs when air contaminates a hydraulic fluid. Air in the hydraulic fluid makes a knocking noise when it compresses and decompresses as it circulates through the system. Symptoms of aeration include foaming of the fluid and erratic actuator movement. Aeration accelerates degradation of the fluid and causes damage to system components through loss of lubrication, overheating, and burning of seals.

### Cavitation

Cavitation is the formation of vapor cavities within a fluid. Cavitation occurs when the volume of fluid demanded by any part of a hydraulic circuit exceeds the volume of fluid being supplied. This causes the absolute pressure in that part of the circuit to fall below the vapor pressure of the hydraulic fluid and results in the formation of vapor cavities within the fluid, which implode when compressed, causing a characteristic knocking noise.

## Demulsifiers

Demulsifiers promote separation of fluid and water in hydraulic systems exposed to water.

## Pour-Point Depressants

A fluid's *pour point* is the temperature at which the fluid ceases to flow under the influence of gravity. In cold weather, fluid with a high pour point makes startup of machinery difficult or impossible. Pour-point depressants reduce the size and cohesiveness of the crystal structures, resulting in a reduced pour point and an increased cold flow at reduced temperatures.

## Fluid-Protective Additives

Fluid-protective additives are employed to protect the base fluid instead of the equipment. Oxidation inhibitors and antifoam inhibitors are examples.

### Oxidation Inhibitors

Oxidation of lubricants is a major issue in industry because high oleic base oils (HOBS) tend to readily oxidize. Until the advent of Stabilized HOBS, a patented processing technology, oxidation has kept most industries from broadly adopting vegetable-based fluids as their preferred factory lubricant. Oxidation inhibitors reduce the quantity of oxygen reacting with HOBS by forming inactive soluble compounds that bond to the lubricant receptors in place of oxygen. Oxidation inhibitors also work by passivating metal-bearing surfaces to retard the oxidation rate.

### Antifoam Inhibitors

Antifoam inhibitors are added to reduce foaming. In many applications, air or other gases may become entrained in the oil. Unless these gases are released, foam is produced. Foaming can result in insufficient oil delivery to bearings, causing premature failure. Foam also may interfere with proper operation of equipment, such as lubricating pumps, and may result in false oil level readings. Under some circumstances, foam may overflow from oil reservoirs.

## Antileak Additives

Antileak additives prevent the leakage of fluid past seals and worn tolerances under high-pressure conditions. These additives:

- Maintain seal and hose flexibility
- Provide positive seal seating
- Reduce oil leaks

- Extend seal and hose life
- Prevent blown or cracked hoses

*Note:* Leaks rank first in the list of most common problems with hydraulic equipment. Any component leaking in the system is increasing the heat load on the system, that is, a cylinder leaking high-pressure fluid past its piston seal. Leaks cause the pressure in the system to drop. When pressure drops, heat is generated.

## ASTM D2272 Rotary Bomb Oxidation Test

The rotary bomb oxidation test (RBOT) is a rapid method of comparing the oxidation life of lubricants in similar formulations. It is used to evaluate the oxidation characteristics of turbine, hydraulic, transformer, and gear oils. The results are reported as the number of minutes to a 25 psi loss. The longer the time, the greater is the stability of the fluid.

## Superior Total Acid Number (TAN)

Hydraulic system and pump manufactures have reported acidic attack of internal metal components when lubricants reach a TAN increase of 2.0 in real-life hydraulic equipment applications.

# Greases

## EP Grease Characteristics

Common ASTM tests for EP grease characteristics are listed on each product's specification page, for example, "Multi-Purpose High-Temp EP Grease, NLGI 2" and "Graphite EP Grease, NLGI 1."

## Apparent Viscosity

At startup, grease has a resistance to motion, implying a high viscosity. However, as grease is sheared between wearing surfaces and moves faster, its resistance to flow reduces. Its viscosity decreases as the rate of shear increases.

In contrast, an oil at constant temperature would have the same viscosity at startup as it has when it is moving. To distinguish between the viscosity of oil and grease, the viscosity of a grease is referred to as *apparent viscosity*. Apparent viscosity is the viscosity of a grease that holds only for the shear rate and temperature at which the viscosity is determined.

## Bleeding, Migration, Syneresis

*Bleeding* is a condition in which a liquid lubricant separates from its thickener. It is induced by high temperatures and also occurs during long storage periods. *Migration* is a form of bleeding that occurs when oil in a grease migrates out of the thickener network under certain circumstances. For example, when grease is pumped through a pipe in a centralized lubrication system, it may encounter resistance to the flow and form a plug. The oil continues to flow, migrating out of the thickener network. As the oil separates from the grease, thickener concentration increases, and plugging gets worse.

If two different greases are in contact, the oils may migrate from one grease to the other and change the structure of the grease. Therefore, it is unwise to mix two greases.

*Syneresis* is a special form of bleeding caused by shrinking or rearrangement of the structure as a result of physical or chemical changes in the thickener.

## Consistency, Penetration

The most important feature of a grease is its *rigidity* or *consistency*. A grease's consistency is its resistance to deformation by an applied force. A grease that is too stiff may not feed into areas requiring lubrication, whereas a grease that is too fluid may leak out. EP grease consistency

depends on the type and amount of thickener used and the apparent viscosity of its HOBS base oil.

The measure of grease consistency is called *penetration*. Penetration depends on whether the consistency has been altered by handling or working. ASTM D217 and ASTM D1403 testing methods measure penetration of unworked and worked greases.

To measure penetration, a cone of given weight is allowed to sink into a grease for five seconds at a standard temperature of 25°C (77°F). The depth, in tenths of a millimeter, to which the cone sinks into the grease is the penetration. A penetration of 100 would represent a solid grease, whereas a penetration of 450 would be a semifluid grease.

## National Lubricating Grease Institute (NLGI) Numbers

The NLGI has established consistency numbers or grade numbers ranging from 000 to 6 corresponding to specified ranges of penetration numbers.

## Contaminants

Greases tend to hold solid contaminants on their outer surfaces. This action protects lubricated surfaces from wear. If the contamination becomes excessive or eventually works its way down to the lubricated surfaces, the reverse occurs: the grease retains abrasive materials at the lubricated surface, and wear occurs.

## Dropping Point

*Dropping point* is an indicator of the heat resistance of EP greases. As grease temperature rises, penetration increases until the grease liquefies and the desired consistency is lost. The dropping point is the temperature at which a grease becomes fluid enough to drip. The dropping point indicates the upper temperature limit at which a grease retains its structure, not the maximum temperature at which a grease may be used.

## Evaporation

The mineral oil in traditional greases evaporates at temperatures above 177°C (350°F). Excessive oil evaporation causes mineral oil greases to harden as a result of increased thickener concentration. Therefore, higher evaporation rates require more frequent relubrication.

## Fretting Wear and False Brinelling

*Fretting* is frictional wear of components at contact points caused by minute oscillations. The oscillations are so minute that grease is displaced from between parts but is not allowed to flow back in. Localized oxidation of wear particles results, and wear accelerates.

In bearings, this localized wear appears as a depression in the race caused by oscillation of the ball or roller. The depression resembles that which occurs during Brinell hardness determination, hence the term *false brinelling*. An example would be fretting wear of automotive wheel bearings when a car is transported by train. The car is secured, but the vibration of the train over the tracks causes minute oscillation resulting in false brinelling of the bearing race.

## Oxidation Stability

Oxidation stability is the ability of an EP grease to resist a chemical union with oxygen. Similar to petroleum-based EP greases, the reaction of raw, nonstabilized vegetable oil–based greases with oxygen produces insoluble gum, sludge, carbon, and lacquer deposits. Prolonged high-

temperature exposure accelerates oxidation in these greases. Enhanced chemical stability ensures long-lasting performance.

## Shear Stability

Grease consistency may change as it is mechanically worked or sheared between wearing surfaces. EP grease's ability to maintain its consistency when worked is its *shear stability* or *mechanical stability*. A grease that softens as it is worked is called *thixotropic*. Greases that harden when worked are called *rheopectic*.

## High-Temperature Effects

*Note:* High temperatures harm greases more than they harm oils.

Grease, by its nature, cannot dissipate heat by convection like a circulating oil. Consequently, without the ability to transfer heat away, excessive temperatures result in accelerated oxidation or even carbonization, where a grease hardens or forms a crust.

*Note:* High temperatures induce softening and bleeding, causing mineral oil greases to flow away from needed areas. The mineral oil in traditional greases can flash, burn, or evaporate at temperatures above 177°C (350°F). High temperatures, above 73–79°C (165–175°F), can dehydrate certain greases, such as calcium soap grease, and cause structural breakdown. The higher evaporation and dehydration rates at elevated temperatures require more frequent mineral oil grease replacement.

Effective grease lubrication under high-temperature operation is a function of EP grease's consistency.

## Low-Temperature Effects

If the temperature of a grease is lowered enough, it will become so viscous that it can be classified as a hard grease. Pumpability suffers, and machinery operation may become impossible owing to torque limitations and power requirements. The temperature at which this occurs depends on the shape of the lubricated part and the power being supplied to it.

## Water Resistance

This is the ability of EP grease to withstand the effects of water with no change in its ability to lubricate. A soap–water lather may suspend the oil in the grease, forming an emulsion that can wash away or, to a lesser extent, reduce lubricity by diluting and changing grease consistency and texture. Rusting becomes a concern if water is allowed to contact iron or steel components.

## Base Oils

The base oil selected in formulating a grease should have the same characteristics as if the equipment is to be lubricated by oil. For instance, lower-viscosity base oils are used for grease applications at lower temperatures or high speeds and light loads, whereas higher-viscosity base oils are used for higher temperatures or low-speed and heavy-load applications.

*Note:* Traditionally, petroleum-based mineral oils or synthetic base oils are used to formulate greases. For petroleum oils in general, naphthenic oils tend to chemically mix better with soaps and additives and form stronger structures than paraffinic oils. Synthetic oils are higher in first cost but are more effective than mineral oils at high- and low-temperature extremes.

## Soap Thickeners

Dispersed in its base oil, a soap thickener gives grease its physical character. Soap thickeners not only provide consistency to grease, but they also affect desired properties such as water and heat resistance and pumpability. They also can affect the amount of an additive, such as a rust inhibitor, required to obtain a desired quality. The soap influences how a grease will flow, change shape, and age as it is mechanically worked at temperature extremes. Each soap type brings its own characteristic properties to a grease.

The principal ingredients in creating a soap are a fatty acid and an alkali. *Fatty acids* can be derived from vegetable fat such as soybean, corn, canola, sunflower, olive, castor, palm, and peanut oils. The most common alkalis used are the hydroxides from earth metals such as aluminium, calcium, lithium, and sodium.

Soap is created when a long-chain fatty acid reacts with a metal hydroxide. The metal is incorporated into the carbon chain, and the resulting compound develops a polarity. The polar molecules form a fibrous network that holds the HOBS. Thus, a somewhat rigid gel-like material called *grease* is developed. Soap concentration is varied to obtain different grease thicknesses.

Furthermore, the viscosity of the base oil affects thickness as well. Because soap qualities are also determined by the fatty acid from which the soap is prepared, not all greases made from soaps containing the same metals are identical. The name of the soap thickener refers to the metal (calcium, lithium, aluminum, etc.) from which the soap is prepared.

## Complex Soaps

The high temperatures generated by modern equipment necessitated an increase in the heat resistance of normal soap-thickened greases. As a result, *complex soap greases* were developed.

The dropping point of a complex grease is at least 38°C (100°F) higher than that of its normal soap-thickened counterpart, and its maximum usable temperature is around 177°C (350°F).

*Note:* In the past, complex soap greases were limited to this temperature because the mineral oil can flash, evaporate, or burn above this temperature.

A *complexing agent* made from a salt of the named metal is the additional ingredient in forming a complex grease. A complex soap is formed by the reaction of a fatty acid and alkali to form a soap and the simultaneous reaction of the alkali with a short-chain organic or inorganic acid to form a metallic salt, the complexing agent. Basically, a complex grease is made when a complex soap is formed in the presence of a base oil.

Common organic acids are acetic and lactic acids, and common inorganic acids are carbonates or chlorides.

### (1) Lithium Complex Grease

Smooth, buttery-textured lithium grease is by far the most popular grease when compared with all others. Traditional petroleum-based lithium soap grease contains lithium 12-hydroxystearate soap. It has a dropping point around 204°C (400°F) and can be used at temperatures up to about 135°C (275°F). It also can be used at temperatures as low as –35°C (–31°F). It has good shear stability and a relatively low coefficient of friction, which permits higher machine operating speeds. It has good water resistance, but not as good as that of calcium or aluminum. Pumpability and resistance to oil separation range from good to excellent. It does not naturally inhibit rust, but additives can provide rust resistance. Antioxidants and antiwear EP additives are also responsive in lithium greases.

Lithium complex grease and lithium soap grease have similar properties, except that the complex grease has superior thermal stability, as

indicated by a dropping point of 260°C (500°F). It is generally considered to be the nearest thing to a true multipurpose grease.

### (2) Aluminum Complex Grease

Food-grade EP greases are premium-grade aluminum complex–based greases designed to meet the stringent demands of food, beverage, pharmaceutical, and cosmetics processing plants. Aluminum complex grease has a maximum usable temperature of almost 100°C (212°F) higher than aluminum soap greases. Aluminum complex grease also has a high resistance to water and chemicals.

*Note:* When made with mineral oils, these greases tend to have shorter life in high-temperature, high-speed applications.

### (3) Calcium Complex Grease

Calcium complex grease is prepared by adding the salt calcium acetate. This salt provides the grease with EP characteristics without using an additive. Dropping points greater than 280°C (536°F) can be obtained, and the maximum usable temperature increases to approximately 177°C (350°F).

*Note:* Petroleum-based calcium complex greases often have poor pumpability in high-pressure centralized systems, where caking and hardening sometimes occur.

### (4) Calcium Sulfonate

Calcium sulfonate–based greases also have a higher dropping point, making them attractive for high-temperature applications. These properties, combined with the fact that calcium sulfonate can be formulated with EP greases for use in H1 food-grade applications, make it an attractive alternative to other grease additives. Calcium sulfonate thickeners have been around for more than 50 years.

## Additives

Solid lubricants such as molybdenum disulfide and graphite are added to grease in certain applications for high-temperature (>315°C [599°F]) and EP applications.

## Lamellar Solids

The most common additive materials are graphite and molybdenum disulfide.

### (1) Graphite

Graphite has a low coefficient of friction and very high thermal stability (2000°C [3632°F] and above). However, practical application is limited to a range of 500–600°C (932–1112°F) due to oxidation.

Furthermore, because graphite relies on adsorbed moisture or vapor to achieve low friction, use may be further limited. At temperatures as low as 100°C (212°F), the amount of water vapor adsorbed may be significantly reduced to the point that low friction cannot be maintained. In some instances, sufficient vapor may be extracted from contaminants in the surrounding environment or may be deliberately introduced to maintain low friction. When necessary, additives composed of inorganic compounds may be added to enable use at temperatures up to 550°C (1022°F).

*Note:* Another concern is that graphite promotes electrolysis. Graphite has a very significant potential of +0.25 V, which can lead to severe galvanic corrosion of copper alloys and stainless steels in saline waters.

### (2) Molybdenum Disulfide (Moly)

Like graphite, molybdenum disulfide has a low coefficient of friction, but unlike graphite, it does not rely on adsorbed vapor or moisture. In fact, adsorbed vapor actually may result in a slight but insignificant increase in friction. Moly also has greater load-carrying capacity, and its manufacturing quality is better controlled. Thermal stability in nonoxidizing environments is acceptable to 1100°C (2012°F), but in air it may be reduced to a range of 350–400°C (662–752°F).

## Dispersions of Powdered Solids

*Dispersions* are mixtures of solid lubricants in grease or fluid lubricants. The most common solids used are graphite and molybdenum disulfide, as discussed earlier. The grease or fluid provides normal lubrication, whereas the solid lubricant increases lubricity and provides EP protection. The addition of molybdenum disulfide to lubricating oils can increase load-carrying capacity, reduce wear, and increase life in roller bearings and also has been found to reduce wear and friction in automotive applications.

*Note:* Caution must be exercised when using these solids with petroleum-based greases and lubricating fluids.

- Petroleum-based greases and oils may prevent good adhesion of the solid to the protected surface.
- Detergent additives in some petroleum-based oils can inhibit the wear-reducing ability of moly and graphite, and some AW additives may actually increase wear.
- Solid lubricants also can affect the oxidation stability of petroleum-based oils and greases. Consequently, the concentration of oxidation inhibitors required must be carefully examined and controlled.

## Types of Common Greases

### (1) Calcium Grease

Calcium or lime grease, the first of the modern production greases, is prepared by reacting mineral oil with fats, fatty acids, a small amount of water, and calcium hydroxide (also known as *hydrated lime*). The water modifies the soap structure to absorb mineral oil. Because of water evaporation, calcium grease is sensitive to elevated temperatures. It dehydrates at temperatures around 79°C (175°F), at which point its structure collapses, resulting in softening and, eventually, phase separation. Greases with soft consistencies can dehydrate at lower temperatures, whereas greases with firm consistencies can lubricate satisfactorily to temperatures up to 93°C (200°F).

Despite the temperature limitations, lime grease does not emulsify in water and is excellent at resisting washout. Also, its manufacturing cost is relatively low. If a calcium grease is prepared from 12-hydroxystearic acid, the result is an anhydrous (waterless) grease. Because dehydration is not a concern, anhydrous calcium grease can be used continuously to a maximum temperature of around 110°C (230°F).

### (2) Sodium Grease

Sodium grease was developed for use at higher operating temperatures than the early hydrated calcium greases. Sodium grease can be used at temperatures up to 121°C (250°F), but it is soluble in water and readily washes out. Sodium is sometimes mixed with other metal soaps, especially calcium, to improve water resistance.

Although it has better adhesive properties than calcium grease, the use of sodium grease is declining owing to its lack of versatility. It cannot compete with water-resistant, more heat-resistant multipurpose greases. It is, however, still recommended for certain heavy-duty applications and well-sealed electric motors.

### (3) Aluminum Grease

Aluminum grease is normally clear and has a somewhat stringy texture, more so when produced from high-viscosity oils. When heated above 79°C (175°F), this stringiness increases and produces a rubberlike substance that pulls away from metal surfaces, reducing lubrication and increasing power consumption. Aluminum grease has good water resistance, good adhesive properties, and inhibits rust without additives, but it tends to be short lived. It has excellent inherent oxidation stability but relatively poor shear stability and pumpability.

## Other Greases

Thickeners other than soaps are available to make greases. Although most of these are restricted to very special applications, two non-soap greases are worthy of mention. One is organic and the other inorganic.

### (1) Polyurea Grease

Polyurea is the most important organic nonsoap thickener. It is a low-molecular-weight organic polymer produced by reacting amines (an ammonia derivative) with isocyanates, which results in an oil-soluble chemical thickener. Polyurea grease has outstanding resistance to oxidation because it contains no metal soaps (which tend to invite oxidation). It effectively lubricates over a wide temperature range of –20 to 177°C (–4 to 350°F) and has a long life.

Water resistance is good to excellent depending on the grade. Polyurea works well with many elastomer seal materials. It is used with all types of bearings but has been particularly effective in ball bearings. Its durability makes it well suited for sealed-for-life bearing applications.

Polyurea complex grease is produced when a complexing agent, most commonly calcium acetate or calcium phosphate, is incorporated into the polymer chain. In addition to the excellent properties of normal polyurea grease, these agents add inherent EP and wear-protection properties that increase the multipurpose capabilities of polyurea greases.

### (2) Organoclay

Organoclay is the most commonly used inorganic thickener. It is a modified clay that is insoluble in oil in its normal form but through complex chemical processes converts to platelets that attract and hold oil. Organoclay thickener structures are amorphous and gel-like rather than the fibrous, crystalline structures of soap thickeners.

This grease has excellent heat resistance because clay does not melt. Maximum operating temperature is limited by the evaporation temperature of its base oil. However, with frequent grease changes, this multipurpose grease can operate for short periods at temperatures up to its dropping point, which is about 260°C (500°F). A disadvantage is that greases made with higher-viscosity oils for high thermal stability will have poor low-temperature performance.

Organoclay grease has excellent water resistance but requires additives for oxidation and rust resistance. Work stability is fair to good. Pumpability and resistance to oil separation are good for this buttery-textured grease.

## Compatibility

Greases are considered incompatible when the physical or performance characteristics of the mixed grease fall below original specifications. In general, greases with different chemical compositions should not be mixed. Mixing greases of different thickeners can form a mix that is too firm to provide sufficient lubrication or, more commonly, too soft to stay in place.

*Note:* Combining greases of different base oils can produce a fluid component that will not provide a continuous lubrication film. Additives can be diluted when greases with different additives are mixed. Mixed greases may become less resistant to heat or have lower shear stability.

When a new brand of grease is introduced, it is recommended that the component part be disassembled and thoroughly cleaned to remove all the old grease. If this is not practical, the new EP grease should be injected until all traces of the prior product are flushed out. Also, the first grease changes should be more frequent than normally scheduled.

## Gear Oils

In gear systems, energy is transmitted from a power source to a terminal point through gears that change speeds, directions, and torque. EP gear oils meet or exceed the key objectives for a high-performance gear lubricant, including:

- Reduction of friction and wear
- Corrosion prevention
- Reduction of operating noise
- Improvement in heat transfer
- Removal of foreign or wear particles (from critical contact areas of gear tooth surfaces)

Gears vary greatly in their design and in their lubrication requirements. Proper lubrication is important to prevent premature wear of gear tooth surfaces. When selecting a lubricant for any gear application, the following issues must be considered:

- Types and materials of gears
- Operating conditions

Operating conditions include:

- Temperature range
- Rolling or sliding speed
- Type of steady load
- Method of lubricant application
- Type of service
- Environment

## Gear Systems

*Enclosed gears* are gears encased in an oil-tight housing. Depending on the operating conditions, enclosed gears usually require an oil with various additives (e.g., rust, oxidation, and foam inhibitors are common). EP additives are also used when loads are severe.

*Worm gears* are special because the action between the worm and the mating bull gear is sliding rather than the rolling, which is common in most gears. The sliding action allows fluid-film lubrication to take place. Another significant difference is that worm gears are usually made of dissimilar materials, which reduces the chance of galling and reduces friction.

*Note:* EP additives usually are not required for worm gears and actually may be detrimental to a bronze worm gear.

In *open-gear* applications, the lubricant must resist being thrown off by centrifugal force or being scraped off by the action of the gear teeth. A highly adhesive lubricant is required for most open-gear applications.

*Note:* Most open-gear lubricants are heavy oils, asphalt-based compounds, or soft greases. Depending on the service conditions, oxidation inhibitors or EP additives may be added. Caution must be exercised when using adhesive lubricants because they may attract and retain dust and dirt, which can act as abrasives.

## Gear Types

### (1) Spur Gears

*Spur gears* are the most common type used. Tooth contact is primarily rolling, with sliding occurring during engagement and disengagement. Some noise is normal, but it may become objectionable at high speeds.

### (2) Rack and Pinion Gears

*Rack and pinion gears* are essentially a variation of spur gears and have similar lubrication requirements.

### (3) Helical Gears

*Helical gears* operate with less noise and vibration than spur gears. At any time, the load on helical gears is distributed over several teeth, resulting in reduced wear. Because of their angular cut, teeth meshing results in thrust loads along the gear shaft. This action requires thrust bearings to absorb the thrust load and maintain gear alignment.

### (4) Herringbone Gears

*Herringbone gears* are essentially two side-by-side opposite-handed helical gears. This design eliminates thrust loads, but alignment is very critical to ensure correct teeth engagement.

### (5) Bevel Gears

*Bevel gears* are used to transmit motion between shafts with intersecting centerlines. The intersecting angle is normally 90 degrees but may be as high as 180 degrees. When the mating gears are equal in size and the shafts are positioned at 90 degrees to each other, they are referred to as *miter gears*. The teeth of bevel gears also can be cut in a curved manner to produce *spiral bevel gears*, which produce smoother and quieter operation than straight-cut bevels.

### (6) Worm Gears

Operation of *worm gears* is analogous to a screw. The relative motion between these gears is sliding rather than rolling. The uniform distribution of tooth pressures on these gears enables use of metals with inherently low coefficients of friction such as bronze wheel gears with hardened-steel worm gears. These gears rely on full fluid-film lubrication and require heavy oil compounded to enhance lubricity and film strength to prevent metal contact.

### (7) Hypoid Gears

*Hypoid gears* are similar to spiral bevel gears except that the shaft centerlines do not intersect. Hypoid gears combine the rolling action and high tooth pressure of spiral bevels with the sliding action of worm gears. This combination and the all-steel construction of the drive and driven gear result in a gear set with special lubrication requirements, including oiliness and AW additives to withstand the high tooth pressures and high rubbing speeds.

### (8) Annular Gears

*Annular gears* have the same tooth design as spur and helical gears, but unlike these gears, the annular gear has an internal configuration. The tooth action and lubrication requirements for annular gears are similar to those of spur and helical gears.

## Gear Wear and Failure

Gear failures can be traced to mechanical problems or lubricant failure. Lubricant-related failures are usually traced to contamination, oil-film collapse, additive depletion, and use of an improper lubricant for the application. The most common failures are due to particle contamination of the lubricant. Dust particles are highly abrasive and can penetrate through the oil film, causing plowing wear or ridging on metal surfaces. Water contamination can cause rust on working surfaces and eventually destroy metal integrity.

To prevent premature failure, gear selection requires careful consideration of the following:

- Gear tooth geometry
- Tooth action
- Tooth pressures
- Construction materials
- Surface characteristics
- Lubricant characteristics
- Operating environment

The first four items relate to design and application; further discussion is beyond the scope of this manual. This manual discusses only those aspects directly related to and affected by lubrication, including wear, scuffing, and contact fatigue. Refer to ANSI/AGMA Standard 1010-E95 and the *ASM Handbook*, Volume 11A, for photographs illustrating the wear modes described in the following discussion.

## Normal Wear

Normal wear occurs in new gears during the initial running-in period. The rolling and sliding actions of the mating teeth create mild wear that appears as a smooth and polished surface.

## Fatigue

### (1) Pitting

*Pitting* occurs when fatigue cracks are initiated on the tooth surface or just below the surface. Usually pits are the result of surface cracks caused by metal-to-metal contact of asperities or defects owing to low lubricant film thickness. High-speed gears with smooth surfaces and good film thickness may experience pitting due to subsurface cracks. These cracks may start at inclusions in the gear materials, which act as stress concentrators and propagate below and parallel to the tooth surface. Pits are formed when these cracks break through the tooth surface and cause material separation.

When several pits join, a larger pit, or *spall*, is formed. Another suspected cause of pitting is hydrogen embrittlement of metal due to water contamination of the lubricant. Pitting also can be caused by foreign-particle contamination of lubricant. These particles create surface stress concentration points that reduce lubricant film thickness and promote pitting.

The following guidelines should be observed to minimize the onset of pitting in gear units:

- Reduce contact stresses through load reduction or by optimizing gear geometry.
- Steel should be properly heat treated to high hardness. Carburizing is preferable.
- Gear teeth should have smooth surfaces produced by grinding or honing.
- Use proper quantities of cool, clean, and dry lubricant with the required viscosity.

### (2) Micropitting

*Micropitting* occurs on surface-hardened gears and is characterized by extremely small pits approximately 10-μm deep. Micropitted metal has a frosted or gray appearance. This condition generally appears on rough surfaces and is exacerbated by use of low-viscosity lubricants. Slow-speed gears are also prone to micropitting due to thin lubricant films. Micropitting may be sporadic and may stop when good lubrication conditions are restored following run-in. Maintaining adequate lubricant film thickness is the most important factor influencing the formation of micropitting. Higher-speed operation and smooth gear tooth surfaces also hinder the formation of micropitting.

The following guidelines should be observed to reduce the onset of micropitting in gear units:

- Use gears with smooth tooth surfaces produced by careful grinding or honing.
- Use high speeds, if possible.
- Use carburized steel with proper carbon content in the surface layers.
- Use the correct amount of cool, clean, and dry lubricant with the highest viscosity permissible for the application.

## Gear Wear

### (1) Adhesion

New gears contain surface imperfections or roughness that is inherent to the manufacturing process. During the initial run-in period, these imperfections are reduced through wear. Smoothing of the gear surfaces is to be expected. Mild wear occurs even when adequate lubrication is provided, but this wear is limited to the oxide layer of the gear teeth. Mild wear is beneficial because it increases the contact areas and equalizes the load pressures on gear tooth surfaces. Furthermore, the smooth gear surfaces increase the film thickness and improve lubrication.

The amount of wear that is acceptable depends on the expected life, noise, and vibration of the gear units. Excessive wear is characterized by loss of tooth profile, which results in high loading, and loss of tooth thickness, which may cause bending fatigue.

Wear cannot be completely eliminated. Speed, lubricant viscosity, and temperature impose practical limits on gear operating conditions. Gears that are highly loaded, operate at slow speeds (i.e., <30 m/min [100 ft/min]), and rely on boundary lubrication are particularly subject to excessive wear. Slow-speed adhesive wear is highly dependent on lubricant viscosity. Higher lubricant viscosities provide significant wear reduction, but viscosities must be carefully selected to prevent overheating.

The following guidelines should be observed to minimize the onset of adhesive wear in gear units:

■ Gear teeth should have smooth surfaces.
■ If possible, the run-in period for new gear units should be restricted to one-half load for the first hours of operation.
■ Use the highest speeds possible. High-load, slow-speed gears are boundary lubricated and are especially prone to excessive wear. For these applications, nitrided gears should be specified.
■ Avoid using lubricants with sulfur-phosphorus additives for very slow-speed gears (<3 m/min [10 ft/min]).
■ Use the required quantity of cool, clean, and dry lubricant at the highest viscosity permissible.

### (2) Abrasion

*Abrasive wear* is caused by particle contaminants in the lubricant. Particles may originate internally as a result of poor quality control during the manufacturing process. Particles also may be introduced from the outside during servicing or through inadequate filters, breathers, or seals. Internally generated particles are particularly destructive because they may become work hardened during compression between the gear teeth.

The following guidelines should be observed to prevent abrasive wear in gear units:

■ Remove internal contamination from new gearboxes. Drain and flush the lubricant before initial startup and again after 50 hours of operation. Refill with the manufacturer's recommended lubricant. Install new filters or breathers.

■ Maintain oil-tight seals and use filtered breather vents, preferably located in clean, nonpressurized areas.
■ Use good housekeeping procedures.
■ Use fine filtration for circulating oil systems. Filtration to 3 µm has proven effective in prolonging gear life.
■ Unless otherwise recommended by the gear manufacturer, change the lubricant in oil-bath systems at least every 2500 hours or every 6 months.
■ When warranted by the nature of the application, conduct laboratory analyses of lubricants. Analyses may include spectrographic, ferrographic, acid number, viscosity, and water content.
■ Use surface-hardened gear teeth, smooth tooth surfaces, and high-viscosity lubricants.

### (3) Polishing

*Polishing wear* is characterized by a mirror-like finish on the gear teeth. Polishing is caused by antiscuff additives that are too chemically reactive. An excessive reaction rate, coupled with continuous removal of surface films by very fine abrasive particles in the lubricant, may result in excessive polishing wear.

The following guidelines should be observed to prevent polishing wear in gear sets:

■ Use less chemically active antiscuff additives such as borate.
■ Remove abrasives from the lubricant by using fine filtration or by frequent oil changes.

## Scuffing

The terms *scuffing* and *scoring* are frequently interchanged. The following definitions are provided to assist in correctly ascertaining the type of damage observed: The *ASM Handbook*,[2] defines *scuffing* as localized damage caused by the occurrence of solid-phase welding between sliding surfaces. It defines *scoring* as the formation of severe scratches in the direction of sliding. The *Handbook* also stipulates that scoring may be caused by local solid-phase welding or abrasion but suggests that minor scoring should be considered as *scratching*.

Gear scuffing is characterized by material transfer between sliding tooth surfaces. Generally, this condition occurs when inadequate lubrication film thickness permits metal-to-metal contact between gear teeth. Without lubrication, direct metal contact removes the protective oxide layer on the gear metal, and the excessive heat generated by friction welds the surfaces at the contact points. As the gears separate, metal is torn and transferred between the teeth. Scuffing is most likely to occur in new gear sets during the running-in period because the gear teeth have not sufficient operating time to develop smooth surfaces.

## Critical Scuffing Temperature: Mineral Oil

Research has shown that for a given mineral oil without antiscuffing or EP additives, there is a critical scuffing temperature that is constant regardless of operating conditions. Evidence indicates that beyond the critical temperature, scuffing will occur. Therefore, the critical-temperature concept provides a useful method for predicting the onset of scuffing. The critical scuffing temperature is a function of the gear bulk temperature and the flash temperature and is expressed as

$$T_c = T_b + T_f$$

where the bulk temperature $T_b$ is the equilibrium temperature of the gears before meshing, and the flash temperature $T_f$ is the instantaneous

temperature rise caused by the local frictional heat at the gear teeth meshing point. The critical scuffing temperature for mineral oils without antiscuffing or EP additives increases directly with viscosity and varies from 150–300°C (300–570°F). However, this increased scuffing resistance appears to be directly attributed to differences in chemical composition and only indirectly to the beneficial effects of increased film thickness associated with higher viscosity.

Examination of the critical-temperature equation indicates that scuffing can be controlled by lowering either of the two contributing factors. The bulk temperature can be controlled by selecting gear geometry and design for the intended application. The flash temperature can be controlled indirectly by gear tooth smoothness and through lubricant viscosity. Smooth gear tooth surfaces produce less friction and heat, whereas increased viscosity provides greater film thickness, which also reduces frictional heat and results in a lower flash temperature. Furthermore, judicious application of lubricant can cool the gears by removing heat.

## Biobased Oils and Synthetics (Antiscuff)

For base oils, synthetics, and lubricants containing antiscuff additives, the critical temperature depends on the operating conditions and must be determined experimentally for each case. Antiscuff additives commonly used include iron sulfide and iron phosphate. These additives react chemically with the protected metal gear surface to form very strong solid films that prevent metal contact under EP and high-temperature conditions.

*Note:* As noted previously in the discussions of oil additives, the beneficial effects of EP additives are enhanced as the temperature increases.

The following guidelines should be observed to prevent scuffing in gear units:

- Specify smooth tooth surfaces produced by careful grinding or honing.
- Protect gear teeth during the running-in period by coating them with iron–manganese phosphate or plating them with copper or silver. During the first 10 hours of run-in, new gears should be operated at one-half load.
- Make sure that the gear teeth are cooled by supplying adequate amounts of cool lubricant. For circulating-oil systems, use a heat exchanger to cool the lubricant.
- Optimize gear tooth geometry. Use small teeth, addendum modification, and profile modification.
- Use accurate gear teeth, rigid gear mountings, and good helix alignment.
- Use nitrided steels for maximum scuffing resistance. Do not use stainless steel or aluminum for gears if there is a risk of scuffing.
- Use high-viscosity lubricants with antiscuff additives such as sulfur, phosphorus, or borate.

# Gear Lubrication (Lubricant Characteristics)

## EP Gear Oils

EP gear oils possess the following characteristics:

### (1) Viscosity

Good viscosity is essential to ensure cushioning and quiet operation. An oil viscosity that is too high will result in excess friction and degradation of oil properties associated with high oil operating temperature. In cold climates, gear lubricants should flow easily at low temperature.

Gear oils should have a minimum pour point that is 5°C (9°F) lower than the lowest expected temperature. The pour point for mineral gear oil is typically –7°C (20°F).

The following equation from the *ASM Handbook*, Volume 11A, provides a method for verifying the required viscosity for a specific gear based on the operating velocity:

$$v40 = 7000/V_{0.5}$$

where

$v40$ = lubricant kinematic viscosity at 40°C (105°F)(in centistokes)

$V$ = pitch line velocity (in feet per minute), given by

$$V = 0.262nd$$

where $n$ is the pinion speed in revolutions per minute (rpm), and $d$ is the pitch diameter (in inches).

### (2) Film Strength

Good film strength helps prevent metal contact and scoring between the gear teeth.

### (3) Lubricity, Oiliness

Lubricity is necessary to reduce friction.

### (4) Adhesion

Adhesion helps prevent loss of lubrication due to throw-off associated with gravity or centrifugal force, especially at high speeds.

### (5) Gear Speed

The now-superseded *Industrial Gear Lubrication Standards*, American Gear Manufacturers Association (AGMA) 250.04, used *center distance* as the primary criterion for gear lubricant selection. The new version of this standard, designated AGMA 9005-D94 Industrial Gear Lubrication, has adopted *pitch-line velocity* as the primary selection criterion.

As noted earlier, gear speed is a factor in the selection of proper oil viscosity. The pitch-line velocity determines the contact time between gear teeth. High velocities are generally associated with light loads and very short contact times. For these applications, low-viscosity oils are usually adequate. In contrast, low speeds are associated with high loads and long contact times. These conditions require higher-viscosity oils. EP additives may be required if the loads are very high.

### (6) Temperature

Ambient and operating temperatures also determine the selection of gear lubricants. Normal gear oil operating temperature ranges from 50 to 55°C (90–100°F) above ambient. Oils operating at high temperature require good viscosity and high resistance to oxidation and foaming. Caution should be exercised whenever abnormally high temperatures are experienced. High operating temperatures are indicative of oils that are too viscous for the application, excess oil in the housing, or an overloaded condition.

All these conditions should be investigated to determine the cause and correct the condition. Oil for gears operating at low ambient temperatures must be able to flow easily and provide adequate viscosity. Therefore, these gear oils must possess a high VI and low pour points.

## Open Gears

In addition to the general requirements, lubrication for open gears must meet the following requirements:

### (1) Drip Resistance

Drip resistance prevents lubricant loss, especially at high temperatures that reduce viscosity.

### (2) Brittle Resistance

Lubricant must be capable of resisting embrittlement, especially at very low temperatures.

## Enclosed Gears

In addition to the general requirements, lubrication for enclosed gears must meet the following requirements:

### (1) Chemical Stability and Oxidation Resistance

Chemical stability and oxidation resistance prevent thickening and formation of varnish or sludge. This requirement is especially significant in high-speed gears because the oil is subjected to high operating oil and air temperatures.

### (2) EP Protection

EP protection provides additional galling and welding protection for heavily loaded gears when the lubricant film thickness fails. EP lubricants are available for mild and severe (hypoid) lubricant applications.

## Types of Gear Lubricants

### Oils

Refer to AGMA 9005-D94 for the specifications for the following lubricants:

### (1) Rust and Oxidation Oils

Economical to use, rust and oxidation oils are common gear lubricants usually formulated to include chemical additives to enhance their performance qualities. Rust and oxidation oils have essential application properties for gears and bearings, good lubrication qualities, and adequate cooling qualities.

*Note:* Rust and oxidation oils are referred to as *R&O gear oils*. Not to be confused with base rapeseed oil (RO). In this manual, use of RO refers to rapeseed oil, and R&O refers to rust and oxidation oil.

### (2) Compounded Gear Lubricants

These oils are a blend of petroleum-based oils with 3% to 10% fatty or synthetic fatty oils. They are particularly useful in worm-gear drives. Except as noted in the AGMA applicable specifications, compounded oils should comply with the same specifications as R&O oils.

### (3) EP Gear Oils

EP gear oils are specially formulated to include chemical additives capable of producing a film that provides EP and antiscuffing protection. EP gear oils have the advantage of stable application over wide temperature range, good oxidation stability at high temperatures, high VIs, and low volatility. Gear oils are available for all types of applications at high-temperature extremes where other lubricants fail.

## Special Compounds and Greases

These lubricants include special greases formulated for boundary-lubricating conditions such as low-speed, low-load applications where high film strength is required. These lubricants usually contain a base oil, a thickener, and a solid lubricant such as molybdenum disulfide or graphite.

The primary disadvantage of using grease is that it accumulates foreign particles such as metal, dirt, and other loose materials that can cause significant damage if adequate maintenance is not provided. Grease also has a tendency to be squeezed out of the gear tooth meshing zone, and it does not provide satisfactory cooling.

## Open Gear Lubricants

Open gear lubricants are generally reserved for low-speed, low-load boundary-lubricating conditions. Because of the open configuration, the lubricants must be viscous and adhesive to resist being thrown off the gear teeth surfaces. The disadvantages of these lubricants are similar to those noted earlier for greases.

## Solid Lubricants

Traditionally, the solid lubricants used most commonly in gear trains are molybdenum disulfide and graphite. Use of these lubricants is reserved for special applications such as high- and low-temperature extremes where other lubricants fail to perform adequately.

## Gear Oil Applications

Spur, helical, and bevel gears have similar load and speed characteristics and similar requirements for antiscuffing and viscosity.

### (1) Spur and Helical Gears

Spur gears and helical gears usually require EP gear oils with R&O inhibitors. Low-viscosity R&O oils, such as turbine oils, are commonly used in high-speed, low-load gear units. For high-speed, low-load gear applications, EP gear oils without antiscuffing and EP agents can be used successfully, provided that the oil viscosity is capable of maintaining the required film thickness. However, low-speed gears are usually heavily loaded, so antiscuffing and EP agents are necessary to ensure adequate protection.

### (2) Hypoid Gears

Hypoid gears combine the rolling action and high tooth pressure of spiral bevel gears with the sliding action of worm gears. These severe operating conditions result in high loads, high sliding speeds, and high friction. Therefore, hypoid gears are very susceptible to scuffing.

### (3) Worm Gears

Worm gears operate under high sliding velocity and moderate loads. The sliding action produces friction that produces higher operating temperatures than those that occur in other gear sets. Normal operating temperature for worm gears may rise to 93°C (200°F) and is not a cause for concern. Lubricants for worm gears must resist the thinning due to high temperatures and the wiping effect of sliding action, and they must provide adequate cooling.

EP gear oils with superior lubricity properties are recommended. EP additives are usually not required for worm gears. However, when EP protection is required, the additive should be selected with caution to prevent damaging the bronze worm wheel.

### (4) Gear Combinations

Many applications use different gears in the same gear housing. For these applications, the lubricant must be suitable for the gears with the most demanding requirements. Generally, the other gears will operate satisfactorily with such high-performance lubricants.

### (5) Gear Shaft Bearings

Gear shaft bearings are frequently lubricated by gear oil. In most instances, this condition is acceptable. Bearings in high-speed, low-load applications may operate satisfactorily with the gear oil. Low-speed, heavily loaded gears usually require a heavier oil.

# Bearings

Bearings can be divided into two subgroups: plain bearings and rolling-contact bearings. Both have their place in the world of machines. Each type has some obvious advantages and disadvantages, but there are subtle properties as well that are often ignored. Each type of bearing can be found in a multiplicity of places, and each can be lubricated with either oil or grease. Some bearings are lubricated by water, and some are lubricated by air (as in the case of a dentist's drill).

## Plain Bearings

Plain bearings consist of two surfaces, one moving in relation to the other. Plain bearings can be the journal type, where both wear surfaces are cylindrical; thrust type, where there are two planar surfaces, one rotating on the other; and various types of sliding bearings, where one surface slides in relation to the other. All depend on a lubricating film to reduce friction. Unless an oil pump is provided to generate the oil film, these bearings rely on shaft motion to generate a hydrodynamic oil wedge.

## Advantages of Plain Bearings

- They have a very low coefficient of friction if designed and lubricated properly.
- They have very high load-carrying capabilities.
- Their resistance to shock and vibration is greater than that of rolling-contact bearings.
- The hydrodynamic oil film produced by plain bearings damps vibration, so less noise is transmitted.
- They are less sensitive to lubricant contamination than rolling-contact bearings.

## Types of Plain Bearings

### (1) Journal (Sleeve Bearings)

These are cylindrical with oil-distributing grooves. The inner surface can be babbitt lined, bronze lined, or lined with other materials that are generally softer than the rotating journal. On horizontal shafts on motors and pumps, oil rings carry oil from the oil reservoir up to the bearing. In the case of very slow-moving shafts, the bearings may be called *bushings*.

### (2) Segmented Journal Bearings

These are similar to journal bearings except that the stationary bearing consists of segments or bearing shoes. Each shoe is individually adjustable. This type of bearing is commonly found in vertical hydrogenerators and large vertical pumping units. This bearing is usually partially immersed in an oil tub.

### (3)Thrust Bearings

These bearings support axial loading and consist of a shaft collar supported by the thrust bearing, many times in segments called *thrust shoes*. The thrust shoes are sometimes allowed to pivot to accommodate formation of the supporting oil wedges. There are many different configurations of thrust bearings aimed at equalizing loading and oil wedges. The bearing is immersed in a tub of oil. On large hydrogenerators and pumps, an oil pump is sometimes used to provide an oil film at startup.

## Plain Bearing Lubrication Selection

The most common lubricants for plain bearings are oils and greases. Traditionally, nonrenewable petroleum-based mineral oils have been used except in extreme hot- and cold-temperature applications, where petroleum-based synthetics provide superior performance.

Oil is used for faster rotational speeds, where the hydrodynamic oil wedge can be formed and maintained. It also is used in high-temperature conditions where grease may melt or degrade.

Grease is used for slower rotational speeds or oscillating movements where the hydrodynamic oil wedge cannot form. It is used in cases of extreme loading, where the bearing operates in boundary-lubrication conditions. The lubricating properties of greases are significantly affected by the base oil and type of thickeners used.

## Selecting Viscosity

Viscosity is the most critical lubricant property for ensuring adequate lubrication of plain bearings. If the viscosity is too high, the bearings will tend to overheat. If the viscosity is too low, the load-carrying capacity will be reduced.

## Lubrication Choices for Spindle Bearings

Most builders of high-speed spindles today offer the same basic spindle model with either permanently grease-lubricated spindle bearings or some form of oil-lubricated spindle bearings (oil mist, oil air, oil jet lubrication). There is a tradeoff concerning the maximum spindle speed if the bearings used are permanently grease lubricated versus oil lubricated. There are also pros and cons to selecting grease- verses oil-lubricated spindle bearings. Factors influencing the decision on the choice of which type of lubrication to use are grouped and discussed next.

### (1) Oil-Lubricated Spindle Bearing Pros

Oil lubrication never runs out. When properly maintained, oil-lubricated spindle bearings can have a longer service lifetime than grease-lubricated bearings. Oil-lubricated bearings can achieve twice the bearing life of those lubricated with grease. Oil lubricated bearings are continuously flushing themselves of wear particles and other contaminants that can damage the bearings. For a given operating speed, oil-lubricated bearings allow the benefits of a larger bearing size (higher load capacity) and a larger shaft size (greater rigidity). The additional load capacity and rigidity may outweigh any of the negative aspects of oil lubrication for certain applications.

### (2) Oil-Lubricated Spindle Bearing Cons

Oil-lubricated bearings will require additional tubing lines to the spindle. Most oil-lubricated spindle startup routines have to establish lubrication several minutes before the spindle is started as the circulating system cycles the oil into bearings. Oil-lubricated spindles increase the complexity of the spindle system. Oil-lubricated spindles also increase the risk for lubrication-related problems. Incorporation of an effective oil scavenging system can be an expensive option if it was not originally designed into the spindle.

### (3) Grease-Lubricated Spindle Bearing Pros

Grease-lubricated bearings clearly offer a simpler, less-expensive system. There are no end-user maintenance requirements, and no peripheral equipment is needed.

### (4) Grease-Lubricated Spindle Bearing Cons

With grease-lubricated bearings, there is a finite amount of oil in the grease; once this oil is used up, the bearings will fail. Once wear particles and debris get into the grease, they remain. Bearings must be kept clean to achieve maximum bearing life.

## Methods Used to Supply Lubricants

Generally, oil additives are not required in plain bearing applications. Some additives and contaminants may cause corrosion, so caution should be exercised when using bearing lubricants containing additives or when contaminants may be present.

## Rolling-Contact Bearings

In rolling-contact bearings, the lubricant film is replaced by several small rolling elements between an inner and outer ring. In most cases, the rolling elements are separated from each other by cages. Basic varieties of rolling-contact bearings include ball, roller, and thrust bearings.

## Advantages of Rolling-Contact Bearings

- At low speeds, ball and roller bearings produce much less friction than plain bearings.
- Certain types of rolling-contact bearings can support both radial and thrust loading simultaneously.
- Rolling bearings can operate with small amounts of lubricant.
- Rolling-contact bearings are relatively insensitive to lubricant viscosity.
- Rolling-contact bearings have low wear rates and require little maintenance.

## Types of Rolling-Contact Bearings

### (1) Ball Bearings

Ball bearings have spherical rolling elements in a variety of configurations. They are able to carry both radial and moderate axial loads. A special type, called *maximum-type ball bearings*, can take an extra 30% radial load but cannot support axial loads.

*(2) Roller Bearings*

Roller bearings have cylindrical rolling elements and can take much higher radial loads than ball bearings but can carry no axial loads.

*(3) Tapered Roller Bearings*

These bearings have truncated cone-shaped rolling elements and are used for very high radial and thrust loads.

*(4) Double-Row Spherical Bearings*

These bearings have a double row of keg-shaped elements. The inner surface of the outer race describes part of a sphere. These bearings can handle thrust in both directions and very high radial loads.

*(5) Ball Thrust Bearings*

These bearings have ball elements between grooved top and bottom races.

*(6) Straight Roller Thrust Bearings*

These bearings have short segments of cylindrical rollers between upper and lower races. The rollers are short to minimize skidding.

*(7) Spherical Thrust Bearings*

These bearings are also called *tapered roller thrust bearings*. The lower race describes part of a sphere. The rolling elements are barrel shaped, and the outside has a larger diameter than the inside.

*(8) Needle Bearings*

These bearings have rollers whose lengths are at least four times their diameter. They are used where space is a factor and are available with or without an inner race.

## Rolling-Contact Conditions

The loads carried by the rolling elements actually cause elastic deformation of the element and race as rotation occurs. The compressive contact between curved bodies results in maximum stresses, called *hertzian contact stresses*, occurring inside the metal under the surfaces involved. The repeated stress cycling causes fatigue in the most highly stressed metal. As a result, normal wear of rolling-contact bearings appears as flaking of the surfaces. Lubrication carries away the excessive heat generated by the repeated stress cycles. While lubrication is necessary, too much lubrication, especially with grease, results in churning action and heating due to fluid friction.

## Rolling-Bearing Lubricant Selection

In most cases, the lubricant type, that is, oil or grease, is dictated by the bearing or equipment manufacturer. In practice, there can be significant overlap in applying these two types of lubricants to the same bearing. Often the operating environment dictates the choice of lubricant. For example, a roller bearing on an output shaft of a gearbox probably will be oil lubricated because it is contained in an oil environment. However, the same bearing with the same rotational speed and loading would be grease lubricated in a pillow-block arrangement.

## Grease

Grease is used for slower rotational speeds, lower temperatures, and low to medium loads. Grease is used in situations where maintenance is more difficult or irregularly scheduled. It can be used in dirty environments if seals are provided.

## Oil

Oil is used for higher rotational speeds and higher operating temperatures. It is used in maximum-loading situations and for bearing configurations where a high amount of heat generated in the bearing can be carried away by the oil. It is also used in dirty conditions when the oil is circulated and filtered.

For moderate speeds, the following viscosities are recommended:

- Ball and cylindrical roller bearings @ 12 cSt
- Spherical roller bearings @ 20 cSt
- Spherical roller thrust bearings @ 32 cSt

In general, oils will be the medium- to high-VI type with R&O inhibitors.

EP oils are required for tapered or spherical roller bearings when operating under heavy loads or shock conditions. Occasionally, EP oils may be required by other equipment or system components.

## Grease-Lubricated Bearings

For greased bearings, multipurpose high-temperature grease (NLGI 2) is the preferable choice. Maintenance problems with these machines revolve around oil and grease changes. Grease-lubricated bearings can have grease cups that provide a reservoir with a threaded top that allows new grease to be injected into the bearing by turning the top a prescribed amount at set intervals of time. In some cases, grease nipples are provided. These receive a prescribed number of strokes from a manual grease gun at specified time intervals.

Having adequate grease in rolling-element bearings is important, but too much grease can cause overheating and bearing failure. Maintenance procedures must be followed to avoid overgreasing. Bearing housings need to be disassembled and all the old grease cleaned out and replaced at appropriate intervals.

## Gears, Gear Drives, and Speed Reducers: General

Lubrication requirements for gear sets are prescribed by the equipment manufacturers based on the operating characteristics and ambient conditions under which the equipment will operate. Often the nameplate data on the equipment will indicate the type of lubricant required.

## Gear Drives

In general, gear lubricants are formulated to comply with ANSI/AGMA 9005-D94 Industrial Gear Lubrication Standard. Gear lubricants complying with the AGMA are also suitable for drive-unit bearings in contact with the gear lubricant.

The AGMA standard is intended for use by gear designers and equipment manufacturers because it requires knowing the pitch-line velocity of the gear set to select a lubricant. Because this information is rarely known, except by the gear manufacturer, the standard provides little assistance for equipment operators trying to select a gear lubricant. The superseded standards, AGMA 250.01 and AGMA 250.02, require that the operators know the centerline distance for the gear sets.

The centerline distance can be calculated or approximated by measuring the distance between the centerline of the driver and driven gear. Although updated standards have been in use for several years, many gear unit manufacturers and lubricant producers continue to publish selection criteria based on the old standard. Therefore, equipment operators may want to save the old standard for reference until manufacturers and producers update all their publications. When the pitch-line velocity is unknown or cannot be obtained in a timely manner, an educated guess may be necessary. A lubricant can be selected by referring to the old standard and subsequently verified for compliance with the latest standard.

Reference to manufacturers' data indicates that an AGMA Grade 3 or AGMA Grade 4 lubricant will cover most winter applications, and an AGMA Grade 5 or AGMA Grade 6 lubricant will cover most summer applications. EP oil should be used for heavily loaded, low-speed equipment.

*Note:* The AGMA provides recommended gear lubricants for continuous and intermittent operation. The intermittent lubricant recommendations are especially important for applications where water flow regulation requires that the gates remain in a fixed position for prolonged periods. Gear lubricants formulated for continuous operation are too thin and may run off during the standing periods, resulting in inadequate lubrication and possible gear tooth damage when the gate moves to a new position.

EP gear oils should be selected for the highest viscosity consistent with the operating conditions. When very low ambient temperatures are encountered, the oil viscosity should not be lowered. A reduced oil viscosity may be too low when the gears reach their normal operating temperature. If possible, oil heaters should be used to warm the oil in cold environments. The heater should be carefully sized to prevent hot spots that may scorch the oil.

# Couplings

Couplings requiring lubrication are usually spring-, chain-, gear-, or fluid-drive type.

## General Lubrication

Lubrication should follow the manufacturers' recommendations. When no suitable recommendations are available, NLGI Grade 1–3 grease may be used for grid couplings. Gear and chain couplings may be lubricated with NLGI Grade 0–3 grease.

## Grease-Lubricated Couplings

### (1) Normal Application

This condition is descriptive of applications where the centrifugal force does not exceed 200 g (0.44 lb), motor speed does not exceed 3600 rpm, hub misalignment does not exceed three-fourths of 1 degree, and peak torque is less than 2.5 times continuous torque. For these conditions, multipurpose high-temperature EP grease (NLGI 2) with a high-viscosity vegetable-based oil (>198 cSt at 40°C [104°F]) should be used.

### (2) Low-Speed Application

This condition includes operating where the centrifugal force does not exceed 10 g (0.2 lb). If the pitch diameter $d$ is known, the coupling speed $n$ can be estimated from the following equation:

Pitch line velocity is a function of the gear's pitch diameter and its rotational speed:

$$PLV = \frac{\pi d_p \omega}{60}$$

where:
   PLV = pitch line velocity (m/s)
   $d_p$ = pitch diameter (m)
   $\omega$ = rotational speed (rpm)

Misalignment and torque are as described for normal conditions, as indicated earlier. For these conditions, an NGLI Grade 0 or Grade 1 grease with a high-viscosity base oil (>198 cSt at 40°C [104°F]) should be used.

### (3) High-Speed Application

This condition is characterized by centrifugal forces exceeding 200 g (0.44 lb), misalignment of less than 0.5 degree, with uniform torque. The lubricant must have good resistance to centrifugal separation. Consult a manufacturer for recommendations.

*(4) High-Torque, High-Misalignment Application*

This condition is characterized by centrifugal forces of less than 200 g (0.44 lb), misalignment greater than 0.75 degree, and shock loads exceeding 2.5 times the continuous torque. Many of these applications also include high temperatures (100°C [212°F]), which limits the number of effective greases with adequate performance capability. In addition to the requirements for normal operation, the grease must have antifriction and AW additives (polydisulfide), EP additives, a Timken load greater than 20.4 kg (40 lb), and a minimum dropping point of 150°C (302°F).

# Oil Classification

Oil is normally classified by viscosity grade, additives, or use. Some oils are classified as non-specialized industrial oils.

## Classification by Viscosity Grade

Classification according to viscosity is the most prevalent method of describing oils. The most common classification systems are those of the Society of Automotive Engineers (SAE), the American Grease Manufacturers Association (AGMA), and the International Standards Organization (ISO). Each organization uses a different kinematic viscosity range numbering system.

## Classification by Additives

Oil may be further classified according to the additives included in the oil to enhance its performance properties as follows:

- R&O inhibited
- Antiwear (AW)
- Extreme pressure (EP)
- Compounded
- Residual

The first three classes are discussed throughout this manual and require no further explanation; they contain the indicated additives.

*Compounded oil* contains from 3% to 10% fatty or synthetic fatty oils. It is also called *steam cylinder oil*. The added fat reduces the coefficient of friction in situations where an extreme amount of sliding friction occurs. A very common application is in worm gear systems.

*Residual compounds* are heavy-grade straight mineral oils or EP oils. These compounds are normally mixed with a diluent to increase ease of application. After application, the diluent evaporates, leaving a heavy adhesive lubricant coating.

Residuals are often used for open gear applications, where tackiness is required to increase adhesion. This type of heavy oil should not be confused with grease. Residual oil with lower viscosity is also used in many closed gear systems. Compounded oil may contain residual oil if the desired viscosity is high.

## Classification According to Use

This system of classification arises because refining additives and types of high-oleic base stock (e.g., corn, soy, sunflower, canola) may be varied to provide desirable qualities for a given application.

Some of the more common uses include:

- Turbine oils
- Lubricants
- Engine oils—automotive, aircraft, marine, commercial
- EP gear oils
- Slideway oils
- Wire-rope lubricants
- Bar and chain lubricants
- Air-tool oils
- Cutting oils—coolants for metal cutting

## Nonspecialized Industrial Oil

This classification includes oils that are not formulated for a specific application and are frequently referred to as *general-purpose oil* in the manufacturer's product literature. These oils are generally divided into two categories: *general purpose* and *EP gear oils*.

General-purpose (GP) lubricants contain R&O additives, AW agents, antifoamants, and demulsifiers. They may be used in mechanical applications where a specialized oil is not required. Their ISO viscosity ranges from about 32 to around 460. These oils are often referred to as R&O oils or lubricants, although they may contain other additives and are not intended exclusively for hydraulic use.

Some of these oils are more highly refined and provide longer life and better performance than others. These are usually referred to as *turbine oils* or *premium grades*.

*Note:* Although used in turbines, the name *turbine oil* does not mean that their use is restricted to turbines but refers to the quality of the oil.

EP gear oils generally have a higher viscosity range, from about ISO Grade 68 to around Grade 1500 and may be regarded as GP lubricants with EP additives. Although commonly used in gear systems, EP gear oils can be used in any application where their viscosity range and additives are required.

*Note:* EP gear oils should not be confused with SAE qualifying gear oils, which are specially formulated for automotive applications.

## Producer Brand Names

Oil producers often identify their products by names that may or may not be connected with standard classifications, for example, premium R&O oil with an ISO viscosity of 68.

*Note:* Regardless of how much information may be implied by the brand name, it is insufficient to select a lubricant. A user must refer to the producer's product information brochures, performance data sheet, or material safety data sheet to determine the intended use, additives, and specifications.

## Oil Producer's Product Data and Specifications

Oil producers publish product information in brochures, pamphlets, handbooks, or on the product container or packaging. Although the amount of information varies, it generally includes the intended use, the additives (AW, EP, R&O, etc.), oil type (i.e., paraffinic, naphthenic, synthetic, compounded, etc.), and the specifications. Some producers may identify the product by its usage classification such as those noted earlier, or they may simply note the machinery class where the product can be used. Often both methods of identification are used.

Intended use designations can be misleading. For example, fact sheets for three different oils by the same producer indicate that the oils can be used for electric motors and general-purpose applications. However, all three are not suitable for every application of this equipment. One oil contains no oxidation inhibitors and is intended for use where the oil is frequently replaced. The second is an R&O oil with the usual antifoaming and demulsifying agents. AW agents are also included. The third is a turbine oil similar to the second except that the refining method and additive package provide greater protection. One turbine viscosity grade (ISO 32) is treated to resist the effects of hydrogen used as a coolant in generators. Failure to notice these differences when evaluating the data can lead to incorrect application of these lubricants.

*Note:* Producers do not usually list additives. Instead, they indicate characteristics such as good AW qualities, good water resistance, or good oxidation resistance. These qualities are not inherent to oil or contained in sufficient quantities to provide the degree of protection necessary. Therefore, the user is safe in assuming that the appropriate agent has been added to obtain the given quality.

*Note:* Product literature also gives the base oil type (i.e., biobased: rapeseed, or petroleum-based: paraffinic, naphthenic, residual compounded, or synthetic). This is important when distinguishing between renewable lubricants and nonrenewable mineral oils and synthetic fluids advertising degradability in their labels.

*Producer specifications* amount to a certification that the product meets or exceeds listed physical characteristics in terms of specific test values. The magnitude of chemical impurities also may be given. Producers vary somewhat in the amount of information in their specifications. However, kinematic viscosity (in centistokes) at 40 and 100°C (104 and 212°F), Saybolt viscosity (SUS) at 37 and 98°C (100 and 210°F), and API gravity, pour point, and flash point are generally listed. Other physical and chemical measurements also may be given if they are considered to influence the intended use.

## Grease Classification

Grease is classified by penetration number and by type of soap or other thickener. Penetration classifications have been established by the NLGI. ASTM D217 and ASTM D1403 are the standards for performing penetration tests. A penetration number indicates how easily a grease can be fed to lubricated surfaces (i.e., pumpability) or how well it remains in place.

*Note:* No method exists to classify soap thickeners. However, the producer will indicate which soap is in it. The type of soap thickener indicates probable water resistance and maximum operating temperature and gives some indication of pumpability.

*Note:* Although these are important factors, they are not the only ones of interest. These simple classifications should be regarded as starting requirements to identify a group of appropriate grease types. The final selection must be made on the basis of other information provided in the producer's specifications. Viscosity of the oil included in a grease also must be considered.

Producers also provide information and specifications for grease in brochures, pamphlets, handbooks, or on the product container or packaging. Grease specifications normally include soap thickener, penetration, included oil viscosity, and dropping point. The producer also may include ASTM test information on wear, loading, lubrication life, water washout, corrosion, oil separation, and leakage.

*Note:* Grease additives are not usually stated except for solid additives such as molybdenum disulfide or graphite or that an EP additive is included. If EP or solid additives are used, the producer will often state this emphatically, and the product name may indicate the additive.

## Principles of Selection: Manufacturer Recommendations

The prime considerations are film thickness and wear. Although film thickness can be calculated, the wear properties associated with different lubricants are more difficult to assess. Lubricants are normally tested by subjecting them to various types of physical stress. However, these tests do not completely indicate how a lubricant will perform in service. Experience has probably played a larger role than any other single criterion. Through a combination of testing and experience, machine manufacturers have learned which classes of lubricants will perform well in their products.

Professional societies have established specifications and classifications for lubricants to be used in a given mechanical application. For example, AGMA has established standard specifications for enclosed and open gear systems. These specifications have been developed from the experience of the association's membership for a wide range of applications. Thus any manufacturer has access to the collective knowledge of many contributors.

It should be noted that the equipment manufacturer's recommendation should not necessarily be considered the best selection. Individual manufacturers may have different opinions based on their experience and equipment design. The concept of *best* lubricant is ambiguous because it is based on opinion.

Despite this ambiguity, the manufacturer is probably in the best position to recommend a lubricant. This recommendation should be followed unless the lubricant fails to perform satisfactorily. When poor performance is evident, the manufacturer should be consulted for additional recommendations.

*Note:* It is especially critical that the manufacturer be consulted if the equipment is still under warranty.

Some manufacturers may provide a list of alternative lubricants that also meet the operating requirements for their equipment. One of the recommended lubricants should be used to avoid compromising the equipment warranty if it is still in effect. Physical qualities (such as viscosity or penetration number), chemical qualities (such as paraffinic or naphthenic oils), and applicable test standards are usually specified.

## Lubricant Producer Recommendations

Many lubricant producers employ product engineers to assist users in selecting lubricants and to answer technical questions. Given a manufacturer's product description, operating characteristics, unusual operating requirements, and lubricant specification, product engineers can identify lubricants that meet the manufacturer's specifications.

Viscosity should be the equipment manufacturer's recommended grade. If a recommendation seems unreasonable, the user should ask for verification or consult a different lubricant producer for a recommendation. These products probably will vary in quality and cost. The application should dictate lubricant selection.

## User Selection

The user should ensure that applicable criteria are met regardless of who makes the lubricant selection. Selection should be in the class recommended by the machinery manufacturer (R&O, EP, AW, etc.) and be in the same base-stock category (paraffinic, naphthenic, or synthetic). Furthermore, physical and chemical properties should be equal to or exceed those specified by the manufacturer. Generally, the user should follow the manufacturer's specification.

If the manufacturer's specifications are not available, determine what lubricant is currently in use. If it is performing satisfactorily, continue to use the same brand. If the brand is not available, select a brand with specifications equal to or exceeding the brand used previously. If the lubricant is performing poorly, obtain the recommendation of a product engineer. If the application is critical, get several recommendations.

Generally, the user will make a selection in either of two possible situations: *substitute a new brand for one previously in use* or *select a brand that meets an equipment manufacturer's specifications*. This will be accomplished by comparing producers' specifications with those of the manufacturer.

Product selection starts by using a *substitution list* maintained by most lubricant producers. A substitution list usually shows the products of major producers and the equivalent or competing product by other producers. Substitution lists are useful, but they have limitations. They may not be subdivided by class of lubricant.

*Note:* It is difficult to do more than compare a lubricant of one producer with one given by the publishing producer. For example, consider three producers called A, B, and C. Producer A's substitution list may compare B's products with A's or C's with A's. However, B and C cannot be compared unless A has a product equivalent to both B and C. A user would need substitution lists from many producers to be able to effectively select more than one option. Many producers claim they do not have a substitution list, or are reluctant to provide one.

A substitution list or chart is valuable because it correlates the array of brand names used by producers. It eliminates producers that do not have the desired product in their line. A substitution list should be regarded as a starting point to quickly identify potential selections.

*Note:* The lists do not suggest or imply that lubricants listed as being equivalent are identical. The lists do indicate that the two lubricants are in the same class, have the same viscosity, and are intended for the same general use.

The chart of interchangeable industrial lubricants lists the following categories:

- General-purpose lubricants
- AW hydraulic fluid
- Spindle oil
- Slideway oil
- EP gear oil
- Worm gear oil
- Cling-type gear shield oil (open gears)
- General-purpose EP lithium-based grease
- Molybdenum disulfide EP grease

Spindle oils and slideway oils are not widely used.

One of the last three classes on the list is a special preparation for open gears, and the other two are classes of grease. General-purpose oils, lubricants, and EP gear oils are best described by comparison with the nonspecialized industrial oils (discussed earlier).

Nonspecialized oils contain a category called *general-purpose oils*. This term differs from the previously described general-purpose oil category in that the additives may not be the same. In some cases, brand names indicate that EP additives have been included. In other cases, AW is indicated but not R&O. This raises the possibility that R&O additives are not present. Hydraulic fluid is a general-purpose oil, but its AW properties are sufficient to pass the Vickers vane test for hydraulic applications when this is required.

The EP gear oils should correspond to those described under nonspecialized industrial oils except that EP additives are included and viscosities may be as high as ISO 2200.

*Note:* The EP classification of EP gear oil should not be confused with SAE qualifying gear oils specially formulated for automotive applications. SAE gear oils are formulated differently.

While grease preparation varies greatly among producers, two traditional types are given: No. 2 Lithium EP and Molybdenum Disulfide EP No. 2. These are the two most widely used industrial greases. The name *Molybdenum Disulfide* designates lubricant type and does not reflect the type of soap, but the soap will usually be lithium.

*Note:* While both types are intended to provide extra protection against wear, one contains EP additives and the other contains molybdenum disulfide.

Lithium greases are the most widely used, but calcium, aluminum, polyurea, and sodium-calcium greases are also used. Furthermore, greases ranging from NLGI Grade 00 to Grade 3 are used.

Cling-type gear shield lubricants are residual oils to which a tackiness agent has been added. They are extremely adhesive and so viscous that solvents are added to permit application. After application, the solvent evaporates, leaving the adhesive viscous material. Some products contain no solvent and must be heated to reduce viscosity for application.

Ultimately, information brochures provided by the producers must be examined to verify the following:

### (1) Viscosity

The product viscosity meets the manufacturer's recommendation or is the same as a previously used lubricant that performed well. When a grease is considered, the viscosity of the included oil should be the same as that of the previous lubricant.

### (2) Intended Use

The product's intended use, as given by the producer, corresponds to the application in which the lubricant will be used.

### (3) Class of Lubricant

The class of lubricant is the same as that recommended by the equipment manufacturer or the same as a previously used lubricant that performed well. If the manufacturer recommended an R&O, AW, or EP oil or a No. 2 lithium grease, that is what should be used.

### (4) Performance Specifications

Performance specifications equal to or better than those recommended by the equipment manufacturer or those of a previously used lubricant that performed well should be used.

### (5) Additives

The product additives perform the required function even though they may not be chemically identical in several possible alternative lubricants.

## Lubricant Consolidation: General

Older machines tend to operate at slow speeds with light loads. These machines also tend to have large clearances and few lubricating points. Lubrication of such older machines is not as critical, comparatively speaking, as for modern machines that operate at higher speeds under heavier loads and with closer mechanical tolerances.

A common maintenance practice is to have inventories of several types of lubricant to service both older and newer versions of similar equipment, that is, speed reducers. This problem is further aggravated by the different types of unrelated equipment operating at a complex facility, that is, turbines, speed reducers, ropes, and chains, each requiring lubrication.

There are operating advantages and cost savings that are often incurred from lubricant consolidation:

- Reduced inventories
- Reduced storage requirements
- Reduced worker safety and health hazards
- Reduced environmental hazards
- Reduced lubricant costs

Consolidation, done properly, is a rational approach to handling the lubrication requirements at a facility while reducing the total number of lubricants in the inventory.

## Manufacturer's Recommendations

Manufacturers may recommend lubricants by brand name or by specifying the lubricant characteristics required for a machine. Depending on the machine, lubricant specifications may be restrictive, or they may be general, allowing considerable latitude. Usually the manufacturer's warranty will be honored only if the purchaser uses the lubricants recommended by the manufacturer.

Voiding the terms of a warranty is normally not advisable. In general, the specified lubricants should be used until the warranty has expired. After warranty expiration, the machine and its lubrication requirements may be included in the consolidation list for the facility. In these instances, equipment owners trade off warranty protection for:

- Advanced equipment protection
- Improvements in worker health and safety
- Compliance with environmental regulations
- Lower yearly energy requirements
- Production-line efficiencies
- Return on investment via cost savings

## Consolidation Considerations

Consolidation of base lubricants requires careful analysis and matching of equipment requirements and lubricant properties. Factors that influence selection of base lubricants include:

- Operating conditions
- Viscosity

- VI
- Pour point
- EP properties
- Oxidation inhibitors
- Rust inhibitors
- Detergent dispersant additives

With a grease, considerations also must include:

- Composition of soap base
- Consistency
- Dropping point
- Pumpability

There are several precautions that must be followed when consolidating lubricants.

### (1) Characteristics

Consideration should be given to the most severe requirements of any of the original and consolidated lubricants. To prevent equipment damage, the selected lubricant also must have these same characteristics. This is also true for greases.

### (2) Special Requirements

Applications with very specific lubricant requirements should not be consolidated.

### (3) Compatibility

Remember that some lubricant additives may not be compatible with certain metals or seals.

## Consolidation Procedure

Consolidation may be accomplished through the services of a lubricant producer or attempted by facility personnel who have knowledge of the equipment operating characteristics and lubricating requirements and an ability to read lubricant producers' product data.

### (1) Consolidation by In-House Personnel

In-house personnel should begin the consolidation process by preparing a spreadsheet identifying equipment, lubricating requirements, lubricant characteristics, and brand names. The equipment should be sorted by type of lubricant (oil, hydraulic fluid, synthetics, degradable, grease) required. Under each type, the properties of each lubricant should be grouped, such as oil viscosity, detergent dispersant requirements, EP requirements, R&O inhibitors, NLGI grade of grease, viscosity of oil component in the grease, pumpability, and so on.

At this stage, viscosity grouping can be done. For instance, if three similar oils have viscosities of 110, 150, and 190 SUS at 100°F, the 150 oil may be used as a final selection. If one of the original oils was R&O inhibited, the final product also should have this property. A second group of oils with viscosities of 280, 330, and 350 SUS at 100°F could be reduced to one oil having a viscosity in the neighborhood of 315 SUS at 100°F.

The goal is to identify the viscosity requirements and range for various equipment and see if a single lubricant can span the range. If the range can be covered, then consolidation is possible.

*Note:* The lubricant viscosity for a machine must comply with the manufacturer's requirements. Obviously, an exact match of viscosity for all equipment cannot be accomplished with the same lubricant when consolidation is the goal. Lubricants with vastly different viscosity requirements must not be consolidated.

### (2) Use Higher-Quality Lubricants

Another alternative for consolidation is to use higher-grade base lubricants that are capable of meeting the requirements of various machines. Although the cost of high-grade base lubricants is greater, this may still be offset by the benefits of consolidation:

- Reduction in the number of different lubricants needed
- Reduction in inventory-management requirements
- Price discounts for purchasing certain lubricants in greater quantity

### (3) Use Multipurpose Lubricants

Multipurpose and other general-purpose lubricants can be applied to a wide range of equipment and help reduce the number of lubricants required. Although some lubricants are not listed as multipurpose, they may be used in this capacity. For example, assume two lubricants by the same producer: one is listed as an R&O turbine oil and the other as a gear oil. Examination of product literature shows that the R&O turbine oil also can be used in bearings, gear sets, compressors, hydraulic systems, machine tools, electric motors, and roller chains, whereas the EP gear oils also can be used in circulating systems, chain drives, plain bearings, antifriction bearings, and slides. These oils may be suitable for use in a consolidation effort.

## Maintenance Schedules

Modern maintenance schedules are computer generated and are frequently referred to as *computer maintenance management systems (CMMSs)*. These systems are essential in organizing, planning and executing required maintenance activities for complex manufacturing facilities. A complete discussion of CMMSs is beyond the scope of this manual, but the following discussion summarizes some key concepts.

### (1) Key Concepts of CMMSs

The primary goals of a CMMS include scheduling resource-optimizing resource availability and reducing the cost of production, labor, materials, and tools. These goals are accomplished by tracking equipment, parts, repairs, and maintenance schedules.

The most effective CMMSs are integrated with a predictive maintenance program. This type of program should not be confused with preventive maintenance (PM), which schedules maintenance and/or replacement of parts and equipment based on manufacturers' suggestions. A PM program relies on established service intervals without regard to the actual operating conditions of the equipment. This type of program is very expensive and often results in excess downtime and premature replacement of equipment.

While a PM program relies on elapsed time, a predictive maintenance program relies on condition monitoring of machines to help determine when maintenance or replacement is necessary. Condition monitoring involves the continuous monitoring and recording of vital characteristics that are known to be indicative of a machine's condition. The most commonly measured characteristic is vibration, but other useful tests include lubricant analysis, thermography, and ultrasonic measurements. The desired tests are conducted on a periodic basis. Each new measurement is compared with previous data to determine if a trend is developing.

This type of analysis is commonly referred to as *trend analysis* or *trending* and is used to help predict failure of a particular machine component and to schedule maintenance and order parts. Trending data can be collected for a wide range of equipment, including pumps, turbines, motors, generators, gearboxes, fans, and compressors. The obvious advantage of condition monitoring is that failure often can be predicted, repairs planned, and downtime and costs reduced.

## Relative Cost of Biobased Lubricants

Cost is one of the factors to be considered when selecting lubricants. This is especially true when making substitutions such as using renewable-based lubricants, fluids, and greases in place of mineral oils and synthetics. Biobased lubricants are slightly more expensive than mineral oil lubricants. Therefore, justification for their use must be based on operating requirements for which suitable mineral oil and synthetic lubricants are not available.

# Lubricating Oil Degradation

A lubricating oil may become unsuitable for its intended purpose as a result of one or several processes. Most of these processes have been discussed in previous parts of this section, so the following discussions are brief summaries.

## Oxidation

Oxidation occurs by chemical reaction of the oil with oxygen. The first step in the oxidation reaction is the formation of hydroperoxides. Subsequently, a chain reaction is started and other compounds such as acid, resins, varnishes, sludge, and carbonaceous deposits are formed.

## Water and Air Contamination

Water may be dissolved or emulsified in oil. Water affects viscosity, promotes oil degradation and equipment corrosion, and interferes with lubrication. Air in oil systems may cause foaming, slow and erratic system response, and pump cavitation.

### (1) Results of Water Contamination in Fluid Systems

- Fluid breakdown, such as additive precipitation and oil oxidation
- Reduced lubricating-film thickness
- Accelerated metal surface fatigue
- Corrosion
- Jamming of components due to ice crystals formed at low temperatures
- Loss of dielectric strength in insulating oils

### (2) Effects of Water on Bearing Life

Studies have shown that the fatigue life of a bearing can be extended dramatically by reducing the amount of water contained in a petroleum-based lubricant.

### (3) Effect of Water and Metal Particles

Oil oxidation is increased in a hydraulic or lubricating oil in the presence of water and particulate contamination. Small metal particles act as catalysts to rapidly increase the neutralization number of the acid level.

### (4) Sources of Water Contamination

- Heat exchanger leaks
- Seal leaks
- Condensation of humid air
- Inadequate reservoir covers
- Temperature drops changing dissolved water to free water

### (5) Forms of Water in Oil

- Free water. Emulsified or droplets
- Dissolved water. Below typical oil saturation levels:
  - Hydraulic. 200–400 ppm (0.02%–0.04%)
  - Lubricating. 200–750 ppm (0.02%–0.075%)
  - Transformer. 30–50 ppm (0.003%–0.005%)

### (6) Results of Dissolved Air and Other Gases in Oils

- Foaming
- Slow system response with erratic operation
- A reduction in system stiffness
- Higher fluid temperatures
- Pump damage due to cavitation
- Inability to develop full system pressure
- Acceleration of oil oxidation

## Loss of Additives

Two of the most important additives in turbine lubricating oil are the rust inhibiting and oxidation inhibiting agents. Without these additives, oxidation of oil and the rate of rusting will increase.

## Accumulation of Contaminants

Lubricating oil can become unsuitable for further service by accumulation of foreign materials in the oil. The source of contaminants may be from within the system or from outside. Internal sources of contamination are rust, wear, and sealing products. Outside contaminants are dirt, weld spatter, metal fragments, and so on, which can enter the system through ineffective seals, dirty oil-fill pipes, and dirty makeup oil.

## Biological Deterioration

Lubricating oils are susceptible to biological deterioration if the proper growing conditions are present. Hydraulic oils are also susceptible to deterioration due to infections from local contamination. Procedures for preventing and coping with local contamination include cleaning and sterilizing, adding biocides, frequent draining of moisture from the system, and avoiding dead legs in pipes.

## Hydraulic Fluid Degradation (Water Contamination)

Given the hygroscopic nature of hydraulic fluid, water contamination is a common occurrence. Water may be introduced by exposure to humid environments, condensation in the reservoir, and when adding fluid from drums that may have been improperly sealed and exposed to rain. Leaking heat exchangers, seals, and fittings are other potential sources of water contamination.

The water saturation level is different for each type of hydraulic fluid. Below the saturation level, water will completely dissolve in the oil. Oil-based lubricants have a saturation level between 100 and 1000 ppm (0.01% and 0.1%). This saturation level will be higher at the higher operating temperatures normally experienced in hydraulic systems.

### (1) Effects of Water Contamination

Hydraulic system operation may be affected when water contamination reaches 1%–2%.

### (2) Reduced Viscosity

If the water is emulsified, the fluid viscosity may be reduced and result in poor system response, increased wear of rubbing surfaces, and pump cavitation.

### (3) Ice Formation

If free water is present and exposed to freezing temperatures, ice crystals may form. Ice may plug orifices and clearance spaces, causing slow or erratic operation.

### (4) Chemical Reactions

(a) **Galvanic Corrosion.** Water may act as an electrolyte between dissimilar metals to promote galvanic corrosion. This condition first occurs and is most visible as rust formations on the inside top surface of the fluid reservoir.

(b) **Additive Depletion.** Water may react with oxidation additives to produce acids and precipitates that increase wear and cause system fouling. AW additives such as zinc dithiophosphate (ZDTP) are commonly used for boundary-lubrication applications in high-pressure pumps, gears, and bearings. However, chemical reaction with water can destroy this additive when the system operating temperature rises above 60°C (140°F). The end result is premature component failure due to metal fatigue.

(c) **Agglomeration.** Water can act as an adhesive to bind small contaminant particles into clumps that plug the system and cause slow or erratic operation. If the condition is serious, the system may fail completely.

(d) **Microbiological Contamination.** Growth of microbes such as bacteria, algae, yeast, and fungi can occur in hydraulic systems contaminated with water. The severity of microbial contamination is increased by the presence of air. Microbes vary in size from 0.2 to 2.0 µm for single cells to up to 200 µm for multicell organisms. Under favorable conditions, bacteria reproduce exponentially. Their numbers may double in as little as 20 minutes. Unless they are detected early, bacteria may grow into an interwoven mass that will clog the system. A large quantity of bacteria also can produce significant waste products and acids capable of attacking most metals and causing component failure.

## Essential Properties of Used Oil

Several important properties of used oil must be retained to ensure continued service, as discussed next.

### (1) Viscosity

New turbine oils are sold under the ISO Viscosity Grade System. Oil manufacturers normally produce lubricating oil with viscosity of ISO VG-22, VG-32, VG-46, VG-68, VG-100, VG-150, VG-220, VG-320, and VG-460. The numbers 22–460 indicate the average oil viscosity in centistoke units at 40°C (104°F) with a range of ±10%. Most hydroelectric power plants use ISO VG-68 or VG-100 oils.

### (2) Oxidation Stability

One of the most important properties of new turbine oil is its oxidation stability. New turbine oils are highly stable in the presence of air or oxygen. In service, oxidation is gradually accelerated by the presence of a metal catalyst in the system (such as iron and copper) and by the depletion of antioxidant additives. Additives control oxidation by attacking the hydroperoxides (the first product of the oxidation step) and breaking the chain reaction that follows.

When oxidation stability decreases, the oil will undergo a complex reaction that eventually produces insoluble sludge. This sludge may settle in critical areas of the equipment and interfere with the lubrication and cooling functions of oil.

Most rust inhibitors used in turbine oils are acidic and contribute to the acid number of the new oil. An increase in acid number above the value for new oil indicates the presence of acidic oxidation products or, less likely, contamination with acidic substances. An accurate determination of the total acid number (TAN) is very important.

*Note:* The TAN test does not strictly measure oxidation stability reserve, which is better determined by the rotating bomb oxidation test (RBOT), ASTM Test Method D2272.

### (3) Freedom From Sludge

Sludge is the by-product of oil oxidation. Given the nature of the highly refined lubricant base stocks used in the manufacture of turbine oils, these oils are very poor solvents for sludge. This is the main reason why the oxidation stability reserve of the oil must be carefully monitored. Only a relatively small degree of oxidation can be permitted; otherwise, there is considerable risk of sludge deposition in bearing housings, seals, and pistons. Filtration and centrifugation can remove sludge from oil as it is formed, but if oil deterioration is allowed to proceed too far, sludge will deposit in parts of the equipment, and system flushing and an oil change may be required.

### (4) Freedom from Abrasive Contaminants

The most deleterious solid contaminants found in turbine oil systems are those left behind when the system is constructed and installed or when it is opened for maintenance and repair. Solid contaminants may also enter the system when units are outdoors, through improperly installed vents, and when units are opened for maintenance. Other means of contamination are from the wearing of metals originating within the system, rust and corrosion products, and dirty makeup oil. The presence of abrasive solids in the oil cannot be tolerated because they will cause serious damage to the system. These particles must be prevented from entering the system by flushing the system properly and using clean oil and tight seals. Once abrasive solids have been detected, they must be removed by filtration or centrifugation or both.

### (5) Corrosion Protection

The corrosion protection provided by a lubricant is of significant importance for turbine systems. New turbine oil contains a rust-inhibitor additive and must meet ASTM Test Method D665. The additive may be depleted by normal usage, removal with water in the oil, absorption on wear particles and debris, or chemical reaction with contaminants.

### (6) Water Separability

Water can enter a turbine lubricating oil system through cooler leaks, by condensation, and, to a lesser degree, via seal leaks. Water in the oil can be in either the dissolved or insoluble form. The insoluble water may be in the form of small droplets dispersed in the oil (emulsion) or in a separate phase (free state) settled at the bottom of the container. Water can react with metals to catalyze and promote oil oxidation. It may deplete rust inhibitors and may also cause rusting and corrosion.

In addition to these chemical effects on the oil, additives, and equipment, water also affects the lubrication properties of the oil. Oil containing large amounts of water does not have the same viscosity and lubricating effect as clean oil. Therefore, turbine lubricating oil should not contain a significant amount of free or dispersed water. Normally, if the oil is in good condition, water will settle to the bottom of the storage tank, where it should be drained off as a routine operating procedure. Water also may be removed by purification systems.

*Note:* If turbine oil develops poor water separability properties (poor demulsibility), significant amounts of water will stay in the system and create problems.

The water separability characteristics of an oil are adequately measured using the ASTM Test Method D1401 procedure. Insoluble water can be removed by filtration and centrifugation.

## Other Properties of Used Oils

Other properties of lubricating oil that are important, but for which direct measurement of their quantitative values is less significant, are described next.

### (1) Color

New turbine oils are normally light in color. Oil will gradually darken in service. This is accepted. However, a significant color change occurring in a short time indicates that something has changed. For example, if oil suddenly becomes hazy, it is probably being contaminated with water. A rapid darkening or clouding may indicate that oil is contaminated or excessively degraded.

### (2) Foaming Characteristics

Foaming characteristics are measured by ASTM Test Method D892. This test will show the tendency of oil to foam and the stability of the foam after it is generated. Foaming can result in poor system performance and cause serious mechanical damage. Most lubricants contain an antifoam additive to break up the foam.

### (3) Water Content

Turbine oil should be clear and bright. Most turbine oil will remain clear up to 75 ppm water at room temperature. A quick and easy qualitative analysis of insoluble water in oil is the *hot-plate test*. A small amount of oil is placed on a hot plate. If the oil smokes, there is no insoluble water. If it spatters, the oil contains free or suspended water.

### (4) Inhibitor Content

The stability of turbine lubricating oil is based on the combination of high-quality base stock with highly effective additives. Therefore, it is very important to monitor the oxidation of the turbine oil. ASTM Test Method D2272 (RBOT) is very useful for approximating the oxidation inhibitor content of the turbine oil. The remaining useful life of the oil can be estimated from this test.

### (5) Wear and Contaminant Metals

Quantitative spectrographic analysis of used-oil samples may be used to detect trace metals (and silica) and identify metal-containing contaminants. System metals such as iron and copper can be accurately identified if the sample is representative and the metals are solubilized or are very finely divided. A high silica level generally indicates dirt contamination.

### (6) Oil Operating Temperature

The recommended oil operating temperature range for a particular application is usually specified by the equipment manufacturer. Exceeding the recommended range may reduce the oil's viscosity, resulting in inadequate lubrication. Subjecting oil to high temperatures also increases the oxidation rate. As noted previously, for every 18°F (10°C) above 150°F (66°C), an oil's oxidation rate doubles, and the oil's life is essentially cut in half.

Longevity is especially critical for turbines in hydroelectric generating units, where the oil life expectancy is several years. Ideally, the oil should operate at between 50 and 60°C (120 and 140°F). Consistent operation above this range may indicate a problem such as misalignment or tight bearings. Adverse conditions of this nature should be verified and corrected.

*Note:* When operating at higher temperatures, the oil's neutralization (acid) number should be checked more frequently than dictated by normal operating temperatures. An increase in the neutralization number indicates that the oxidation inhibitors have been consumed and the oil is beginning to oxidize. The lubricant manufacturer should be contacted for recommendations on the continued use of the oil when the operating temperatures for a specific lubricant are unknown.

# Lubricant Storage and Handling

Lubricants are frequently purchased in large quantities and must be stored safely. The amount of material stored should be minimized to reduce the potential for contamination, deterioration, and health and explosion hazards associated with lubricant storage. Although lubricant storage receives due attention, equipment that has received a lubricant coating and stored is frequently forgotten. Stored equipment should be inspected on a periodic basis to ensure that damage is not occurring.

## Oil

Oil is stored in active oil reservoirs, where it is drawn as needed, and in oil drums for replenishing used stock. Each mode has its own storage requirements.

### (1) Filtered and Unfiltered Oil Tanks

Most hydroelectric power plants use bulk-oil storage systems consisting of filtered (clean) and unfiltered (dirty) oil tanks to store the oil for the thrust bearings, guide bearings, and governors. Occasionally, the filtered oil tank can become contaminated by water condensation, dust, or dirt. To prevent contamination of the bearing or governor oil reservoirs, the filtered oil should be filtered again during transfer to the bearing or governor reservoir. If this is not possible, the oil from the filtered tank should be transferred to the unfiltered oil tank to remove any settled contaminants. The filtered oil storage tank should be drained and thoroughly cleaned periodically. If the area where the storage tanks are located is dusty, a filter should be installed in the vent line. If water contamination is persistent or excessive, a water-absorbent filter, such as silica gel, may be required.

### (2) Oil Drums

If possible, oil drums should be stored indoors, away from sparks, flames, and extreme heat. The storage location must ensure that the proper temperature, ventilation, and fire-protection requirements are maintained. Tight oil drums breathe in response to temperature fluctuations, so standing water on the lid may be drawn into the drum as it inhales. Proper storage is especially important when storing lubricants because of their hygroscopic nature. To prevent water contamination, place a convex lid over drums stored outdoors.

Alternatively, the drums should be set on their side with the bungs parallel to the ground. The bungs on the drums should be tightly closed except when oil is being drawn out. If a tap or pump is installed on the drum, the outlet should be wiped clean after drawing oil to prevent dust from collecting.

## Grease

Grease should be stored in a tightly sealed container to prevent dust, moisture, or other contamination. Excessive heat may cause the grease to bleed and oxidize. Store grease in clean areas where it will not be exposed to potential contaminants and away from excessive heat sources such as furnaces or heaters. The characteristics of some greases may change with time. A grease may bleed, change consistency, or pick up contaminants during storage.

To reduce the risk of contamination, the amount of grease in storage should not exceed a one-year supply. Before purchasing grease supplies, the manufacturer or distributor should be consulted for information about the maximum shelf life and other storage requirements for the specific grease.

## Safety and Health Hazards

Safety considerations related to lubricants include knowledge of handling and the potential hazards. With this information, the necessary precautions can be addressed to minimize the risk to personnel and equipment.

## Material Safety Data Sheets

When handled properly, most lubricants are safe, but when handled improperly, some hazards may exist. Occupational Safety and Health Administration (OSHA) Communication Standard 29 CFR 1910.1200 requires that lubricant distributors provide a Material Safety Data Sheet (MSDS) at the time lubricants are purchased.

The MSDS provides essential information on the potential hazards associated with a specific lubricant and should be readily accessible to all personnel responsible for handling lubricants. The lubricant's MSDS should provide information on any hazardous ingredients, physical and chemical characteristics, fire and explosion data, health hazards, and precautions for safe use.

## Risk of Fire, Explosion, and Health Hazards

### (1) Nonrenewable Oils

Although nonrenewable lubricating oils are not highly flammable, there are many documented cases of fires and explosions. The risk of an explosion depends on the spontaneous ignition conditions for the oil vapors. These conditions can be produced when oils are contained in enclosures such as crankcases, reciprocating compressors, and large gearboxes.

### (2) Nonrenewable Lubricants

Typically, nonrenewable hydraulic systems are susceptible to explosion hazards. A leaking hose under high pressure can atomize hydraulic fluid, which can ignite if it contacts a hot surface. Synthetic fluids are less flammable than mineral oils. In normal circumstances, synthetic fluid will not support combustion once the ignition source has been removed.

### (3) Lubricants

Use of fire-resistant lubricants significantly reduces the risk of explosion. Low-volatility fluids do not produce harmful vapors and therefore have high flash points and fire points, typically 300–600°F.

## Nonrenewable Oil Health Hazards

Traditionally, nonrenewable lubricants also present health hazards when in contact with skin. Health hazards associated with nonrenewable lubricants include:

- Toxicity. Some additives contained in mineral oils may be toxic.
- Dermatitis. This may be caused by prolonged contact with neat or soluble cutting oil.
- Acne. This is caused mainly by neat cutting and grinding oils.
- Cancer. This may be caused by some mineral oil constituents.

MSDSs for products should be reviewed carefully by personnel to ensure that the proper handling procedures are used.

## Development of Environmental Regulations

### (1) Legislation Passed by Congress Is Termed an Act of Congress

The responsibility for developing rules or regulations to implement the requirements of the acts is given to various agencies of the federal government, such as the Environmental Protection Agency (EPA). The proposed regulations developed by these agencies are published daily in the *Federal Register*. After publication, the public is permitted to review and comment on the proposed regulations. All comments are evaluated after the specified review time (30 days, 60 days, etc.) has passed. The comments may or may not result in changes to the proposed regulations, which are published in the *Federal Register* as the final rules.

The final rules from the *Federal Register* are compiled annually in the *Code of Federal Regulations* (CFR). The CFR is divided into 50 titles, numbered 1–50, which represent broad areas subject to federal regulation.

*Title 40, Protection of the Environment,* contains regulations for protection of the environment. References to the CFR are made throughout this section. Copies of the CFR can be obtained from the Superintendent of Documents, U.S. Government Printing Office, Washington, DC 20402.

The general format for identifying a specific regulation in the CFR involves the use of a combination of numbers and letters. For example,

40 CFR 112.20, Facility Response Plans, indicates that the regulation is found in Title 40 of the CFR. It is further identified as Part 112. A part covers a 12–31 specific regulatory area and can range in length from a few sentences to hundreds of pages. The number 20 that follows the decimal point indicates a given section where the specific information is found. A section also may range in length from a few sentences to many pages. The section number may be followed by a series of letters and numbers in parentheses to further identify individual paragraphs.

## Water Quality Regulations

The EPA has developed water pollution regulations under legal authority of the Federal Water Pollution Control Act, also known as the Clean Water Act. These regulations are found in 40 CFR Subchapter D, Water Programs, and encompass Parts 100–149. Prominent parts of the regulation addressing oil pollution of the water are 40 CFR 110, Discharge of Oil; 40 CFR 112, Oil Pollution Prevention; and 40 CFR 113, Liability Limits for Small Onshore Storage Facilities.

### (1) Reportable Oil Discharge

40 CFR 110 requires the person in charge of a facility that discharges harmful oil to report the spill to the National Response Center (800-424-8802). The criteria for harmful oil discharges include:

1. Discharges that violate applicable water quality standards
2. Discharges that cause a film or sheen on or discoloration of the surface of the water or adjoining shorelines (Sheen means an iridescent appearance on the surface of the water.)
3. Discharges that cause a sludge or emulsion to be deposited beneath the surface of the water or on adjoining shorelines

### (2) Spill Prevention Control and Countermeasures (SPCC) Plan

40 CFR 112 requires regulated facilities that have discharged or could reasonably discharge harmful oil into navigable U.S. waters or adjoining shorelines to prepare and implement an SPCC plan. The regulation applies to non-transportation-related facilities and provides that:

- The facility's total above-ground oil storage capacity is greater than 5000 liters (1320 gallons).
- The above-ground storage capacity of a single container is in excess of 2500 liters (660 gallons).
- The total underground storage capacity of the facility is greater than 160,000 liters (42,000 gallons).

### (3) General Requirements

40 CFR 112.7 provides guidelines for preparing and implementing an SPCC plan. The SPCC plan is to follow the sequence outlined in the section and includes a discussion of the facility's conformance with the appropriate guidelines. Basic principles to embody in an SPCC plan include:

- Practices devoted to the prevention of oil spills, such as plans to minimize operational errors and equipment failures that are the major causes of spills. Operational errors can be minimized by training personnel in proper operating procedures and increasing operator awareness of the imperative nature of spill prevention. Equipment failures can be minimized through proper construction, preventive maintenance, and frequent inspections.
- Plans to contain or divert spills or use equipment to prevent discharged oil from reaching navigable waters. When it is impracticable to implement spill-

containment measures, the facility must develop and incorporate a spill contingency plan into the SPCC plan.

■ Plans to remove and dispose of spilled oil.

### (4) Specific Requirements

■ (40 CFR 112.3) Time Limits. Prepare the SPCC within six months from startup. Implement the plan within 12 months from startup, including carrying out spill prevention and containment measures. Extensions may be authorized due to nonavailability of qualified personnel or delay in construction or equipment delivery beyond the control of the owner or operator.

■ (40 CFR 112.3) Certification. A registered professional engineer must certify the SPCC plan and amendments.

■ (40 CFR 112.3) Plan Availability. A complete copy of the SPCC plan must be maintained at an attended facility or at the nearest field office if the facility is not attended at least eight hours per day.

■ (40 CFR 112.7) Training. Employee training on applicable pollution-control laws, rules and regulations, proper equipment operation, and maintenance to prevent oil discharge must be available, and spill-prevention briefings should be conducted to ensure adequate understanding of the contents of the SPCC plan.

■ (40 CFR 112.5) Plan Review. The SPCC plan should be reviewed at least once every three years.

■ (40 CFR 112.4) Amendments. Certified amendments to the SPCC plan are required when

• (40 CFR 112.4) The EPA regional administrator requires amendment after a facility has discharged more

than 3785 liters (1000 gallons) of oil into navigable waters in a single spill event or discharged oil in harmful quantities into navigable waters in two spill events within any 12-month period.

• (40 CFR 112.5) There is a change in design, construction, operation, or maintenance that affects the potential for an oil spill.

• (40 CFR 112.5) The required three-year review indicates more effective field-proven prevention and control technology will significantly reduce the likelihood of a spill.

## Facility Response Plans

40 CFR 112.20 requires facility response plans to be prepared and implemented if a facility, because of its location, could reasonably be expected to cause substantial harm to the environment by discharging oil into navigable waters or on adjoining shorelines. This regulation applies to facilities that transfer oil over water to or from vessels and have a total oil storage capacity greater than 160,000 liters (42,000 gallons) or the facility's total oil storage capacity is at least 3.78 million liters (1 million gallons) with conditions.

## Liability Limits

40 CFR 113 establishes size classifications and associated liability limits for small onshore oil storage facilities with fixed capacity of 160,000 liters (1000 barrels or 42,000 gallons) or less that discharge oil into U.S. waters and removal of the discharge is performed by the U.S. government.

*Notes*

1. Edited by Brett A. Miller, Roch J. Shipley; Ronald J. Parrington, and Daniel P. Dennies, *American Society for Metals Handbook,* ASM International, Volume 11A, 2021, doi https://doi.org/10.31399/asm.hb.v11A.9781627083294.

2. Edited by Brett A. Miller, Roch J. Shipley; Ronald J. Parrington, and Daniel P. Dennies, *American Society for Metals Handbook,* ASM International, Volume 11A, 2021, doi https://doi.org/10.31399/asm.hb.v11A.9781627083294.

# Practice Exam

1. The American Petroleum Institute (API) has defined the categories of lubricant base stocks as:

    A. I, II, II+, III, III+, IV, V
    B. Groups I, II, III, and synthetics
    C. Groups I, II, III, IV, V
    D. Mineral, synthetic, semisynthetic

2. The _____ chemical reaction of the oil with hydrogen changes the polar compounds slightly but retains them in the oil.

    A. hydrofinishing
    B. hydrolysis
    C. hydrogenation
    D. saturation

3. _____ offers methods to further aid in the removal of aromatics, sulfur, and nitrogen from the lube base stocks.

    A. Hydrofinishing
    B. Hydrolysis
    C. Hydrogenation
    D. Hydroprocessing

4. Two of the _____ methods use the feedstock from the solvent refining process are hydrofinishing and hydrotreating.

    A. synthesis
    B. hydrolysis
    C. hydrogenation
    D. hydroprocessing

5. _____ are chemical compounds added to lubricating oils to impart specific properties to the finished oils.

    A. Mistifiers
    B. Chelating agents
    C. Additives
    D. Organics

6. Pour-point depressants function by inhibiting the formation of _____ that would prevent oil flow at low temperatures.

    A. ice
    B. wax
    C. high-molecular-weight polymers
    D. sludge

7. The concentration of organic acids in the oil increases when _____

    A. the oil is mixed with another oil.
    B. the oil is mixed with water.
    C. the oil is heated.
    D. the oil is cooled.

8. _____ additives are used in many lubricating oils to reduce friction, wear, scuffing, and scoring under boundary-lubrication conditions, that is, when full lubricating films cannot be maintained.

    A. Antiwear
    B. Extreme pressure
    C. Antioxidant
    D. Anticorrosion

9. At high temperatures or under heavy loads where more severe sliding conditions exist, _____ additives are required to reduce friction, control wear, and prevent severe surface damage.

   A. antiwear
   B. extreme pressure
   C. antioxidant
   D. anticorrosion

10. _____ agents are usually compounds containing sulfur, chlorine, or phosphorus, either alone or in combination.

   A. Antiwear
   B. Extreme pressure
   C. Antioxidant
   D. Anticorrosion

11. The principal thickeners used in greases are _____

   A. clay based.
   B. aluminum.
   C. metallic soaps.
   D. animal fat.

12. _____ is a combination of a conventional metallic soap-forming material with a complexing agent.

   A. Complex grease
   B. Organic-based grease
   C. Organoclay-based grease
   D. Synthetic grease

13. Additives and modifiers commonly used in lubricating greases are _____ lubricating oils.

   A. much the same as similar materials added to
   B. specific and very different from those in
   C. only the same for extreme-duty
   D. only the same for food-grade (H1)

14. Molybdenum disulfide is used in many greases for applications in which _____

   A. the grease has to be black.
   B. high vibrations exist.
   C. food-grade grease is needed.
   D. there are low speeds.

15. _____ impart specific properties to grease.

   A. Base oils
   B. Thickeners
   C. Bearings
   D. Conditions

16. The penetration point is a measure of the relative hardness or softness of a grease and may indicate _____

   A. flow and dispensing properties.
   B. low-temperature flow.
   C. dropping point.
   D. contamination.

17. The _____ of a grease is the temperature at which material falls from the orifice of a test cup under prescribed test conditions.

   A. flow and dispensing properties
   B. low-temperature flow
   C. dropping point
   D. water washout contamination

18. A _____ is produced by combining or building individual units into a unified entity.

    A. synthetic oil
    B. biobased base oil
    C. grease
    D. conspiracy

19. _____ have outstanding flow characteristics at low temperatures, and their stability at high temperatures marks the preferred use of these lubricants.

    A. Group III base oils
    B. Soy oil base oils
    C. Polyglutarimides
    D. Synthetic base oils

20. The single most important physical characteristic of a hydraulic fluid is _____

    A. viscosity.
    B. antifoaming.
    C. pump wear protection.
    D. API classification.

21. _____ means that entrained air is released from an oil.

    A. Antifoam
    B. Air separation
    C. Air release
    D. Anti-entrapment

22. The $nD_m$ is commonly referred to as the _____

    A. bearing speed factor.
    B. pitch-line velocity.
    C. Avogedro's value.
    D. REO Speed Wagon.

23. The _____ is determined by multiplying the rotational speed in revolutions per minute $n$ by the pitch diameter in millimeters $D_m$.

    A. bearing speed factor
    B. pitch-line velocity
    C. Avogedro's value
    D. REO Speed Wagon

24. Under _____ conditions, the effect of load on film thickness is not as great as the effect of speed or oil viscosity.

    A. boundary
    B. mixed-film
    C. hydrodymanic
    D. EHL

25. Most open gears and wire ropes, many drive chains and rolling-element bearings, and some cylinders, bearings, and enclosed gears are lubricated by _____ methods.

    A. spray
    B. all-loss
    C. boundary
    D. viscoelastic

26. What is the advantage of an engine oil with a high viscosity index?

    A. Less viscous drag during starting
    B. Provides thinner oil films
    C. Stabilizes oil consumption
    D. Reduced oxidation rates

27. Many additives can have a few functions. The _____ can have the ability to neutralize the acidic end products of fuel combustion and oil oxidation.

    A. antifoam agents
    B. chelating agents
    C. detergents
    D. polyalphaolefins

28. Filter ratings are important to any system. Which factor is used to calculate filter efficiency?

    A. Media type
    B. Micron rating
    C. Beta ratio
    D. Flow rate

29. When determining the grease quantity for bearings, the following formula is used: $G = 0.005DB$. What does the $D$ represent in this formula?

    A. Bearing outside diameter (mm)
    B. Bearing inside diameter (mm)
    C. Bearing width (mm)
    D. Bearing roller diameter (mm)

30. A centralized system for rotating equipment needs to be lubricated within the steel industry over a large area. The system must be designed to have good filtration (contamination is a challenge), reduce oil usage (by as much as 30%), and lower operating costs. Which would be the best type of design for this system?

    A. Wick oilers
    B. Oil mist lubrication
    C. Constant-level oilers
    D. Splash lubrication

31. When establishing a lube route, it is important to identify the _____

    A. lubrication points.
    B. lubricants in storage.
    C. lubrication costs.
    D. lubricants not in use.

32. Storage and handling of lubricants are critical. Which of the following is *not* a best practice when receiving lubricants at your site?

    A. Keep lubricants in clean, cool, dry storage at all times.
    B. Place identification on lubricants when in storage.
    C. Use any available containers for lubricant transfers.
    D. Control transfer of lubricant from the delivery vehicle.

33. Automatic grease systems need to be maintained. Which of the following is the *most appropriate* maintenance technique?

    A. Inspections can be done once per year.
    B. No calibration is required for these grease guns.
    C. Inspection is not required for the applicators.
    D. Visual inspections can detect anomalies.

34. As per the Stribeck curve, which of the following is in the correct order?

    A. Boundary, mixed film, elastohydrodynamic, hydrodynamic
    B. Boundary, mixed film, hydrodynamic, elastohydrodynamic
    C. Mixed film, boundary, elastohydrodynamic, hydrodynamic
    D. Mixed film, boundary, hydrodynamic, elastohydrodynamic

35. Which of the following is a characteristic required of biodegradable lubricants?

   A. Viscosity of the biodegradable products
   B. Speed at which the products biodegrade
   C. Operating temperature of the products
   D. Energy produced to create the products

36. When selecting a lubricating oil, which is *not* a factor to be considered?

   A. Cost
   B. Operating environment
   C. Load
   D. Temperature

37. Fire-resistant fluids are required for certain applications. Which of the following is *not* a consideration when selecting a fire-resistant fluid?

   A. Cost
   B. Function
   C. Load
   D. Risk

38. Industrial gears can experience different types of action between the gear teeth depending on the type of gear. Which of the following gears would experience sliding at right angles to the lines of contact?

   A. Helical gears
   B. Spur gears
   C. Herringbone gears
   D. Worm gears

39. When selecting the right type of grease for a chassis, it is important that the product contains _____

   A. antifoaming additives.
   B. antiwear additives.
   C. viscosity improvers.
   D. detergents.

40. The flash point of a lubricant is the lowest temperature at which the lubricant must be heated before its vapor when mixed with air will

   _____

   A. ignite and continue to burn.
   B. ignite but not continue to burn.
   C. spark and fully combust.
   D. spark spontaneously.

41. When taking an oil sample, which of the following should be used?

   A. Sample from a nonturbulent area.
   B. Sample when the machine is hot.
   C. Sample tubing can be reused.
   D. Sample from the bottom of the sump.

42. Oxidation degrades the quality of an oil. Which of the following is *not* a contributing factor to oxidation?

   A. High temperature
   B. Load speed
   C. Wear metal
   D. Oil aeration

43. The NLGI grade of a grease relates to which of its properties?

   A. Thickener
   B. Base oil
   C. Consistency
   D. Additive

**44.** Coupling greases must be able to withstand _____

   **A.** centripetal forces.
   **B.** centrifugal forces.
   **C.** low-torque loads.
   **D.** low rotational speeds.

**45.** Additive depletion can occur in several ways. Which of the following is *not* a mode of additive depletion?

   **A.** Decomposition
   **B.** Contamination
   **C.** Adsorption
   **D.** Separation

**46.** Food-grade lubricants must provide the same functions as oils but be approved by the USDA. Which category best describes H1 food-grade oils?

   **A.** Possibility of incidental food contact
   **B.** No possibility of food contact
   **C.** Edible oils
   **D.** Nonedible oils

**47.** Polyalkylene glycols are also known as PAGs. Which base oil group do they fall under?

   **A.** Group II
   **B.** Group III
   **C.** Group IV
   **D.** Group V

**48.** Hydrodynamic lubrication is usually achieved during _____

   **A.** startup.
   **B.** shutdown.
   **C.** operation.
   **D.** the highest load.

**49.** Pitting is a form of wear characterized by the presence of _____

   **A.** fractures.
   **B.** surface cavities.
   **C.** extensive grooves.
   **D.** scratches.

**50.** What should be avoided when handling full drums of lubricant at your site?

   **A.** Use of appropriate lifting gear
   **B.** Working under suspended loads
   **C.** Bending knees when lifting
   **D.** Use of two people to move drums

# Glossary of Lubrication Terms

**Abrasion**  A general wearing away of a surface by constant scratching, usually due to the presence of foreign matter such as dirt, grit, or metallic particles in the lubricant. It may also cause a breakdown of the material (such as the tooth surfaces of gears). Lack of proper lubrication may result in abrasion.

**Abrasive wear**  Wear between two surfaces in relative motion due to particles (three-body) or surface roughness (two-body).

**Abrasive wear** (or *cutting wear*)  Comes about when hard-surface asperities or hard particles that have embedded themselves into a soft surface plow grooves into the opposing harder surface (e.g., a journal).

**Absolute filtration rating**  The diameter of the largest hard spherical particle that will pass through a filter under specified test conditions. This is an indication of the largest opening in the filter elements.

**Absolute viscosity**  A term used interchangeably with viscosity to distinguish it from either kinematic viscosity or commercial viscosity. Absolute viscosity is the ratio of shear stress to shear rate. It is a fluid's internal resistance to flow. The common unit of absolute viscosity is the poise. Absolute viscosity divided by fluid density equals kinematic viscosity. It is occasionally referred to as *dynamic viscosity*. Absolute viscosity and kinematic viscosity are expressed in fundamental units. Commercial viscosity such as Saybolt viscosity is expressed in arbitrary units of time, usually seconds.

**AC fine test dust (ACFTD)**  A test contaminant used to assess both filters and the contaminant sensitivity of all types of tribologic mechanisms.

**Accumulator**  A container in which fluid is stored under pressure as a source of fluid power.

**Acid number**  The quantity of base, expressed in milligrams of potassium hydroxide, that is required to neutralize the acidic constituents in 1 gram of sample.

**Acid sludge**  The residue left after treating petroleum oil with sulfuric acid for the removal of impurities. It is a black viscous substance containing the spent acid and impurities.

**Acid treating**  A refining process in which unfinished petroleum products, such as gasoline, kerosene, and lubricating oil stocks, are contacted with sulfuric acid to improve their color, odor, and other properties.

**Acidity**  In lubricants, acidity denotes the presence of acid-type constituents whose concentration is usually defined in terms of total acid number. The constituents vary in nature and may or may not markedly influence the behavior of the lubricant.

**Activated alumina**  A highly porous material produced from dehydroxylated aluminum hydroxide; is used as a desiccant and a filtering medium.

**Actuator**  A device used to convert fluid energy into mechanical motion.

**Additive**  Any material added to a base stock to change its properties, characteristics, or performance.

**Additive**  A chemical substance added to a petroleum product to impart or improve certain properties. Common petroleum product additives are antifoaming agents, antiwear additives, corrosion inhibitors, demulsifiers, detergents, dispersants, emulsifiers, extreme-pressure additives, oiliness agents, oxidation inhibitors, pour-point depressants, rust inhibitors, tackiness agents, and viscosity index (VI) improvers.

**Additive level**  The total percentage of all additives in an oil (expressed in percent of mass [weight] or percent of volume).

**Additive stability** The ability of additives in a fluid to resist changes in their performance during storage or use.

**Adhesion** The attraction or joining of two materials, such as a lubricating grease and a metal.

**Adhesive wear** Wear caused by metal-to-metal contact; characterized by local welding and tearing of the surface.

**Adsorbent filter** A filter medium intended primarily to hold soluble and insoluble contaminants on its surface by molecular adhesion.

**Adsorption** Adhesion of the molecules of gases, liquids, or dissolved substances to a solid surface, resulting in relatively high concentration of the molecules at the place of contact, for example, the plating out of an antiwear additive on metal surfaces.

**Adsorptive filtration** The attraction to and retention of particles in a filter medium by electrostatic forces or by molecular attraction between the particles and the medium.

**Age hardening** An increase in consistency (hardening) over time (also see *Thixotropy*).

**AGMA** An acronym for American Gear Manufacturers Association, an organization serving the gear industry.

**AGMA lubricant numbers** AGMA specification covering gear lubricants. The viscosity ranges of the AGMA numbers (or grades) conform to the International Standards Organization (ISO) viscosity classification system (see *ISO viscosity classification system*).

**Air bleeder** A device for removal of air from a hydraulic fluid line.

**Air entrainment** A device that converts compressed gas into mechanical force and motion. It usually provides rotary mechanical motion.

**Air motor** The incorporation of air in the form of bubbles as a dispersed phase in a bulk liquid. Air may be entrained in a liquid through mechanical means and/or by release of dissolved air due to a sudden change in the environment. The presence of entrained air is usually readily apparent from the appearance of the liquid (i.e., bubbly, opaque, etc.), whereas dissolved air can only be determined by analysts.

**Air/oil system** A lubrication system in which small measured quantities of oil are introduced into an air–oil mixing device that is connected to a lube line that terminates at a bearing or other lubrication point. The air velocity transports the oil along the interior walls of the lube line to the point of application. These systems provide positive air pressure within the bearing housing to prevent the ingress of contaminants, provide cooling airflow to the bearing, and perform the lubrication function with a continuous flow of minute amounts of oil.

**Air-gap solenoid** A solenoid that is sealed to prevent leakage of a liquid into the plunger cavity.

**Almen EP lubricant tester** A journal bearing machine used for determining the load-carrying capacity or extreme pressure (EP) properties of gear lubricants.

**Aluminum alloy** White particles that indicate wear of an aluminum component such as a casing wall.

**Ambient temperature** Temperature of the area or atmosphere around a process (not the operating temperature of the process itself).

**Amp** Ampere.

**Analytical ferrography** The magnetic precipitation and subsequent analysis of wear debris from a fluid sample. This approach involves passing a volume of fluid over a chemically treated microscope slide that is supported over a magnetic field. Permanent magnets are arranged in such a way as to create varying field strengths over the length of the substrate. These varying strengths cause wear debris to precipitate in a distribution with respect to size and mass over the ferrogram. Once rinsed and fixed to the substrate, this debris deposit serves as an excellent medium for optical analysis of the composite wear particulates.

**Anhydrous**   A lubricating grease without water (as determined by ASTM D128).

**Anhydrous**   Devoid of water.

**Aniline point**   The minimum temperature for complete miscibility of equal volumes of aniline and a sample under test ASTM Method D611. A product of high aniline point will be low in aromatics and naphthenes and therefore high in paraffins. Aniline point is often specified for spray oils, cleaning solvents, and thinners, where effectiveness depends on aromatics content. In conjunction with API gravity, the aniline point may be used to calculate the net heat of combustion for aviation fuels.

**Antifoam agent**   An additive used to suppress the foaming tendency of petroleum products in service; may be a silicone oil to break up surface bubbles or a polymer to decrease the number of small entrained bubbles.

**Antifriction bearing**   A rolling-contact bearing in which the rotating or moving member is supported or guided by means of ball or roller elements; does not mean without friction.

**Antioxidant (oxidation inhibitor)**   An additive to retard oxidation. Antioxidants are critical to any application that is not sealed from the atmosphere. If you don't have a nitrogen blanket on your expansion tank or reservoir, it is crucial that your fluid contain an antioxidant. Oxidation leads to sludge formation that, left unchecked, could cause blockages and lead to complete system failure.

**Antistatic additive**   An additive that increases the conductivity of a hydrocarbon fuel to hasten the dissipation of electrostatic charges during high-speed dispensing, thereby reducing the fire/explosion hazard.

**Antiwear agents**   Additives or their reaction products that form thin, tenacious films on highly loaded parts to prevent metal-to-metal contact.

**API**   An acronym for American Petroleum Institute, a trade association of petroleum producers, refiners, marketers, and transporters, organized for the advancement of the petroleum industry by conducting research, gathering and disseminating information, and maintaining cooperation between government and the industry on all matters of mutual interest.

**API engine service categories**   Gasoline and diesel engine oil quality levels established jointly by API, SAE, and ASTM and sometimes called SAE or API/SAE categories; formerly called API Engine Service Classifications.

**API gravity**   A gravity scale established by the API and in general use in the petroleum industry, the unit being called the *API degree*. This unit is defined in terms of specific gravity.

**Apparent viscosity**   The ratio of shear stress to rate of shear of a non-Newtonian fluid, such as lubricating grease or a multigrade oil, calculated from Poiseuille's equation and measured in poises. The apparent viscosity changes with changing rates of shear and temperature and therefore must be reported as the value at a given shear rate and temperature (ASTM D1092).

**Appearance**   A general term relating to characteristics that are observable only by visual inspection (see *Bloom*, *Bulk appearance*, *Color*, *Luster*, and *Texture*).

**Aqueous decontamination** Removal of a chemical or biological hazard with a water-base solution.

**Aromatic** Derived from or characterized by the presence of a benzene ring.

**ARP** An acronym for aeronautical recommended practice.

**Ash** Metallic deposits formed in the combustion chamber and other engine parts during high-temperature operation.

**Ash** A measure of the amount of inorganic material in lubricating oil; determined by burning the oil and weighing the residue. Results are expressed as percent by weight.

**Ash (sulfated)** The ash content of an oil determined by charring the oil, treating the residue with sulfuric acid, and evaporating to dryness; expressed as percent by mass.

**Asperities** Microscopic projections on metal surfaces resulting from normal surface-finishing processes. Interference between opposing asperities in sliding or rolling applications is a source of friction and can lead to metal welding and scoring. Ideally, the lubricating film between two moving surfaces should be thicker than the combined height of the opposing asperities.

**ASTM D2670, Pin and V-Block Test** ASTM Test Method D2670 is for measuring the antiwear properties of liquid lubricants. The load is applied to the jaws and maintained by a toothed wheel. The wear is a function of the number of the teeth that need to be engaged to keep the load constant for a fixed time.

**ASTM D5302, Sequence VE Gasoline Engine Test** ASTM Test Method D5302 has been correlated with vehicles used in stop-and-go service prior to 1988, particularly with regard to sludge and valve-train wear.

**ASTM D5533, Sequence IIIF Gasoline Engine Test** ASTM Test Method D5533 has been correlated with vehicles used in high-temperature service prior to 1988, particularly with regard to oil thickening and valve-train wear.

**atm** Unit of measure of atmospheric pressure (atmosphere).

**Atmospheric pressure** Pressure exerted by the atmosphere at any specific location. (Sealevel pressure is approximately 14.7 pounds per square inch [psi] absolute.)

**Atomic absorption spectroscopy** Measure of the radiation absorbed by chemically unbound atoms by analyzing the transmitted energy relative to the incident energy at each frequency. The procedure consists of diluting the fluid sample with methyl isobutyl ketone (MIBK) and directly aspirating the solution. The actual process of atomization involves reducing the solution to a fine spray, dissolving it, and finally vaporizing it with a flame. The vaporization of the metal particles depends on their time in the flame, the flame temperature, and the composition of the flame gas. The spectrum occurs because atoms in the vapor state can absorb radiation at certain well-defined characteristic wavelengths. The wavelength bands absorbed are very narrow and differ for each element. In addition, the absorption of radiant energy by electronic transitions from ground to excited state is essentially an absolute measure of the number of atoms in the flame and therefore is the concentration of the element in a sample.

**Atomization** The conversion of a liquid into a spray of very fine droplets.

**Autoignition** Minimum temperature that a substance must be heated to without application of flame or spark to cause the substance to ignite.

**Automatic transmission fluid (ATF)** Fluid for automatic hydraulic transmissions in motor vehicles.

**Babbitt** A soft, white, nonferrous alloy bearing material composed principally of copper, antimony, tin, and lead.

**Back-pressure** The pressure encountered on the return side of a system.

**Background contamination** The total of the extraneous particles that are introduced in the process of obtaining, storing, moving, transferring, and analyzing a fluid sample.

**Bacteria** Microorganisms often composed of a single cell.

**Bactericide** Additive to inhibit bacterial growth in the aqueous component of fluids; prevents foul odors.

**Baffle** A device to prevent direct fluid flow or impingement on a surface.

**Ball bearing** An antifriction rolling-type bearing containing rolling elements in the form of balls.

**Barrel** A unit of liquid volume of petroleum oils equal to 42 U.S. gallons or approximately 35 Imperial gallons.

**Base** A material that neutralizes acids; an oil additive containing colloidally dispersed metal carbonate used to reduce corrosive wear.

**Base number** The amount of acid, expressed in terms of the equivalent number of milligrams of potassium hydroxide (KOH), required to neutralize all basic constituents present in 1 g of sample.

**Base oil** A base stock or blend of base stocks used in an API-licensed engine oil.

**Base-oil credit** In lubricant cost calculations, the value of the base fluid displaced by the additive package.

**Base stock** The base fluid, usually a refined petroleum fraction or a selected synthetic material, into which additives are blended to produce finished lubricants.

**Bases** Compounds that react with acids to form salts plus water. Alkalis are water-soluble bases used in petroleum refining to remove acidic impurities. Oil-soluble bases are included in lubricating oil additives to neutralize acids formed during the combustion of fuel or oxidation of the lubricant.

**Batch** Any quantity of material handled or considered as a *unit* in processing; that is, any sample taken from the same batch will have the same properties and/or qualities.

**Bearing** A support or guide by means of which a moving part such as a shaft or axle is positioned with respect to the other parts of a mechanism.

**Bellows seal** A type of mechanical seal that uses bellows for providing secondary sealing and spring-type loading.

**Bernouilli's theory** If no work is done on or by a flowing frictionless liquid, its energy due to pressure and velocity remains constant at all points along the streamline.

**Beta rating** A method of comparing filter performance based on efficiency. This is done using the multipass test, which counts the number of particles of a given size before and after fluid passes through a filter.

**Beta ratio** The ratio of the number of particles greater than a given size in the influent fluid to the number of particles greater than the same size in the effluent fluid under specified test conditions (see *Multipass test*).

**Bevel gear** A straight-toothed gear with the teeth cut on sloping faces with the gear shafts at an angle (normally a right angle).

**Biocides** Additive designed to inhibit the growth of microorganisms in liquids.

**Biodegradation** The chemical breakdown of materials by living organisms in the environment. The process depends on certain microorganisms, such as bacteria, yeast, and fungi, that break down molecules for sustenance. Certain chemical structures are more susceptible to microbial breakdown than others; vegetable oils, for example, will biodegrade more rapidly than petroleum oils. Most petroleum products typically will completely biodegrade in the environment within two months to two years.

**Bitumen** Also called *asphalt* or *tar*, bitumen is the brown or black viscous residue from the vacuum distillation of crude petroleum. It also occurs in nature as asphalt lakes and tar sands. It consists of high-molecular-weight hydrocarbons and minor amounts of sulfur and nitrogen compounds.

**Black oils** Lubricants containing asphaltic materials, which impart extra adhesiveness, that are used for open gears and steel cables.

**Bleeding** The separation of some of the liquid phase from a grease.

**Blending** The process of mixing fluid lubricant components for the purpose of obtaining desired physical properties (see *Compounding*).

**Bloom** The typical blue or green surface color of a grease when viewed by reflected daylight.

**Blow-by** Passage of unburned fuel and combustion gases past the piston rings of internal combustion engines, resulting in fuel dilution and contamination of the crankcase oil.

**Boiling point** The temperature at which a substance boils or is converted into vapor by bubbles forming within the liquid; varies with pressure.

**Boiling range** For a mixture of substances, such as a petroleum fraction, the temperature interval between the initial and final boiling points.

**Bomb oxidation** A test for the oxidation stability of a product obtained by sealing it in a closed container with oxygen under pressure. The drop in pressure of the oxygen is a measure of the amount of oxidation that has occurred.

**Boundary lubrication** Lubrication between two rubbing surfaces without the development of a full fluid lubricating film. It occurs under high loads and requires the use of antiwear or extreme-pressure additives to prevent metal-to-metal contact.

**Brinelling** Permanent deformation of bearing surfaces where the rollers (or balls) contact the races. Brinelling results from excessive load or impact on stationary bearings. It is a form of mechanical damage in which metal is displaced or upset without attrition.

**Bubble point** The differential gas pressure at which the first steady stream of gas bubbles is emitted from a wetted filter element under specified test conditions.

**Bulk appearance** Appearance of an undisturbed grease surface; described as *bleeding*, free oil on the surface (or in the cracks of a cracked grease); *cracked*, surface cracks; *grainy*, composed of small granules or lumps of constituent thickener; and *smooth*, relatively free of irregularities.

**Bulk modulus (of elasticity)** A ratio of normal stress to a change in volume. A term used in determining the compressibility of a fluid. Data for petroleum products can be found in the International Critical Tables.

**Burst pressure rating** The maximum specified inside-out differential pressure that can be applied to a filter element without outward structural or filter-medium failure.

**Bypass valve (relief valve)** A valve mechanism that ensures system fluid flow when a preselected differential pressure across the filter element is exceeded; the valve allows all or part of the flow to bypass the filter element.

**Cams** Eccentric shafts used in most internal combustion engines to open and close valves.

**Capillary viscometer** A viscometer in which the oil flows through a capillary tube.

**Carbon (deposit)** Solid black residue in piston grooves that can interfere with piston ring movement leading to wear and/or loss of power.

**Carbon residue** Coked material remaining after an oil has been exposed to high temperatures under controlled conditions.

**Carbon type** The distinction between paraffinic, naphthenic, and aromatic molecules; in relation to lubricant base stocks, the predominant type present.

**Carbonyl iron powder** A contaminant that consists of up to 99.5% pure iron spheres.

**Carcinogen** A cancer-causing substance; certain petroleum products are classified as potential carcinogens (OSHA criteria). Suppliers are required to identify such products as potential carcinogens on package labels and Material Safety Data Sheets.

**Cartridge seal** A completely self-contained assembly including seal, gland, sleeve, mating ring, etc., usually needing no installation measurement.

**Case drain filter** A filter located in a line conducting fluid from a pump or motor housing to a reservoir.

**Case drain line** A line conducting fluid from a component housing to a reservoir.

**Catalyst** A substance that initiates or increases the rate of a chemical reaction without itself being used up in the process.

**Catastrophic failure** Sudden, unexpected failure of a machine resulting in considerable cost and downtime.

**Caustic** A highly alkaline substance such as sodium hydroxide.

**Cavitation** In a heat-transfer system, failure of the material to flow to the suction of the system pump for any reason.

**Cavitation** Formation of an air or vapor pocket (or bubble) due to lowering of pressure in a liquid, often as a result of a solid body, such as a propeller or piston, moving through the liquid; also, the pitting or wearing away of a solid surface as a result of the violent collapse of a vapor bubble. Cavitation can occur in a hydraulic system as a result of low fluid levels that draw air into the fluid, producing tiny bubbles that expand followed by rapid implosion, causing metal erosion and eventual pump destruction.

**Cavitation erosion** A material-damaging process that occurs as a result of vaporous cavitation. *Cavitation* refers to the occurrence or formation of gas- or vapor- filled pockets in flowing liquids due to the hydrodynamic generation of low pressure (below atmospheric pressure). This damage results from the hammering action when cavitation bubbles implode in the flow stream. Ultrahigh pressures caused by the collapse of the vapor bubbles produce deformation, material failure, and, finally, erosion of the surfaces.

**Cellulose medium** A filter material made from plant fibers. Because cellulose is a natural material, its fibers are rough in texture and vary in size and shape. Compared with synthetic media, these characteristics create a higher restriction to the flow of fluids.

**Centi** Hundredth.

**Centipoise (cP) and centistoke (cSt)** A centipoise is 1/100th of the unit of absolute viscosity (the poise); for example, the viscosity of water at 20°C is approximately 1 cP. The centipoise is derived from one kinematic unit of viscosity (the centistoke) by multiplying the latter by the density of the liquid; that is, 1 cP equals 1 cSt times the density of the liquid. These units are part of the metric system, commonly used in Europe and becoming adopted in the United States and Canada.

**Cetane index** A value calculated from the physical properties of a diesel fuel to predict its cetane number.

**Cetane number** A measure of the ignition quality of a diesel fuel, as determined in a standard single-cylinder test engine, which measures ignition delay compared with primary reference fuels. The higher the cetane number, the easier a high-speed, direct-injection engine will start and the less white smoking and diesel knock after startup.

**Cetane number improver** An additive (usually an organic nitrate) that boosts the cetane number of a fuel.

**cfm** Abbreviation for cubic feet per minute.

**Channeling** The phenomenon observed among gear lubricants and greases when they thicken due to cold weather or other causes to such an extent that a groove is formed through which the part to be lubricated moves without actually coming in full contact with the lubricant. A term used in percolation filtration; may be defined as a preponderance of flow through certain portions of a clay bed.

**Chemical stability** The tendency of a substance or mixture to resist chemical change.

**Chip control (grit control, last-chance) filter** A filter intended to prevent only large particles from entering a component immediately downstream.

**Chlorinated wax** Certain solid hydrocarbons treated with chlorine gas to form straight-chain hydrocarbons with a relatively high chlorine component. Chlorinated waxes are used primarily as polyvinyl chloride plasticizers, extreme-pressure additives for lubricants, and formulation components for many cutting fluids.

**Chromatography** An analytical technique whereby a complex substance is adsorbed on a solid or liquid substrate and progressively eluted by a flow of a substance (the eluant) in which the components of the substance under investigation are differentially soluble. The eluant can be a liquid or a gas. When the substrate is filter paper and the eluant is a liquid, a chromatogram of colored bands can be developed by use of indicators. For gas chromatography, electronic detectors are normally used to indicate passage of the various components from the system.

**Circulating header system** A lubrication system having isolated lube zones wherein the lube pump runs continuously and circulates oil through the header and a return filter and back to tank during the idle period. When lubrication is required, a normal open solenoid valve in the return loop is actuated, allowing pump pressure to build. The zone valves are then sequentially opened to provide lubricant to the individual zones. Oil dispensed to the friction points is not reused; therefore, the system is a terminating type.

**Circulating oil** A lubrication system wherein the oil pump runs continuously and circulates oil to the friction points on a continuous basis. The oil is drained back to tank, filtered, cooled as required, and reused.

**Circulating system** A lubricating system in which oil is recirculated from a central sump to the parts requiring lubrication and then returned to the sump.

**Clay filtration** A refining process using Fuller's earth (activated clay), bauxite, or another mineral to absorb minute solids from lubricating oil, as well as remove traces of water, acids, and polar compounds.

**Clean**   100 particles >10 μm/mL—in regard to the cleanliness of an oil sample bottle.

**Cleanable filter**   A filter element that, when loaded, can be restored by a suitable process to an acceptable percentage of its original dirt capacity.

**Cleanliness level**   A measure of relative freedom from contaminants.

**Clearance bearing**   A journal bearing in which the radius of the bearing surface is greater than the radius of the journal surface.

**Cleveland open-cup test**   A flash-point test in which the surface of the sample is completely open to the atmosphere and which is therefore relatively insensitive to small traces of volatile contaminants.

**Cloud point**   The temperature at which a cloud of wax crystals appears when a lubricant or distillate fuel is cooled under standard conditions; indicates the tendency of the material to plug filters or small orifices under cold weather conditions.

**cm**   Abbreviation for centimeter.

**Coalescor**   A separator that divides a mixture or emulsion of two immiscible liquids using the interfacial tension between the two liquids and the difference in wetting of the two liquids on a particular porous medium.

**Coefficient of friction**   Coefficient of *static* friction is the ratio of the tangential force initiating sliding motion to the load perpendicular to that motion. Coefficient of *kinetic* friction (usually called *coefficient of friction*) is the ratio of the tangential force sustaining sliding motion at constant velocity to the load perpendicular to that motion.

**Cohesion**   Molecular attraction between grease particles contributing to the resistance to flow of the grease.

**Coking**   The undesirable accumulation of carbon (coke) deposits in an internal combustion engine or a refinery plant. The process of distilling a petroleum product to dryness.

**Cold cranking simulator (CCS)**   An intermediate shear-rate viscometer that predicts the ability of an oil to permit a satisfactory cranking speed to be developed in a cold engine.

**Collapse**   An inward structural failure of a filter element that can occur due to an abnormally high pressure drop (differential pressure) or resistance to flow.

**Color**   The predominant hue of lubricating grease (such as amber, brown or perhaps green, red, or blue for dyed grease) and intensity (light, medium, or dark) when viewed to eliminate bloom.

**Complex grease**   A lubricating grease thickened by a complex soap consisting of a normal soap and a complexing agent.

**Complex soap**   A soap crystal or fiber formed usually by co-crystallization of two or more compounds. Complex soaps can be a normal soap (such as metallic stearate or oleate) or incorporate a complexing agent that causes a change in grease characteristics; usually recognized by an increase in dropping point.

**Compound**   Chemically speaking, a distinct substance formed by the combination of two or more elements in definite proportions by weight and possessing physical and chemical properties different from those of the combining elements.

**Compound**   In petroleum processing, generally connotes fatty oils and similar materials foreign to petroleum added to lubricants to impart special properties.

**Compounding** (see *Blending*) The mixing or otherwise combining of lubricant components with other components for the purpose of securing chemical and/or physical properties not usually obtainable by blending of fluid lubricant components alone.

**Compressed air** Air at any pressure greater than atmospheric pressure.

**Compressibility** A compound that enhances some property of or imparts some new property to a base fluid. In some hydraulic fluid formulations, the additive volume may constitute as much as 20% of the final composition. The more important types of additives include antioxidants, antiwear additives, corrosion inhibitors, viscosity index improvers, and foam suppressants.

**Compressor** A device that converts mechanical force and motion into pneumatic fluid power.

**Consistency (hardness)** The resistance of a lubricating grease to deformation under load; usually indicated by ASTM Cone Penetration, ASTM D217 (IP 50), or ASTM D1403.

**Contaminant capacity** (dirt, ACFTD) The weight of a specified artificial contaminant that must be added to the influent to produce a given differential pressure across a filter under specified conditions; used as an indication of relative service life.

**Contaminant failure** Any loss of performance due to the presence of contamination. Two basic types of contamination failure: lock and control.

**Contaminant lock** A particle- or fiber-induced jam caused by solid contaminants.

**Contamination control** A broad subject that applies to all types of material systems (including both biological and engineering). It is concerned with planning, organizing, managing, and implementing all activities required to deter-

mine, achieve, and maintain a specified contamination level.

**Copper-strip corrosion** The gradual eating away of copper surfaces as a result of oxidation or other chemical action; caused by acids or other corrosive agents.

**Corrosion** The wearing away and/or pitting of a metal surface due to chemical attack (see *Fretting*).

**Corrosion inhibitor** Helps prevent oxidation of metal by displacing water from metal surfaces. It plates the metal with a polarized effect to give an internal "umbrella," helping to resist iron oxide formation. Most systems at some point in time will have some water contamination. Whether from leaky heat exchangers or drawn from humid air, moisture venting through the expansion tank or oil reservoir can lead to corrosion inside the tank. Corrosion inhibitors virtually eliminate this risk.

**Corrosive wear** Wear caused by chemical reactions.

**Coupling** A straight connector for fluid lines.

**Cracking pressure** The pressure at which a pressure-operated valve begins to pass fluid.

**Crankcase oil** Lubricant used in the crankcase of an internal combustion engine.

**Crown** The top of a piston in an internal combustion engine above the fire ring; exposed to direct flame impingement.

**Cutting oil** A lubricant used in machining operations for lubricating the tool in contact with the workpiece and to remove heat. The fluid can be petroleum based, water based, or an emulsion of the two. The term *emulsifiable cutting oil* normally indicates a petroleum-based concentrate to which water is added to form an emulsion, which is the actual cutting fluid.

**Cycle** A single complete operation consisting of progressive phases starting and ending at the neutral position.

**Cylinder oil** A lubricant for independently lubricated cylinders, such as those of steam engines and air compressors; also for lubrication of valves and other elements in the cylinder area. Steam cylinder oils are available in a range of grades with high viscosities to compensate for the thinning effect of high temperatures; of these, the heavier grades are formulated for superheated and high-pressure steam, and the less heavy grades, for wet, saturated, or low-pressure steam. Some grades are compounded for service in excessive moisture; see *Compounding*. Cylinder oils lubricate on a once-through basis.

**Defoaming agents** During startup, air can become trapped in a system. Pumping creates air bubbles (foaming), which can lead to pump cavitations, possibly damaging pumps and other system components. My company's proprietary additive package contains defoaming agents to help keep air from foaming in the oil.

**Degradation** The progressive failure of a machine or lubricant.

**Dehydrator** A separator that removes water from a system fluid.

**Delamination wear** A complex wear process in which a machine surface is peeled away or otherwise removed by forces of another surface acting on it in a sliding motion.

**Demulsibility** A measure of a fluid's ability to separate from water.

**Demulsifier** An additive that promotes oil–water separation in lubricants that are exposed to water or steam.

**Denaturants** Toxic or noxious components used in fuel ethanol to make it unfit for use as a beverage.

**Density** Mass per unit volume.

**Density (specific gravity)** A material's weight or compactness. It changes as temperatures fluctuate. Fluids with a higher density typically transfer heat more efficiently than fluids that are less dense. Density is sometimes expressed as specific gravity, a dimensionless value that has no unit of measure. It refers to the ratio of a fluid's weight in comparison with water. For example, a material with a specific gravity of less than 1 would be considered less dense than water and therefore would float on water.

**Deplete** The depletion of additives expressed as an approximate percentage.

**Deposits** Oil-insoluble materials that result from oxidation of oil and contamination from external sources and settle out in system components as sludge and varnish.

**Depth filter media** Porous materials that primarily retain contaminants within a tortuous path, performing the actual process of filtration.

**Dermatitis** Inflammation of the skin. Repeated contact with petroleum products can be a cause.

**Desorption** Opposite of absorption or adsorption. In filtration, it relates to the downstream release of particles previously retained by a filter.

**Detergent** In lubrication, either an additive or a compounded lubricant having the property of keeping insoluble matter in suspension, thus preventing its deposition where it would be harmful. A detergent also may redisperse deposits already formed.

**Detergent oil** A lubricating oil possessing special sludge-dispersing properties usually conferred on the oil by the incorporation of special additives. Detergent oils hold formed sludge particles in suspension and thus promote cleanliness, especially in internal combustion engines. However, detergent oils do not contain detergents such as those used for cleaning of laundry or dishes. Also, detergent oils do not clean already dirty engines but rather keep in suspension the sludge that petroleum oil forms so that the engine remains cleaner for longer. The formed sludge particles are either filtered out by oil filters or drained out when the oil is changed.

**Detergent/dispersant** An additive package that combines a detergent with a dispersant.

**Detonation** Uncontrolled burning of the last portion (end gas) of an air–fuel mixture in the cylinder of a spark ignition engine. Also known as *knock* or *ping*.

**Dewaxing** Removal of wax from a base oil in order to reduce the pour point.

**Dielectric strength** A measure of the ability of an insulating material to withstand electric stress (voltage) without failure. Fluids with high dielectric strength (usually expressed in volts or kilovolts) are good electrical insulators (ASTM D877).

**Differential pressure indicator** An indicator that signals the difference in pressure between two points, typically between the upstream and downstream sides of a filter element.

**Differential pressure valve** A valve whose primary function is to limit differential pressure.

**Dilution of engine oil** Contamination of crankcase oil by unburned fuel, leading to reduced viscosity and flash point; may indicate component wear or fuel system maladjustment.

**Directional control servo valve** A directional control valve that modulates flow or pressure as a function of its input signal.

**Directional control valve** A valve whose primary function is to direct or prevent flow through selected passages.

**Dirt capacity** The weight of a specified artificial contaminant that must be added to the influent to produce a given differential pressure across a filter under specified conditions; used as an indication of relative service life.

**Dispensibility** The ease with which a grease can be delivered through its dispensing system to the point of application (see *Pumpability* and *Feedability*).

**Dispersant** In lubrication, a term usually used interchangeably with detergent. An additive, usually nonmetallic (*ashless*), that keeps fine particles of insoluble materials in a homogeneous solution. Hence particles are not permitted to settle out and accumulate.

**Dissolved air** Air that is dispersed in a fluid to form a mixture.

**Dissolved water** Water that is dispersed in a fluid to form a mixture.

**Distillation** The basic test used to characterize the volatility of a gasoline or distillate fuel.

**Distillation range** A measurement of the temperatures within which a liquid distills or boils; usually expressed as a percentage of the material that boils between two temperatures. A higher initial boiling point indicates a more thermally stable fluid as well as lower vapor pressures. A narrower boiling range is also more desirable.

**Double seal** Two mechanical seals designed to permit a liquid or gas barrier fluid between the seals mounted back-to-back or face-to-face.

**Drag** The resistance to movement caused by oil viscosity.

**Dropping point** The temperature at which grease becomes soft enough to form a drop that falls from the orifice of the test apparatus of ASTM D566 (IP 132) and ASTM D2265.

**Dropping point** In general, the temperature at which a grease passes from a semisolid to a liquid state. This change in state is typical of greases containing conventional soap thickeners. Greases containing thickeners other than conventional soaps may, without change in state, separate oil.

**Dry film lubricant** Low-shear-strength lubricants that shear in one particular plane within their crystal structure (such as graphite, molybdenum disulfide, and certain soaps).

**Dry lubrication** The situation when moving surfaces have no liquid lubricant between them.

**Dry sump** An engine design in which oil is not retained in a pan beneath the crankshaft, thus permitting splash lubrication. There may be a remote sump from which oil is recirculated, or there may be a total-loss system.

**Dual-line system** A positive-displacement terminating (oil or grease) lubrication system that employs two main lines supplied from a pump connected to a four-way (reverser) valve. Pressure in one main line (while the other is open to the tank) causes the measuring piston(s) in the dual-line valve(s) to stroke in one direction dispensing lubricant to one group of lube points. Switching the four-way (reverser) valve directs pump flow to the second main line and opens the first main line to the tank. This allows pressure to build in the second main line causing the dual-line valve(s) measuring piston(s) to stroke back to their original position dispensing lubricant to a second group of lube points. The system is a parallel type, and each dual-line valve operates independently of any other in the system.

**Duplex filter** An assembly of two filters with valving for selection of either or both filters.

**Dust capacity** The weight of a specified artificial contaminant that must be added to the influent to produce a given differential pressure across a filter at specified conditions; used as an indication of relative service life.

**Dynamic seal** A seal that moves due to axial or radial movement of the unit.

**Elastohydrodynamic lubrication (EHD)** A lubricant regime characterized by high unit loads and high speeds where the mating parts (usually in rolling-element bearings) deform elastically causing an increase in lubricant viscosity and load-carrying capacity.

**Elastomer** A rubber or rubber-like material, both natural and synthetic, used in making a wide variety of products such as seals and hoses. In oil seals, an elastomer's chemical composition is a factor in determining its compatibility with a lubricant.

**Electrical insulating oil** A high-quality oxidation-resistant oil refined to give long service as a dielectric and coolant for electrical equipment, most commonly transformers. An insulating oil must resist the effects of elevated temperatures, electrical stress, and contact with air, which can lead to sludge formation and loss of insulation properties. It must be kept dry because water is detrimental to dielectric strength—the minimum voltage required to produce an electric arc through an oil sample, as measured by test method ASTM D877.

**Emission spectrometer** Works on the basis that atoms of metallic and other particular elements emit light at characteristic wavelengths when they are excited in a flame, arc, or spark. Excited light is directed through an entrance slit in the spectrometer. This light penetrates the slit, falls on a grate, and is dispersed and reflected. The spectrometer is calibrated by a series of standard samples containing known amounts of the elements of interest. By exciting these standard samples, an analytical curve can be established that gives the relationship between the light intensity and its concentration in the fluid.

**Emissions (mobile sources)** The combustion of fuel leads to the emission of exhaust gases that may be regarded as pollutants. Water and $CO_2$ are not included in this category but CO, $NO_x$, and hydrocarbons are subject to legislative control. All three are emitted by gasoline engines; diesel engines also emit particulates that are controlled.

**Emissions (stationary sources)** Fuel composition can influence emissions of sulfur oxides and particulates from power stations. Local authorities control the sulfur content of heavy fuel oils used in such applications.

**Emulsifier** Additive that promotes the formation of a stable mixture, or emulsion, of oil and water.

**End point** Highest vapor temperature recorded during a distillation test of a petroleum stock.

**Engine deposits** Hard or persistent accumulation of sludge, varnish, and carbonaceous residues due to blow-by of unburned and partially burned fuel or the partial breakdown of the crankcase lubricant. Water from the condensation of combustion products, carbon, residues from fuel or lubricating oil additives, dust, and metal particles also contribute.

**Entrained air** A mechanical mixture of air bubbles having a tendency to separate from the liquid phase.

**EP (extreme-pressure) lubricants** Lubricants that impart to rubbing surfaces the ability to carry appreciably greater loads than would be possible with ordinary lubricants without excessive wear or damage.

**EP (extreme-pressure) agent** Lubricant additive that prevents sliding metal surfaces from seizing under extreme-pressure conditions.

**EPA complex model** Scheduled for implementation January 1, 1997, this model is more restrictive than the simple model and contains limits on Reid Vapor Pressure, oxygen, olefins, benzene, sulfur, and T-90. In addition, it will include requirements on aromatic content and T-50 temperatures.

**EPA simple model** Used to define reformulated gasoline effective January 1, 1995, this model includes Reid Vapor Pressure and oxygen content requirements to reduce volatile organic compound emissions. It caps oxygen, benzene, sulfur, olefins, and T-90 content at levels equal to or lower than a refiner's 1990s baseline.

**Erosion** Wearing away of a surface by an impinging fluid or solid.

**Evaporative loss** Loss of a portion of a lubricant due to volatization (evaporation). Test methods include ASTM D972 and ASTM D2595.

**Exhaust gas recirculation (EGR)** System to reduce automotive emission of nitrogen oxides ($NO_x$). It routes exhaust gases into the carburetor or intake manifold, where they dilute the air–fuel mixture and reduce peak combustion temperatures, thereby reducing the tendency for $NO_x$ to form.

**Externally pressurized seal** A seal that has pressure acting on the seal parts from an external independent source of supply.

**Extreme-pressure (EP) property** The property of a grease that, under high applied loads, reduces scuffing, scoring, and seizure of contacting surfaces. Common laboratory tests are the Timken OK load test (ASTM D2509 and ASTM D2782) and the four-ball load wear index test (ASTM D2596 and ASTM D2783).

**Fabrication integrity point** The differential gas pressure at which the first stream of gas bubbles is emitted from a wetted filter element under standard test conditions.

**Face seal** A device that prevents leakage of fluids along rotating shafts. Sealing is accomplished by a stationary primary seal ring bearing against the face of a mating ring mounted on a shaft. Axial pressure maintains the contact between the seal ring and the mating ring.

**False brinelling** False brinelling of needle roller bearings is actually a fretting corrosion of the surface because the rollers are the inside diameter of the bearing. Although its appearance is similar to that of brinelling, false brinelling is characterized by attrition of the steel, and the load on the bearing is less than that required to produce the resulting impression. It is the result of a combination of mechanical and chemical actions that are not completely understood and occurs when a small relative motion or vibration is accompanied by some loading in the presence of oxygen.

**FAS** Abbreviation for free alongside.

**Fat** An animal or vegetable oil that will combine with an alkali to saponify and form a soap.

**Fatigue** Cracking, flaking, or spalling of a surface due to stresses beyond the endurance limit of the material.

**Fatigue chunks** Thick three-dimensional particles exceeding 50 μm indicating severe wear of gear teeth.

**Fatigue life** The theoretical number of revolutions (or hours of operation) a bearing will last under a given constant load and speed before the first evidence of fatigue develops on one or more of the components.

**Fatigue platelets** Normal particles between 20 and 40 μm found in gearbox and rolling-element bearing oil samples observed by analytical ferrography. A sudden increase in the size and quantity of these particles indicates excessive wear.

**Fatigued** A structural failure of a filter medium due to flexing caused by cyclic differential pressures.

**Feedability** The ease with which a grease flows within a dispensing pump.

**Ferrography** An analytical method for assessing machine health by quantifying and examining ferrous wear particles suspended in the lubricant or hydraulic fluid.

**Fiber** Grease soap thickeners that occur in fibrous form. Most soap-thickener fibers are microscopic, and the grease appears smooth; if the fiber bundles are large enough to be seen, the grease appears fibrous. Although the most common fibrous grease contains sodium soap thickener, not all sodium base greases are fibrous (see *Appearance* and *Texture*).

**Fiber grease** A grease with a distinctly fibrous structure that is noticeable when portions of the grease are pulled apart.

**Fibril** A tiny fiber barely visible even with an electron microscope. Fibers may be formed by bundles of fibrils that collect together.

**Filler** Materials such as talc, pigment, and carbon black that increase a lubricant's bulk or density. Fillers can have various effects (or no effect) on the lubricating properties of a grease.

**Film strength** The ability of a lubricant film to withstand the effects of load, speed, and temperature without breaking down or rupturing.

**Filter efficiency** Method of expressing a filter's ability to trap and retain contaminants of a given size.

**Fire point** The temperature at which a fluid will sustain a fire if ignited by an outside ignition source. It is quite common for heat-transfer systems to be operated at temperatures above the fire point of the fluid because ignition sources should always be far removed from any heat-transfer system.

**Fire-resistant fluid** Lubricant used especially in high-temperature or hazardous hydraulic applications. Three common types of fire-resistant fluids are (1) water–petroleum oil emulsions, in which the water prevents burning of the petroleum constituent, (2) water–glycol fluids, and (3) nonaqueous fluids of low volatility, such as phosphate esters, silicones, and halogenated hydrocarbon fluids.

**Fixed displacement pump** A pump in which the displacement per cycle cannot be varied.

**Flash point** Minimum temperature at which a fluid will support instantaneous combustion (a flash) but before it will burn continuously (fire point). Flash point is an important indicator of the fire and explosion hazards associated with a petroleum product.

**Floc point** The temperature at which wax or solids separate in an oil.

**Flow-control valve** A valve whose primary function is to control flow rate.

**Fluid friction** Occurs between the molecules of a gas or liquid in motion and is expressed as shear stress. Unlike solid friction, fluid friction varies with speed and area.

**Fluid velocity** The measured speed at which a fluid moves through the inside of a tube.

**Foam** An agglomeration of gas bubbles separated from each other by a thin liquid film that is observed as a persistent phenomenon on the surface of a liquid.

**Foam inhibitor** A substance introduced in a very small proportion to a lubricant or a coolant to prevent the formation of foam due to aeration of the liquid and to accelerate the dissipation of any foam that may form.

**Foaming** A frothy mixture of air and a petroleum product (e.g., lubricant, fuel oil) that can reduce the effectiveness of the product and cause sluggish hydraulic operation, air binding of oil pumps, and overflow of tanks or sumps. Foaming can result from excessive agitation, improper fluid levels, air leaks, cavitation, or contamination with water or other foreign mate-

rials. Foaming can be inhibited with an anti-foaming agent. The foaming characteristics of a lubricating oil can be determined by blowing air through a sample at a specified temperature and measuring the volume of foam, as described in test method ASTM D892.

**FOB** Abbreviation for free on board.

**Food-grade lubricants** Lubricants acceptable for use in meat, poultry, and other food-processing equipment, applications, and plants. The lubricant types in food-grade applications are broken into categories based on the likelihood they will contact food. The U.S. Department of Agriculture (USDA) created the original food-grade designations H1, H2, and H3, which are the current terminology used. The approval and registration of a new lubricant into one of these categories depend on the ingredients used in the formulation.

**Force-feed lubrication** A system of lubrication in which the lubricant is supplied to the bearing surface under pressure.

**FOT** Abbreviation for free on truck.

**Four-ball tester** Frequently used to describe either of two similar laboratory machines, the four-ball wear tester and the four-ball EP tester. These machines are used to evaluate a lubricant's antiwear qualities, frictional characteristics, or load-carrying capabilities. It derives its name from the four ½-inch steel balls used as test specimens. Three of the balls are held together in a cup filled with lubricant, while the fourth ball is rotated against them.

**Free air** Air at ambient temperature, pressure, relative humidity, and density.

**Free water** Water droplets or globules in a system's fluid that tend to accumulate at the bottom or top of the system fluid depending on the fluid's specific gravity.

**Fretting** Wear characterized by the removal of fine particles from mating surfaces. Fretting is caused by vibratory or oscillatory motion of limited amplitude between contacting surfaces (see *False brinelling*).

**Fretting corrosion** Can take place when two metals are held in contact and subjected to repeated small sliding relative motions. Other names for this type of corrosion include *wear oxidation*, *friction oxidation*, *chafing*, and *brinelling*.

**Friction** The resisting force encountered at the common boundary between two bodies when, under the action of an external force, one body moves or tends to move relative to the surface of the other.

**Frost** Field of micropits; form of microadhesive wear.

**FTIR** Acronym for fourier-transform infrared spectroscopy, a test where infrared light absorption is used for assessing levels of soot, sulfates, oxidation, nitro-oxidation, glycol, fuel, and water contaminants.

**Fuel dilution** The amount of raw, unburned fuel that ends up in the crankcase of an engine. It lowers an oil's viscosity and flash point, creating friction-related wear almost immediately by reducing film strength.

**Fuel economy** The amount of fuel required to move a machine over a given distance.

**Fuel ethanol** Ethanol (ethyl alcohol, $C_2H_5OH$) with impurities, including water but excluding denaturants.

**Full fluid-film lubrication** Presence of a continuous lubricating film sufficient to completely separate two surfaces, as distinct from boundary lubrication. Full fluid-film lubrication is normally hydrodynamic lubrication, whereby the oil adheres to the moving part and is drawn into the area between the sliding surfaces, where it forms a pressure barrier.

**FZG** Acronym for Forschungstelle für Zahnräder und Getriebau.

**FZG four-square gear oil test** Used in developing industrial gear lubricants to meet equipment manufacturers' specifications. The FZG test equipment consists of two gear sets arranged in a four-square configuration and driven by an electric motor. The test gear set is run in the lubricant at gradually increased load stages until failure, which is the point at which a 10-mg weight loss by the gear set is recorded. Also called *Niemann four-square gear oil test*.

**Galling** A form of wear in which seizing or tearing of the gear or bearing surface occurs (see *Adhesive wear*).

**Gaseous fuels** Liquefied or compressed hydrocarbon gases (e.g., propane, butane, or natural gas) that are finding increasing use in motor vehicles as replacements for gasoline and diesel fuel.

**Gasoline** A volatile mixture of liquid hydrocarbons containing small amounts of additives and suitable for use as a fuel in spark-ignition internal combustion engines.

**Gasoline–ethanol blend** A spark-ignition automotive engine fuel containing denatured fuel ethanol in a base gasoline. It may be leaded or unleaded.

**Gear** A machine part that transmits motion and force by means of successively engaging projections called *teeth*. The smaller gear of a pair is called the *pinion*; the larger, the *gear*. When the pinion is on the driving shaft, the gear set acts as a speed reducer; when the gear drives, the set acts as a speed multiplier. The basic gear type is the spur gear, or straight-tooth gear, with teeth cut parallel to the gear axis. Spur gears transmit power in applications using parallel shafts.

In this type of gear, the teeth mesh along their full length, creating a sudden shift in load from one tooth to the next, with consequent noise and vibration. This problem is overcome by the helical gear, which has teeth cut at an angle to the center of rotation so that the load is transferred progressively along the length of the tooth from one edge of the gear to the next. When the shafts are not parallel, the most common gear type used is the bevel gear, with teeth cut on a sloping gear face rather than parallel to the shaft. The spiral bevel gear has teeth cut at an angle to the plane of rotation, which, like the helical gear, reduces vibration and noise. A hypoid gear resembles a spiral bevel gear except that the pinion is offset so that its axis does not intersect the gear axis; it is widely used in automobiles between the engine driveshaft and rear axle. Offset of the axes of hypoid gears introduces additional sliding between the teeth, which, when combined with high loads, requires a high-quality EP oil. A worm gear consists of a spirally grooved screw moving against a tooth wheel; in this type of gear, where the load is transmitted across sliding rather than rolling surfaces, compounded or EP oils are usually necessary to maintain effective lubrication.

**Gear oil**　A high-quality oil with good oxidation stability, load-carrying capacity, rust protection, and resistance to foaming for service in gear housings and enclosed chain drives. Specially formulated industrial EP gear oils are used where highly loaded gear sets or excessive sliding action (as in worm gears) is encountered.

**Gearbox (gear housing)**　A casing for gear sets that transmit power from one rotating shaft to another. A gearbox has a number of functions: it is precisely bored to control gear and shaft alignment, it contains the gear oil, and it protects the gears and lubricant from water, dust, and other environmental contaminants. Gearboxes are used in a wide range of industrial, automotive, and home machinery. Not all gears are enclosed

in gearboxes; some are open to the environment and are commonly lubricated by highly adhesive greases.

**Gravity (specific)**　In petroleum products, the mass/volume relationship, expressed as: g/cc.

**Gravity separation**　A method of separating two components from a mixture. Under the influence of gravity, separation of immiscible phases (gas-solid, liquid-solid, liquid-liquid, and solid-solid) allows the denser phase to settle out.

**Grease**　A lubricant composed of an oil or oils thickened with a soap, soaps, or other thickener to a semisolid or solid consistency.

**Grease fitting**　A small fitting that connects a grease gun and the component to be lubricated. The fitting is installed by a threaded connection, leaving a nipple to which the grease gun attaches.

**Grease gun**　A tool (normally hand powered) that is used for lubrication tasks. By squeezing the trigger of the gun, grease is applied through an aperture to a specific point.

**Gross additive treating cost (GATC)**　The cost of additive in one-volume units of finished product not including base-fluid credit or shipping costs.

**Gross delivered treating cost (GDTC)**　The cost of additive in a one-volume unit of finished product including shipping cost but not base-fluid credit.

**H1 lubricant**　Food-grade lubricant used in food processing environments where there is some possibility of incidental food contact. Lubricant formulations may only be composed of one or more approved base stocks, additives, and thickeners (if grease) listed in "Guidelines of Security," *Code of Federal Regulations* (CFR) Title 21, §178.3570.

**H2 lubricant** Lubricants used on equipment and machine parts in locations where there is no possibility that the lubricant or lubricated surface contacts food. Because there is no risk of contacting food, these lubricants do not have a defined list of acceptable ingredients. They cannot, however, contain intentionally heavy metals such as antimony, arsenic, cadmium, lead, mercury, or selenium. Also, the ingredients must not include substances that are carcinogens, mutagens, teratogens, or mineral acids.

**H3 lubricant** Also known as *soluble* or *edible oil*, these lubricants are used to clean and prevent rust on hooks, trolleys, and similar equipment.

**Heat capacity (specific heat)** This is a measure of a fluid's capacity to carry heat. It indicates how many Btus it would take to increase the temperature of 1 lb of fluid by 1F. Heat capacity is used along with other properties to determine efficiency as quantified by the heat-transfer coefficient.

**Heavy ends** The portions of a petroleum distillate fraction that are highest boiling and therefore distill at higher temperatures if the temperature is raised progressively.

**Homogenization** The intimate mixing of a grease to produce a uniform dispersion of components.

**Hybrid bearing** A bearing that consists of metal rings and ceramic balls.

**Hydraulic fluid** Fluid serving as the power-transmission medium in a hydraulic system. The most commonly used fluids are petroleum oils, synthetic lubricants, oil–water emulsions, and water–glycol mixtures. The principal requirements of a premium hydraulic fluid are proper viscosity, high viscosity index, antiwear protection (if needed), good oxidation stability, adequate pour point, good demulsibility, rust inhibition, resistance to foaming, and compatibility with seal materials. Antiwear oils are frequently used in compact, high-pressure, and high-capacity pumps that require extra lubrication protection.

**Hydraulic motor** A device that converts hydraulic fluid power into mechanical force and motion by transfer of flow under pressure; usually provides rotary mechanical motion.

**Hydraulic oil** An oil specially suited for use as either the specific gravity or the API gravity of a liquid.

**Hydraulic pump** A device that converts mechanical force and motion into hydraulic fluid power by means of producing flow.

**Hydraulic system** A system designed to transmit power through a liquid medium, permitting multiplication of force in accordance with Pascal's law, which states that "a pressure exerted on a confined liquid is transmitted undiminished in all directions and acts with equal force on all equal areas." Hydraulic systems have six basic components: (1) a reservoir to hold the fluid supply; (2) a fluid to transmit the power; (3) a pump to move the fluid; (4) a valve to regulate pressure; (5) a directional valve to control the flow, and (6) a working component—such as a cylinder and piston or a shaft rotated by pressurized fluid—to turn hydraulic power into mechanical motion. Hydraulic systems offer several advantages over mechanical systems: they eliminate complicated mechanisms such as cams, gears, and levers; are less subject to wear; are usually more easily adjusted for control of speed and force; are easily adaptable to both rotary and liner transmission of power; and can transmit power over long distances and in any direction with small losses.

**Hydrodynamic lubrication** A system of lubrication in which the shape and relative motion of the sliding surfaces causes the formation of a fluid film having sufficient pressure to separate the surfaces.

**Hydrofinishing** A process for treating raw extracted base stocks with hydrogen to saturate them for improved stability.

**Hydrogenation** In refining, the chemical addition of hydrogen to a hydrocarbon in the presence of a catalyst; a severe form of hydrogen treating. Hydrogenation may be either destructive or nondestructive. In the former case, hydrocarbon chains are ruptured (cracked), and hydrogen is added where the breaks have occurred. In the latter, hydrogen is added to a molecule that is unsaturated with respect to hydrogen. In either case, the resulting products are highly stable. Temperatures and pressures in the hydrogenation process are usually greater than in hydrofining.

**Hydrolytic stability** Ability of additives and certain synthetic lubricants to resist chemical decomposition (hydrolysis) in the presence of water.

**Hydrophobic** Having antagonism for water; not capable of uniting or mixing with water.

**Hydrophobic** Compounds that repel water.

**Hypoid gear lubricant** A gear lubricant having extreme-pressure characteristics for use with a hypoid type of gear, as in the differential of an automobile.

**Hypoid gears** Gears in which the pinion axis intersects the plane of the ring gear at a point below the ring-gear axle and above the outer edge of the ring gear or above the ring-gear axle and below the outer edge of the ring gear.

**ILMA** Acronym for the Independent Lubricant Manufacturers Association, a trade association of businesses engaged in compounding, blending, formulating, packaging, marketing, and distributing lubricants.

**ILSAC** Acronym for the International Lubricant Standardization and Approval Committee, a joint committee of AAMA (American Automobile

Manufacturers Association) and JAMA (Japan Automobile Manufacturers Association) members that assists in the development of new minimum oil performance standards.

**Incompatibility** When a mixture of two greases shows physical properties or service performance that is markedly inferior to that of either of the greases before mixing, they are incompatible. Performance or properties inferior to one of the products and superior to the other may not be due to incompatibility.

**Incompatible fluids** Fluids that when mixed in a system will have a deleterious effect on that system, its components, or its operation.

**Induction period** In an oxidation test, the time period during which oxidation proceeds at a constant and relatively low rate. It ends at the point where oxidation rate increases sharply.

**Induction period** (grease oxidation) The time during which grease oxidation proceeds at a relatively slow rate (see ASTM D942).

**Industrial lubricant** Any petroleum or synthetic-base fluid or grease commonly used in lubricating industrial equipment, such as gears, turbines, and compressors.

**Infrared analysis** A form of absorption spectroscopy that identifies organic functional groups present in a used oil sample by measuring their light absorption at specific infrared wavelengths; absorbance is proportional to concentration. The test can indicate additive depletion, the presence of water, hydrocarbon contamination of a synthetic lubricant, oxidation, nitration, and glycol contamination from coolant. Fourier-transform infrared (FTIR) permits the generation of complex curves from digitally represented data.

**Infrared spectroscopy** An analytical method using infrared absorption for assessing the properties of used oil and certain contaminants suspended therein. See *FTIR*.

**Inhibitor** Additive that improves the performance of a petroleum product by controlling undesirable chemical reactions, that is, oxidation inhibitor, rust inhibitor, etc.

**Inorganic thickener** See *Nonsoap thickener*.

**Inside-mounted seal** A mechanical seal located inside the seal chamber with the pumped product's pressure at its outside diameter.

**Insolubles** Contaminants found in used oils due to dust, dirt, wear particles, or oxidation products; often measured as pentane or benzene insolubles to reflect insoluble character.

**Insolubles** Grease components that are insoluble in the prescribed reagents in an analytical procedure (such as ASTM D128).

**Intercooler** A device that cools a gas between the compressive steps of a multistage compressor.

**Ion exchange** A transfer of ions between two electrolytes or between an electrolyte solution and a complex. The term normally denotes the processes of purification, separation, and decontamination of aqueous and other ion-containing solutions with an insoluble (usually resinous) solid.

**JIC** Acronym for joint industry conference.

**Joule** A unit of work, energy, or heat. 1 J (joule) = 1 Nm (newton-meter).

**Journal** That part of a shaft or axle that rotates or angularly oscillates in or against a bearing or about which a bearing rotates or angularly oscillates.

**Journal bearing** A sliding type of bearing having either rotating or oscillatory motion and in conjunction with which a journal operates. In a full or sleeve-type journal bearing, the bearing surface is 360 degrees in extent. In a partial bearing, the bearing surface is less than 360 degrees in extent, that is, 150 or 120 degrees, etc.

**Karl Fischer reagent method** The standard laboratory test to measure the water content of mineral base fluids. In this method, water reacts quantitatively with the Karl Fischer reagent. This reagent is a mixture of iodine, sulfur dioxide, pyridine, and methanol. When excess iodine exists, electric current can pass between two platinum electrodes or plates. The water in the sample reacts with the iodine. When the water is no longer free to react with iodine, an excess of iodine depolarizes the electrodes, signaling the end of the test.

**Kinematic viscosity** Measure of a fluid's resistance to flow under gravity at a specific temperature (usually 40 or 100°C).

**km** Abbreviation for kilometer.

**Laminar flow** A flow situation in which fluid moves in parallel lamina or layers.

**Lands** The circumferential areas between the grooves of a piston.

**Lead** Commonly used name for tetraethyl or tetramethyl lead, an additive used in gasoline to improve octane ratings. Elemental lead is commonly used in sleeve bearing and bushing alloys.

**Light ends** Low-boiling volatile materials in a petroleum fraction. They are often unwanted and undesirable, but in gasoline, the proportion of light ends deliberately included is used to assist low-temperature starting.

**Lip seal** An elastomeric or metallic seal that prevents leakage in dynamic and static applications by a scraping or wiping action at a controlled interference between itself and the mating surface.

**Lithium grease** The most common type of grease today, based on lithium soaps.

**Load–wear index (LWI)** Measure of the relative ability of a lubricant to prevent wear under applied loads; calculated from data obtained from the four-ball EP method. Formerly called *mean Hertz load*.

**Lubricant** Any substance interposed between two surfaces in relative motion for the purpose of reducing the friction and/or the wear between them.

**Lubricating grease** A solid to semifluid dispersion of a thickening agent in a liquid lubricant containing additives (if used) to impart special properties.

**Lubrication** Control of friction and wear by the introduction of a friction-reducing film between moving surfaces in contact. May be a fluid, solid, or plastic substance (see *Boundary lubrication*, *Elastohydrodynamic lubrication*, *Hydrodynamic lubrication*)

**Lubricator** A device that adds controlled or metered amounts of lubricant into a pneumatic system.

**Lubricity** Ability of an oil or grease to lubricate; also called *film strength*.

**Luster** Reflected light intensity from a grease; its sheen or brilliance. Luster is either *bright*, reflects light with a relatively strong intensity; or *dull*, reflects light with a relatively weak intensity. A high water content or certain thickeners and fillers may give a grease a characteristic dull luster.

**LVI** An abbreviation for *low viscosity index* (VI), typically below 40 VI units.

**m** Abbreviation for meter.

**Magnetic plug** Strategically located in the flow stream to collect a representative sample of wear debris circulating in a system, for example, engine swarf, bearing flakes, and fatigue chunks.

The rate of buildup of wear debris reflects degradation of critical surfaces.

**Magnetic seal** A seal that uses magnetic material (instead of springs or a bellows) to provide the closing force that keeps the seal faces together.

**Magnetic separator** A separator that uses a magnetic field to attract and hold ferromagnetic particles.

**Manifold** A filter assembly containing multiple ports and integral relating components that services more than one fluid circuit.

**Material Safety Data Sheet (MSDS)** A publication containing health and safety information on a hazardous product (including petroleum). The OSHA Hazard Communication Standard requires that an MSDS be provided by manufacturers to distributors or purchasers prior to or at the time of product shipment. An MSDS must include the chemical and common names of all ingredients that have been determined to be health hazards if they constitute 1% or greater of the product's composition (0.1% for carcinogens). An MSDS also includes precautionary guidelines and emergency procedures.

**Maximum bulk/use temperature** All fluids have a maximum recommended temperature to which they can be heated. Heating a fluid beyond this point will result in thermal degradation or *cracking*. This is characterized by low boilers (or *light ends*) coming off, lowering the flashpoint, increasing vapor pressure, and allowing carbon buildup.

**Maximum film temperature** Heating elements and pipe walls get much hotter than the maximum bulk/use temperature of a system. A fluid's film temperature is always higher than its maximum bulk temperature and, if exceeded, will thermally degrade the fluid.

**Metal deactivators** Some metals used in the construction of heat-transfer systems can actually react with the oil, causing premature breakdown. Metal deactivators ensure compatibility with any system, even those with copper lines, heat exchangers, or fittings.

**Metal oxides** Oxidized ferrous particles that are very old or have been produced recently by conditions of inadequate lubrication. Trend is important.

**Metalworking lubricant** Any lubricant, usually petroleum based, that facilitates the cutting or shaping of metal. Basic types of metalworking lubricants are cutting and tapping fluids and drawing compounds.

**MIL** Abbreviation for military.

**Mineral seal oil** A distillation fraction between kerosene and gas oil, widely used as a solvent oil in gas adsorption processes, as a lubricant for the rolling of metal foil, and as a base oil in many specialty formulations. Mineral seal oil takes its name not from any sealing function but from the fact that it originally replaced oil derived from seal blubber for use as an illuminant for signal lamps and lighthouses.

**Mixed film** A type of lubrication that features a combination of full-film and thin-film elements.

**Mold (release) lubricant** A compound, often of petroleum origin, for coating the interiors of molds for glass and ceramic products. The mold lubricant facilitates removal of the molded object from the mold, protects the surface of the mold, and reduces or eliminates the need for cleaning it.

**Moly** Abbreviation for molybdenum disulfide, a solid lubricant and friction reducer that is colloidally dispersed in some oils and greases.

**Molybdenum disulfide** A black lustrous powder ($MoS_2$) that serves as a dry-film lubricant in certain high-temperature and high-vacuum applications. It is also used in the form of paste to prevent scoring when assembling press-fit parts and as an additive to impart residual lubrication properties to oils and greases. Molybdenum disulfide is often called *moly* or *molysulfide*.

**Morphology** Pertaining to structure and form.

**Motor** A device that converts fluid power into mechanical force and motion. It usually provides rotary mechanical motion.

**Motor bearing** A bearing that supports the crankshaft in an internal combustion engine. It is a support or guide by means of which a moving part is positioned with respect to the other parts of a mechanism.

**Motor oil** Oil that is used to lubricate the moving components of an internal combustion engine.

**Multigrade oil** Engine or gear oil that meets the requirements of more than one SAE viscosity grade classification and that can be used over a wider temperature range than a single-grade oil.

**Multipass test** Filter performance test in which the contaminated fluid is allowed to recirculate through the filter for the duration of the test. Contaminant is usually added to the test fluid during the test. The test is used to determine the beta ratio of an element.

**Naphthenic** A type of petroleum fluid derived from naphthenic crude oil containing a high proportion of closed-ring methylene groups.

**NAS** Acronym for National Aerospace Standard.

**NASA** Acronym for National Aeronautics and Space Administration.

**Needle bearing** A rolling type of bearing containing rolling elements that are relatively long compared with their diameter.

**Net additive treating cost (NATC)** The cost of additive in one unit of finished product including base fluid credit but not shipping costs.

**Net delivered treating cost (NDTC)** The cost of additive in one unit of finished product including base fluid credit and shipping costs.

**Neutral oil** The basis of most commonly used automotive and diesel lubricants; they are light overhead cuts from vacuum distillation.

**Neutralization number** A measure of the acidity or alkalinity of an oil. The number is the mass in milligrams of acid (HCl) or base (KOH) required to neutralize 1 g of oil.

**Newtonian behavior** A lubricant exhibits Newtonian behavior if its shear rate is directly proportional to the shear stress. This constant proportion is the viscosity of the liquid.

**Newtonian flow** Occurs in a liquid system where the rate of shear is directly proportional to the shearing force, as with straight-grade oils that do not contain a polymeric viscosity modifier. When rate of shear is not directly proportional to the shearing force, flow is non-Newtonian, as it is with oils containing viscosity modifiers.

**Newtonian fluid** A fluid with a constant viscosity at a given temperature regardless of the rate of shear. Single-grade oils are Newtonian fluids. Multigrade oils are non-Newtonian fluids because viscosity varies with shear rate.

**Nitration** The process whereby nitrogen oxides attack petroleum fluids at high temperatures, often resulting in a viscosity increase and deposit formation.

**Nitrous oxide** A chemical compound made up of nitrogen and oxygen, $N_2O$. It is a liquid that turns into a gas when injected into an engine.

**NLGI** Acronym for the National Lubricating Grease Institute, a trade association whose main interest is grease and grease technology. The NLGI is best known for its system of rating greases by penetration.

**NLGI automotive grease classifications** Automotive lubricating grease quality levels established jointly by the SAE, ASTM, and NLGI. There are several categories in two classifications: chassis lubricants and wheel bearing lubricants. Quality and performance levels within each category are defined by ASTM tests.

**NLGI consistency grades** Simplified system established by the NLGI for rating the consistency of grease.

**NLGI number** A scale for comparing the consistency (hardness) range of greases (numbers are in order of increasing consistency). Based on the ASTM D217 worked penetration at 25°C (77°F).

**Nominal filtration rating** An arbitrary micrometer value indicated by a filter manufacturer. Given the lack of reproducibility, this rating has deprecated.

**Non-Newtonian behavior** The property of some fluids and many plastic solids (including grease) of exhibiting a variable relationship between shear stress and shear rate.

**Non-Newtonian fluid** Fluids, such as a grease or a polymer-containing oil (e.g., multigrade oil), in which shear stress is not proportional to shear rate.

**Nonsoap thickener** Specially treated or synthetic materials (not including metallic soaps) dispersed in liquid lubricants to form greases. Sometimes called *synthetic thickener*, *inorganic thickener*, or *organic thickener*.

**Nonwoven medium** A filter medium composed of a mat of fibers.

**Normal paraffin** A hydrocarbon consisting of molecules in which any carbon atom is attached to no more than two other carbon atoms; also called *straight-chain paraffin* and *linear paraffin*.

**Obliteration** A synergistic phenomenon of both particle silting and polar adhesion. When water and silt particles coexist in a fluid containing long-chain molecules, the tendency for valves to undergo obliteration increases.

**Octane number** A measure of a fuel's ability to prevent detonation in a spark-ignition engine; measured in a standard single-cylinder variable-compression-ratio engine by comparison with primary reference fuels. Under mild conditions, the engine measures research octane number (RON); under severe conditions, motor octane number (MON). Where the law requires posting of octane numbers on dispensing pumps, the antiknock index (AKI) is used. This is the arithmetic average of RON and MON ($R$ + $M$)/2. It approximates the road octane number, which is a measure of how an average car responds to the fuel.

**Octane requirement (OR)** The lowest octane number reference fuel that will allow an engine to run knock free under standard conditions of service.

**Octane requirement increase (ORI)** As deposits accumulate in the combustion chamber, the ORI of an engine increases, usually reaching an equilibrium value after 10,000 to 30,000 km. ORI is a measure of the increase, which may be in the range of three to ten numbers.

**Oil** A liquid of vegetable, animal, mineral, or synthetic origin that provides the basis for increased wear resistance as well as an income to those in the business of providing value.

**Oil analysis** The routine activity of analyzing lubricant properties and suspended contaminants for the purpose of monitoring and reporting timely, meaningful, and accurate information on lubricant and machine conditions.

**Oil consumption** The amount of lubricating fluid that is consumed by a machine, production line, plant, or company over a given period of time.

**Oil consumption ratio** Annual oil purchases divided by machine charge volume. For example, if you purchased 10,000 gallons of oil in one year and the total amount of oil that all your machine holds is 4200 gallons, your consumption ratio is 2.4.

**Oil drain** A large bolt or plug that secures the drain hole in the oil pan. It is generally fitted with a gasket or O-ring to prevent leakage.

**Oil filter** A device that removes the inherent or introduced impurities from the oil that lubricates an internal combustion engine.

**Oil flushing** A fluid circulation process that is designed to remove contamination and decomposition from a lubrication-based system.

**Oil-mist lubrication** A method of lubricant delivery in which oil is piped throughout a machine to desired locations and dispensed with a spray nozzle. Oil-mist systems are employed to cool and lubricate many machine parts at once.

**Oil-mist system** A device that delivers lubricant to multiple machine parts at once via a setup that includes piping and a spray nozzle.

**Oil oxidation** Occurs when oxygen attacks petroleum fluids. The process is accelerated by heat, light, metal catalysts, and the presence of water, acid, or solid contaminants. It leads to increased viscosity and deposit formation.

**Oil ring** A loose ring, the inner surface of which rides on a shaft or journal and dips into a reservoir of lubricant from which it carries the lubricant to the top of a bearing by its rotation with the shaft.

**Oiler** A device for once-through lubrication. Three common types of oilers are drop feed, wick feed, and bottle feed; all depend on gravity to induce a metered flow of oil to the bearing. The drop-feed oiler delivers oil from the bottom of a reservoir to a bearing one drop at a time; flow rate is controlled by a needle valve at the top of the reservoir. In a wick-feed oiler, the oil flows through a wick and drops from the end of the wick into the bearing; feed is regulated by chaining the number of strands, by raising or lowering the oil level, or by applying pressure to the wick. In a bottle-feed oiler, a vacuum at the top of the jar keeps the fluid from running out; as tiny bubbles of air enter, the vacuum is reduced, and a small amount of oil enters the bearing or is added to a reservoir from which the bearing is lubricated.

**Oiliness**  The property of a lubricant that produces low friction under conditions of boundary lubrication. The lower the friction, the greater is the oiliness.

**Oiliness agent**  A material that forms an adsorbed film to reduce friction. An additive, usually polar in nature, used to improve the lubricity of a mineral oil. Now usually called a *boundary-lubrication additive.*

**Open gear**  A gear that is exposed to the environment rather than being housed in a protective gearbox. Open gears are generally large, heavily loaded, and slow moving. They are found in such applications as mining and construction machinery, punch presses, plastic and rubber mills, tube mills, and rotary kilns. Open gears require viscous, adhesive lubricants that bond to the metal surfaces and resist runoff. Such lubricants are often called *gear shields.* Top-quality lubricants for such applications are specially formulated to protect the gears against the effects of water and other contaminants.

**Organic thickener**  See *Nonsoap thickener.*

**OSHA**  Acronym for the Occupational Safety and Health Administration.

**Outside-mounted seal**  A mechanical seal with its seal head mounted outside the seal chamber that holds the fluid to be sealed. Outside seals have the pumped fluid's pressure at their ID.

**Oxidation**  The process of combining with oxygen. All petroleum products are subject to oxidation to some degree. The reaction increases with rise in temperature. Oxidation produces oil-insoluble oxidized materials, which result in viscosity increase and deposits. Occurs when oxygen attacks petroleum fluids. The process is accelerated by heat, light, metal catalysts, and the presence of water, acids, or solid contaminants. It leads to increased viscosity and deposit formation.

**Oxidation inhibitor**  A chemical additive that minimizes the formation of harmful acids and varnish-forming compounds that form when a fluid is subjected to air at elevated temperatures.

**Oxidation inhibitor**  Substance added in small quantities to a petroleum product to increase its oxidation resistance, thereby lengthening its service or storage life; also called *antioxidant.* An oxidation inhibitor may work in one of these ways: (1) by combining with and modifying peroxides (initial oxidation products) to render them harmless, (2) by decomposing the peroxides, or (3) by rendering an oxidation catalyst inert.

**Oxidation stability**  The resistance of lubricants to chemically react with oxygen. The absorption and reaction of oxygen may lead to deterioration of lubricants.

**Oxidative degradation**  Oxidative degradation is the reaction of oxygen (in air) with the fluid by a free-radical mechanism to form larger molecules that end up as polymers or solids. These thicken the fluid and increase its viscosity. A more viscous fluid will be more difficult to pump and have poorer heat-transfer characteristics, as well as an increased chance of coke formation.

Oxidation is also accompanied by an increase in the acidity (TAN) of the fluid. As with all chemical reactions, oxidation occurs more rapidly as the temperature is increased. At room temperature, the reaction rate is hardly measurable. However, it can become a factor in the life of the fluid in certain types of heat-transfer systems. At temperatures encountered in systems in use in the plastics extrusion and die casting industries, as an example, oxidation is the main cause of fluid degradation.

**Oxygenate** An oxygen-containing, ashless organic compound such as alcohol or ether that can be used as a fuel or fuel supplement.

**Oxygenated fuels** Fuels for internal combustion engines that contain oxygen combined in the molecule (e.g., alcohols, ethers, and esters). Term also applies to blends of gasoline with oxygenates (e.g., gasohol, which contains 10% by volume anhydrous ethanol in unleaded gasoline).

**Oxygenated gasoline** Required annually from September 15 to March 15 for use in carbon monoxide nonattainment areas. Oxygenated gasoline is defined as a spark-ignition engine fuel meeting ASTM D4814 specifications and blended to include a minimum of 2.0% mass oxygen and a maximum of 1.0% volume benzene.

**Ozone and carbon monoxide nonattainment areas** Any area of the continental United States that does not meet the 1990 Clean Air Act requirements for carbon monoxide or ground-level ozone pollutants.

**PAG synthetic fluid** Polyalkaline glycol (PAG) has excellent oxidative and thermal stability, very high viscosity index, excellent film strength, and an extremely low tendency to leave deposits on machine surfaces. The low-deposit-forming tendency is really due to two properties: the oil's ability to dissolve deposits and the fact that the oil burns clean. So, when they are exposed to a very hot surface or subjected to microdieseling by entrained air, PAGs are less likely to leave residue that will form deposits. PAGs also may be the only type of base oil with significantly lower fluid friction, which may allow for energy savings. The other unique property of PAGs is their ability to absorb a great deal of water and maintain lubricity. There are actually two different types of PAGs; one demulsifies and the other absorbs water. The most common applications for PAGs are compressors and critical gearing applications. The negatives of PAGs are their very high cost and the potential to be somewhat hydrolytically unstable.

**PAO synthetic fluid** Polyalphaolefins, often called *synthetic hydrocarbons*, are probably the most common type of synthetic base oil used today. They are moderately priced, provide excellent performance, and have few negative attributes. PAO base oil is similar to mineral oil. The advantage comes from the fact that it is built rather than extracted and modified, making it more pure. Practically all the oil molecules are the same shape and size and are completely saturated. The potential benefits of PAOs are improved oxidative and thermal stability, excellent demulsibility and hydrolytic stability, a high VI, and a very low pour point. Most of the properties make PAOs a good selection for temperature extremes—both high operating temperatures and low start-up temperatures. Typical applications for PAOs are engine oils, gear oils, and compressor oils. The negative attributes of PAOs are the price and poor solubility. The low inherent solubility of PAOs creates problems for formulators when it comes to dissolving additives. Likewise, PAOs cannot suspend potential varnish-forming degradation by-products, although they are less prone to create such material.

**Paraffin** Any hydrocarbon identified by saturated straight (normal) or branched (iso) carbon chains; also called an *alkane*. The generalized paraffinic molecule can be symbolized by the formula $C_nH_{2n+2}$. Paraffins are relatively nonreactive and have excellent oxidation stability. In contrast to naphthenic oils, paraffinic lubricating oils have relatively high wax content and pour point and generally have a high viscosity index (VI). Paraffinic solvents are generally lower in solvency than naphthenic or aromatic solvents.

**Paraffinic** A type of petroleum fluid derived from paraffinic crude oil and containing a high proportion of straight-chain saturated hydrocarbons; often susceptible to cold-flow problems.

**Parallel systems** Lubrication systems where the dispensing devices are connected to the main line in parallel. Each dispensing device operates independent of any other in the system.

**Particle counter** An instrument that detects and counts particles found in a fluid such as oil.

**Particle counting** A microscopic technique that enables the visual counting of particles in a known quantity of fliud. The count identifies the number of particles present greater than a particular micron size per unit volume of fluid, often stated as particles > 10 μm/mL.

**Particulates** Particles made up of a wide range of natural materials (e.g., pollen, dust, and resins) combined with human-made pollutant (e.g., smoke particles and metallic ash); in sufficient concentrations, particulates can be a respiratory irritant.

**Pascal** Unit of pressure in the metric (SI) system.

**Pascal's law** A pressure applied to a confined fluid at rest is transmitted with equal intensity throughout the liquid, and that pressure is considered to act at right angles to each surface contacted by the fluid.

**PCB** Acronym for polychlorinated biphenyl, a class of synthetic chemicals consisting of a homologous series of compounds beginning with monochlorobiphenyl and ending with decachlo-robiphenyl. PCBs do not occur naturally in petroleum but have been found as contaminants in used oil. PCBs have been legally designated as a health hazard, and any oil so contaminated must be handled in strict accordance with state and federal regulations.

**PCV system** An abbreviation for positive crankcase ventilation system, this is a system that prevents the vapors of a crankcase from being discharged directly into the atmosphere.

**PCV valve** A positive crankcase ventilation valve is a one-way valve that ensures the continual flow and evacuation of gases from the crankcase into the engine.

**Penetration** A measure of consistency (hardness) based on an inverse penetration measurement (the softer the consistency, the higher is the penetration number), based on ASTM D217 (IP 50) and similar standardized methods.

**Percentage permanent viscosity loss (PPVL)** Measure of the PVL related to the viscosities of fresh oil; equals PVL divided by fresh oil viscosity multiplied by 100. Difference between the viscosity of an oil measured at low and high shear stresses divided by viscosity measured at low shear stress multiplied by 100.

**Permanent viscosity loss (PVL)** Difference between the viscosity of fresh oil and that of the same oil after engine operation or special test conditions of polymer degradation.

**Petrochemical** Any chemical substance derived from crude oil or its products or from natural gas. Some petrochemical products may be identical to others produced from other raw materials such as coal and producer gas.

**pH** Measure of alkalinity or acidity in water and water-containing fluids. pH can be used to determine the corrosion-inhibiting characteristic in water-based fluids. Typically, pH > 8.0 is required to inhibit corrosion of iron and ferrous alloys in water-based fluids.

**Phenol** A white, crystalline compound ($C_6H_5OH$) derived from benzene; used in the manufacture of phenolic resins, weed killers, plastics, and disinfectants; also used in solvent extraction, a petroleum-refining process. Phenol is a toxic material; skin contact must be avoided.

**Phosphate ester** Any of a group of synthetic lubricants having superior fire resistance. A phosphate ester generally has poor hydrolytic stability, poor compatibility with mineral oil, and a relatively low viscosity index. It is used as a fire-resistant hydraulic fluid in high-temperature applications.

**Pinion** The smaller of two mating or meshing gears; can be either the driving or the driven gear.

**Pitting** Surface cavities; may be related to fatigue, overload, or corrosion.

**Plain bearing** A relatively simple and inexpensive bearing typically made of two parts. A rotary plain bearing can be just a shaft running through a hole. A simple linear bearing can be a pair of flat surfaces designed to allow motion.

**Plastic flow** Surface deformation of metal as a result of yielding under heavy load.

**Plasticity** The ability of a solid material to permanently deform without rupturing under the application of force (plastic flow differs from fluid flow in that the shearing stress must exceed a yield point before any flow occurs).

**Pleated filter** A filter element whose medium consists of a series of uniform folds and has the geometric form of a cylinder, cone, disc, plate, etc. Synonymous with *convoluted* and *corrugated*.

**PNA (polynuclear aromatic)** Any of numerous complex hydrocarbon compounds consisting of three or more benzene rings in a compact molecular arrangement. Some types of PNAs are formed in fossil fuel combustion and other heat processes, such as catalytic cracking.

**Poise** Measurement unit of a fluid's resistance to flow, that is, viscosity, defined by the shear stress (in dynes per square centimeter) required to move one layer of fluid along another over a total layer thickness of 1 cm at a velocity of 1 cm/sec. This viscosity is independent of fluid density and directly related to flow resistance.

**Polar compound** A chemical compound whose molecules exhibit electrically positive characteristics at one extremity and negative characteristics at the other. Polar compounds are used as additives in many petroleum products. Polarity gives certain molecules a strong affinity for solid surfaces; as lubricant additives (oiliness agents), such molecules plate out to form a tenacious friction-reducing film. Some polar molecules are oil soluble at one end and water soluble at the other end; in lubricants, they act as emulsifiers, helping to form stable oil–water emulsions. Such lubricants are said to have good metal-wetting properties. Polar compounds with a strong attraction for solid contaminants act as detergents in engine oils by keeping contaminants finely dispersed.

**Polishing (bore)** Excessive smoothing of the surface finish of the cylinder bore or cylinder liner in an engine to a mirror-like appearance, resulting in depreciation of ring-sealing and oil-consumption performance.

**Polyalkylene glycol** Mixture of condensation polymers of ethylene oxide and water. They are any of a family of colorless liquids with high molecular weight that are soluble in water and in many organic solvents. They are used in detergents and as emulsifiers and plasticizers. PAG-based lubricants are used in diverse applications where petroleum oil–based products do not provide the desired performance—and because they are fire resistant and will not harm workers or the environment.

**Polyglycols** Polymers of ethylene or propylene oxides used as a synthetic lubricant base. Properties include very good hydrolytic stability, high viscosity index (VI), and low volatility; used particularly in water emulsion fluids.

**Polymer** A substance formed by the linkage (polymerization) of two or more simple molecules called *monomers* to form a single larger molecule having the same elements in the same proportions as the original monomers; that is, each monomer retains its structural identity. A polymer may be liquid or solid; solid polymers may consist of millions of repeated linked units. A polymer made from two or more similar monomers is called a *copolymer*; a copolymer composed of three different types of monomers is a *terpolymer*. Natural rubber and synthetic rubbers are examples of polymers. Polymers are commonly used as VI improvers in multigrade oils and as tackifiers in lubricating greases.

**Polymerization** The chemical combination of similar-type molecules to form larger molecules.

**Polyolester** A synthetic lubricant base formed by reacting fatty acids with a polyol (such as a glycol) derived from petroleum. Properties include good oxidation stability at high temperatures and low volatility. Used in formulating lubricants for turbines, compressors, jet engines, and automotive engines.

**Polyolefin** A polymer derived by polymerization of relatively simple olefins. Polyethylene and polyisoprene are important polyolefins.

**Pore** A small channel or opening in a filter medium that allows passage of fluid.

**Porosity** The ratio of pore volume to total volume of a filter medium expressed as a percent.

**Positive crankcase ventilation (PCV)** System for removing blow-by gases from the crankcase and returning them through the carburetor intake manifold to the combustion chamber where the recirculated hydrocarbons are burned. A PC valve controls the flow of gases from the crankcase to reduce hydrocarbon emissions.

**Pour point** Lowest temperature at which an oil or distillate fuel is observed to flow when cooled under conditions prescribed by test method ASTM D97. The pour point is 3°C (5°F) above the temperature at which the oil in a test vessel shows no movement when the container is held horizontally for five seconds.

**Pour-point depressant** An additive that retards the adverse effects of wax crystallization and lowers the pour point.

**Pour stability** The ability of a pour-depressed oil to maintain its original ASTM pour point when subjected to long-term storage at low temperature approximating winter conditions.

**Pour-point suppressants** Prevent insoluble wax molecules in oil from building a honeycomb (lattice-like structure) at colder temperatures; particularly useful for paraffinic oils. Give useful ability to pour at lower temperatures.

**Power unit** A combination of pump, pump drive, reservoir, controls, and conditioning components that may be required for an application.

**ppm** Parts per million (1/ppm = 0.000001). Generally by weight. 100 ppm = 0.01%; 10,000 ppm = 1%.

**Predictive maintenance** A type of condition-based maintenance emphasizing early prediction of failure using nondestructive techniques such as vibration analysis, thermography, and wear debris analysis.

**Preignition** Ignition of the fuel–air mixture in a gasoline engine before the spark plug fires. Often caused by incandescent fuel or lubricant deposits in the combustion chamber, it wastes power and may damage the engine.

**Pressure**   Force per unit area, usually expressed in pounds per square inch (psi).

**Pressure-control valve**   A valve whose primary function is to limit system pressure.

**Pressure drop**   Resistance to flow created by the element (medium) in a filter. Defined as the difference in pressure upstream (inlet side of the filter) and downstream (outlet side of the filter).

**Pressure gauge**   Pressure differential above or below atmospheric pressure.

**Pressure indicator**   An indicator that signals pressure conditions.

**Pressure-line filter**   A filter located in a line conducting working fluid to a working device or devices.

**Pressure switch**   An electric switch operated by fluid pressure.

**Pressure, absolute**   The sum of atmospheric and gauge pressures.

**Proactive maintenance**   A maintenance strategy for stabilizing the reliability of machines or equipment. Its central theme involves directing corrective actions aimed at failure root causes, not active failure symptoms, faults, or machine wear conditions. A typical proactive maintenance regiment involves three steps: (1) setting a quantifiable target or standard relating to a root cause of concern (e.g., a target fluid cleanliness level for a lubricant), (2) implementing a maintenance program to control the root-cause property to within the target level (e.g., routine exclusion or removal of contaminants), and (3) routine monitoring of the root-cause property using a measurement technique (e.g., particle counting) to verify that the current level is within the target.

**Process oil**   An oil that serves as a temporary or permanent component of a manufactured product. Aromatic process oils have good sol-

vency characteristics; their applications include proprietary chemical formulations, ink oils, and extenders in synthetic rubbers. Naphthenic process oils are characterized by low pour points and good solvency properties. Paraffinic process oils are characterized by low aromatic content and light color.

**psi**   Abbreviation for pounds per square inch.

**Pumpability**   The low-temperature, low-shear-stress shear-rate viscosity characteristics of an oil that permit satisfactory flow to and from the engine oil pump and subsequent lubrication of moving components.

**Pumpability**   The flow characteristics of a grease that permit satisfactory delivery from lines, nozzles, and fittings of grease-dispensing systems and subsequent lubrication of moving components.

**Pusher seal**   A mechanical seal in which the secondary seal is pushed along the shaft or sleeve to compensate for misalignment and face wear.

**Quenching oil**   Also called *heat-treating oil*, a high-quality oxidation-resistant petroleum oil used to cool metal parts during their manufacture; often preferred to water because the oil's slower heat transfer lessens the possibility of cracking or warping of the metal. A quenching oil must have excellent oxidation and thermal stability and should yield clean parts, essentially free of residue. In refining terms, a quenching oil is an oil introduced into high-temperature vapors of cracked (see *Cracking pressure*) petroleum fractions to cool them.

**Quick-disconnect coupling**   A coupling that can quickly join or separate a fluid line without the use of tools or special devices.

**R&O (rust and oxidation) inhibited**   A term applied to highly refined industrial lubricating oils formulated for long service in circulating lubrication systems, compressors, hydraulic systems, bearing housing, gearboxes, and so on. The finest R&O oils are often referred to as *turbine oils*.

**Rate of shear**  The difference between the velocities along the parallel faces of a fluid element divided by the distance between the faces.

**Rated flow**  The maximum flow that the power supply system is capable of maintaining at a specific operating pressure.

**Rated pressure**  The qualified operating pressure that is recommended for a component or a system by the manufacturer.

**Refining**  A series of processes for converting crude oil and its fractions into finished petroleum products. Following distillation, a petroleum fraction may undergo one or more additional steps to purify or modify it. These refining steps include thermal cracking, catalytic cracking, polymerization, alkylation, re-forming, hydrocracking, hydroforming, hydrogenation, hydrogen treating, hydrofining, solvent extraction, dewaxing, deoiling, acid treating, clay filtration, and deasphalting. Refined lubricating oils may be blended with other lube stocks, and additives may be incorporated, to impart special properties.

**Refrigeration compressor**  A special type of compressor typically used for refrigeration, heat pumping, and air-conditioning. They are made to turn low-pressure gases into high-pressure and high-temperature gases. The three main types of refrigeration compressors are screw compressors, scroll compressors, and piston compressors.

**Refrigerator oil**  The lubricant added to the working fluid in an expansion-type cooling unit that serves to lubricate the pump mechanism.

**Remaining useful life**  An opinion (based on data, observations, history, records, exposure, etc.) of the number of years before a fluid, system, or component will require replacement or reconditioning.

**Rerefining**  A process of reclaiming used lubricant oils and restoring them to a condition similar to that of virgin stocks by filtration, clay adsorption, or more elaborate methods.

**Reservoir**  A container for storage of liquid in a fluid power system.

**Reservoir filter**  A filter installed in a reservoir in series with a suction or return line; also known as a *sump filter*.

**Residual dirt capacity**  The dirt capacity remaining in a service-loaded filter element after use but before cleaning, measured under the same conditions as the dirt capacity of a new filter element.

**Return line**  A location in a line conducting fluid from a working device to a reservoir.

**Return line filtration**  Filters located upstream of the reservoir but after fluid has passed through the system's output components (cylinders, motors, etc.).

**Reynold's number**  A numerical ratio of the dynamic forces of mass flow to the shear stress due to viscosity. Flow usually changes from laminar to turbulent between Reynold's numbers 2000 and 4000.

**Rheology**  The study of the deformation and flow of matter in terms of stress, strain, temperature, and time. The rheologic properties of a grease are commonly measured by penetration and apparent viscosity.

**Ridging**  In gear teeth, a form of plastic flow characterized by a rippled appearance on the surface.

**Ring lubrication**  A system of lubrication in which the lubricant is supplied to the bearing by an oil ring.

**Ring sticking** Freezing of a piston ring in its groove in a piston engine or reciprocating compressor because of heavy deposits in the piston ring zone.

**Rings** Circular metallic elements that ride in the grooves of a piston and provide compression sealing during combustion; also used to spread oil for lubrication.

**Roller bearing** An antifriction bearing comprising rolling elements in the form of rollers.

**Rolling and peening** In gear teeth, a form of plastic flow that gives the surface a hammered appearance; metal may be rolled over the teeth tips.

**Rolling oil** An oil used in hot or cold rolling of ferrous and nonferrous metals to facilitate feed of the metal between the work rolls, improve the plastic deformation of the metal, conduct heat from the metal, and extend the life of the work rolls. Because of the pressures involved, a rolling oil may be compounded or contain EP additives. In hot rolling, the oil also may be emulsifiable.

**Rotary seal** A mechanical seal that rotates with a shaft and is used with a stationary mating ring.

**Rotating equipment** Equipment that moves liquids, solids, or gases through a system of drivers (turbines, motors, engines), driven components (compressors, pumps), transmission devices (gears, clutches, couplings), and auxiliary equipment (lube and seal systems, cooling systems, buffer gas systems).

**Rotating pressure vessel oxidation test (RPVOT)** Measures an oil's oxidation stability. The oil sample is placed in a vessel containing a polished copper coil. The vessel is then charged with oxygen and placed in a bath at a constant temperature of 150°C. Stability is expressed in terms of the time it takes to achieve a pressure drop of 25.4 psi from maximum pressure.

**Rust inhibitor** A type of corrosion inhibitor used in lubricants to protect surfaces against rusting.

**Rust preventative** Compound for coating metal surfaces with a film that protects against rust; commonly used to preserve equipment in storage.

**SAE port** A straight thread port used to attach tube and hose fittings. It employs an O-ring compressed in a wedge-shaped cavity. A standard of the Society of Automotive Engineers (SAE) J514 and ANSI/B116.1.

**SAE viscosity** The viscosity classification of a motor oil according to the system developed by the Society of Automotive Engineers and now in general use. Winter grades are defined by viscosity measurements at low temperatures and have W as a suffix, whereas summer grades are defined by viscosity at 100°C and have no suffix. Multigrade oils meet both a winter and a summer definition and have designations such as SAE 10W-30, etc.

**Saponification** The formation of a metallic salt (soap) due to the interaction of fatty acids, fats, or esters generally with an alkali.

**Saponification number** The number of milligrams of potassium hydroxide (KOH) that combine with 1 g of oil under conditions specified by ASTM D94. Saponification number is an indication of the amount of fatty saponifiable material in compounded oil. Caution must be used in interpreting test results if certain substances, such as sulfur compounds or halogens, are present in the oil because these also react with KOH, thereby increasing the apparent saponification number.

**Saturation level** The amount of water that can dissolve in a fluid.

**Scoring** Distress marks on sliding metallic surfaces in the form of long, distinct scratches in the direction of motion. Scoring is an advanced stage of scuffing.

**Scratching**   Fine abrasive furrows in the direction of sliding.

**Seal**   A device designed to prevent the movement of fluid from one area to another or to exclude contaminants.

**Seal assembly**   A group of parts or a unitized assembly that includes sealing surfaces, provisions for initial loading, and a secondary sealing mechanism that accommodates the radial and axial movements necessary for installation and operation.

**Seal chamber**   The area between the seal chamber bore and a shaft in which a mechanical seal is installed.

**Seal face**   It is either of the two lapped surfaces in a mechanical seal assembly forming the primary seal.

**Seal face width**   The radial distance from the inside edge to the outside edge of the sealing face.

**Seal swell agent**   These are typically ester-based and provide a controlled swell of certain types of polymers (typically chlorinated rubber) that are used for gasket and O-ring applications.

**Seal swell (rubber swell)**   The swelling of rubber (or other elastomer) gaskets or seals when exposed to petroleum, synthetic lubricants, or hydraulic fluids. Seal materials vary widely in their resistance to the effect of such fluids. Some seals are designed so that a moderate amount of swelling improves sealing action.

**SEM**   Acronym for scanning electron microscope.

**Servovalve**   A valve that modulates output as a function of an input command.

**Settling tank**   A tank in which liquid is stored until particles suspended in the liquid sink to the bottom.

**Severe sliding**   Large ferrous particles that are produced by sliding contacts. Trend is important to determine whether abnormal wear is taking place.

**Shear rate**   Rate at which adjacent layers of fluid move with respect to each other; usually expressed as reciprocal seconds. See ASTM D1092.

**Shear stability**   The resistance of a grease to changes in consistency (hardness) during mechanical working.

**Shear stability index (SSI)**   Measure of a viscosity modifier's contribution to an oil's percentage kinematic viscosity loss when the oil is subjected to engine operation or special test conditions.

**Shear stress**   The stress (force per unit area) tending to cause shearing. The ratio of shear stress to the shear rate in a fluid is its viscosity.

**Shearing**   Relative slipping or sliding between one part of a substance and an adjacent part. Shearing in a solid involves cutting or breaking of the crystal structure; in a fluid or plastic, shearing does not necessarily destroy the continuous nature of the substance.

**Silt**   Contaminant particles 5 μm and less in size.

**Sintered medium**   A metallic or nonmetallic filter medium processed to cause diffusion bonds at all contacting points.

**Sleeve bearing**   A journal bearing, usually a full journal bearing.

**Sloughing off**   The release of contaminants from the upstream side of a filter element to the upstream side of the filter enclosure.

**Sludge** A thick, dark residue, normally of mayonnaise consistency, that accumulates on nonmoving engine interior surfaces; generally removable by wiping unless baked to a carbonaceous consistency. Its formation is associated with insoluble overloading of the lubricant.

**Slumpability** See *Feedability*.

**Soap** See *Thickener, Complex soap, Saponification*.

**Solvent** A material with a strong capability to dissolve a given substance. The most common petroleum solvents are mineral spirits, xylene, toluene, hexane, heptane, and naphthas. Aromatic-type solvents have the highest solvency for organic chemical materials, followed by naphthenes and paraffins. In most applications, the solvent disappears, usually by evaporation, after it has served its purpose. The evaporation rate of a solvent is very important in manufacture.

**Solvent extraction** A refining process used to separate components (unsaturated hydrocarbons) from lube distillates in order to improve the oil's oxidation stability, viscosity index, and response to additives. The oil and the solvent extraction media are mixed in an extraction tower, resulting in the formation of two phases: a heavy phase consisting of the undesirable unsaturates dissolved in the solvent and a lighter phase consisting of a high-quality oil with some solvent dissolved in it. The phases are separated, and the solvent is recovered from each by distillation.

**Solvent refining** A process for extracting lubricant base stocks from stripped heavy gas, oil, or other heavy, stripped crude stream using selective solvents such as furfural or phenol.

**Spalling** Severe damage characterized by large pits, cavities, and related cracks; related to overload and fatigue.

**Spectrographic Oil Analysis Program (SOAP)** Procedures for extracting fluid samples from operating systems and analyzing them spectrographically for the presence of key elements.

**Spin-on filter** A throwaway type bowl and element assembly that mates with a permanently installed head.

**Splash lubrication** A system of lubrication in which parts of a mechanism dip into and splash the lubricant onto themselves and/or other parts of the mechanism.

**Spur gear** The simplest variation of gears. It consists of a cylinder or disk with the teeth projecting radially. Each tooth edge is straight and aligned parallel to the axis of rotation. Such gears can be meshed together correctly only if they are fitted to parallel axles.

**Squeeze-film lubrication** The development of fluid pressure sufficient to support a load between surfaces thickly coated or flooded with lubricant and rapidly moving toward each other. Because of viscosity (or apparent viscosity), the lubricant cannot immediately flow away from the area of contact. Squeeze-film lubrication occurs between gear teeth and between wrist pins and their bushings, for example.

**Static friction** Force just sufficient to initiate relative motion between two bodies under load. The value of the static friction at the instant relative motion begins is termed *break-away friction*.

**Static seal** A seal between two surfaces that have no relative motion.

**Stationary seal** A mechanical seal in which the flexible members do not rotate with the shaft.

**Statistical process control (SPC)** The use of control charts to track and eliminate variables in repetitive manufacturing processes in order to ensure that the product is of consistent and predictable quality. If a chart reveals only chance variations that are inherent to the system, the process is said to be in a state of *statistical control*. If the chart reveals variations traceable to changes in equipment, procedures, or workers, the process is said to be *out of control*. Statistical process control differs from statistical quality control in that the former monitors manufacturing-process parameters and the latter monitors product-quality parameters.

**Stick-slip motion** Erratic, noisy motion characteristic of some machineways due to the starting friction encountered by a machine part at each end of its back-and-forth (reciprocating) movement. This undesirable effect can be overcome with a way lubricant that reduces starting friction.

**Straight mineral oil** Petroleum oil containing no additives. Straight mineral oils include such diverse products as low-cost once-through lubricants and thoroughly refined white oils. Most high-quality lubricants, however, contain additives.

**Straight oil** A mineral oil containing no additives.

**Strainer** A coarse filter element (pore size more than approximately 40 µm).

**Sulfated ash** Ash content of fresh, compounded lubricating oil as determined by ASTM D874; indicates level of metallic additives in the oil.

**Sulfonate** A hydrocarbon in which a hydrogen atom has been replaced with the highly polar ($SO_2OX$) group, where X is a metallic ion or alkyl radical. Petroleum sulfonates are refinery by-products of the sulfuric acid treatment of white oils. Sulfonates have important applications as emulsifiers and chemical intermediates in petrochemical manufacture, and substituted sulfonates are widely used as corrosion inhibitors. Synthetic sulfonates can be manufactured from special feedstocks rather than from white oil base stocks.

**Sulfur** A common natural constituent of petroleum products. While certain sulfur compounds are commonly used to improve the extreme-pressure or load-carrying properties of an oil, high sulfur content in a petroleum product may be undesirable because it can be corrosive and create an environmental hazard when burned. For these reasons, sulfur limitations are specified in the quality control of fuels, solvents, etc.

**Sulfurized oil** Oil to which sulfur or sulfur compounds have been added.

**Surface filter media** Porous materials that primarily retain contaminants on the influent face, performing the actual process of filtration.

**Surface filtration** Filtration that primarily retains contaminants on the influent surface.

**Surfactant** Surface-active agent that reduces interfacial tension of a liquid. A surfactant used in a petroleum oil may increase the oil's affinity for metals and other materials.

**Surge** A momentary rise of pressure in a circuit.

**SUS (SSU)** Saybolt universal seconds, a measure of lubricating oil viscosity in the oil industry. The measuring apparatus is filled with a specific quantity of oil or other fluid, and its flow time through a standardized orifice is measured in seconds. Fast-flowing fluids (low viscosity) will have low values; slow-flowing fluids (high viscosity) will have high values.

**Suspension agent** Some fluids develop carbon and other particulate matter after years of use. Even new systems have things like weld slag and metal shavings that can become trapped in instrument lines or cause problems in other areas. Suspension agents help ensure that particulates are held in suspension and easily filtered or caught in strainers.

**Swarf** The cuttings and grinding fines that result from metalworking operations.

**Syncrude** Unconventional crudes such as those derived from tar sands, oil shale, and coal liquefaction.

**Synthetic grease** A grease that contains a liquid lubricant that is not a mineral oil.

**Synthetic hydrocarbon** Oil molecule with superior oxidation quality tailored primarily out of paraffinic materials.

**Synthetic lubricant** Lubricating fluid made by chemically reacting materials of a specific chemical composition to produce a compound with planned and predictable properties. A lubricant produced by chemical synthesis rather than by extraction or refinement of petroleum to produce a compound with planned and predictable properties.

**Synthetic oils** Oils produced by synthesis (chemical reaction) rather than by extraction or refinement. Many (but not all) synthetic oils offer immense advantages in terms of high-temperature stability and low-temperature fluidity but are more costly than mineral oils. Major advantage of all synthetic oils is their chemical uniformity.

**Synthetic thickener** See *Nonsoap thickener*.

**System pressure** The pressure that overcomes the total resistances in a system. It includes all losses as well as useful work.

**Tacky** A descriptive term applied to lubricating oils and greases that appear particularly sticky or adhesive.

**TAN (total acid number)** Acids are formed when a fluid comes in contact with oxygen. TAN levels are a means to show the extent of which a fluid has been oxidized. New fluids typically have a TAN less than 0.05; most fluids should be changed at and have a condemning limit of a TAN of 1.0. The rate of oxidation is minimal under 200°F, but as the temperature climbs, the effects of oxidation are exponential. It is an industry-accepted standard to assume that the rate of oxidation doubles for each 15-degree increase above 200°F.

**TBN** Acronym for total base number.

**Temporary shear stability index (TSSI)** Measure of the viscosity modifier's contribution to an oil's percentage viscosity loss under high-shear conditions. Temporary shear loss results from the reversible lowering of viscosity in high-shear areas of the engine, an effect that can positively influence fuel economy and cold-cranking speed.

**Temporary viscosity loss (TVL)** Measure of the decrease in dynamic viscosity under high shear rates compared with dynamic viscosity under low shear.

**Texture** The texture of a grease is observed when a small portion of it is pressed together and then slowly drawn apart. Texture can be described as *brittle*, ruptures or crumbles when compressed; *buttery*, separates in short peaks with no visible fibers; *long fibers*, stretches or strings out into a single bundle of fibers; *resilient*, withstands a moderate compression without permanent deformation or rupture; *short fibers*, short breakoff with evidence of fibers; *stringy*, stretches or strings out into long, fine threads but with no evidence of fiber structure.

**Thermal conductivity** The rate that heat transfer occurs through conduction. Fluids with low conductivity will transfer heat slower than a fluid with a higher rate of conductivity.

**Thermal degradation** Thermal degradation or thermal cracking is the breaking of carbon–carbon bonds in the fluid molecules by heat to form smaller fragments that are free radicals. The reaction may either stop at that point, in which case smaller molecules than previously existed are formed, or the fragments may react with each other to form polymeric molecules larger than previously existed in the fluid. In heat-transfer terminology, the two types of degradation products are known as *low boilers* and *high boilers*. If thermal degradation occurs at extreme temperatures greater than 400°C (752°F), the effect is not only to break carbon–carbon bonds but also to separate hydrogen atoms from carbon atoms and form coke. In this case, fouling of the heat-transfer surfaces is very rapid, and the system will soon cease to operate. The effect of the low boilers is to decrease the flash point and viscosity of the fluid as well as to increase its vapor pressure. The effect of the high boilers is to increase the viscosity of the fluid as long as it remains in solution. However, once the solubility limit is exceeded, the fluids begin to form solids that can foul the heat-transfer surfaces.

**Thermal expansion** Heat-transfer fluids change in volume based on temperature. Volume increases when heated and decreases when cooled. This information is helpful when trying to size an expansion tank; typically, an expansion tank should be one-third full when cold and about one-half full when hot.

**Thermal stability** Ability of a fuel or lubricant to resist oxidation under high-temperature operating conditions.

**Thickener** The structure within a grease of extremely small, uniformly dispersed particles in which the liquid is held by surface tension and/or other internal forces.

**Thin-film lubrication** A condition of lubrication in which the film thickness of the lubricant is such that the friction between the surfaces is determined by the properties of the surfaces as well as by the viscosity of the lubricant.

**Thixotropy** Thixotropy of a lubricating grease is manifested by a decrease in consistency (softening) as a result of shearing and an increase in consistency after shearing is stopped. That property of a lubricating grease that is manifested by a softening in consistency as a result of shearing followed by a hardening in consistency starting immediately after the shearing is stopped.

**Three-body abrasion** A particulate wear process by which particles are pressed between two sliding surfaces.

**Timken EP test** Measure of the extreme-pressure properties of a lubricating oil. The test uses a Timken machine, which consists of a stationary block pushed upward by means of a lever arm system against the rotating outer race of a roller bearing, which is lubricated by the product under test. The test continues under increasing load (pressure) until a measurable wear scar is formed on the block.

**Total base number (TBN)** The quantity of acid, expressed in terms of the equivalent number of milligrams of potassium hydroxide, that is required to neutralize all basic constituents present in 1 g of sample (ASTM D974). See *Base number*.

**Tribology** Science of the interactions between surfaces moving relative to each other, including the study of lubrication, friction, and wear.

**Turbine oil** A top-quality rust- and oxidation-inhibited (R&O) oil that meets the rigid requirements traditionally imposed on steam turbine lubrication. Quality turbine oils are also distinguished by good demulsibility, a requisite of effective oil–water separation. Turbine oils are widely used in other exacting applications for which long service life and dependable lubrication are mandatory such as compressors, hydraulic systems, gear drives, and other equipment. Turbine oils can also be used as heat-transfer fluids in open systems, where oxidation stability is of primary importance.

**Turbulent flow** Flow in which the velocity at any point varies on an erratic basic. It occurs when flow velocity exceeds a limiting value or when tube configuration irregularities preclude laminar flow.

**Ultraclean** 1 particle > 10 μm/mL.

**Unbalanced seal**  A mechanical seal arrangement wherein the full hydraulic pressure of the seal chamber acts to close the seal faces.

**Vacuum distillation**  A distillation method that involves reducing the pressure above a liquid mixture to be distilled to less than its vapor pressure (usually less than atmospheric pressure). This causes evaporation of the most volatile liquid(s), those with the lowest boiling points. This method works on the principle that boiling occurs when a liquid's vapor pressure exceeds the ambient pressure. It can be used with or without heating the solution.

**Vacuum pump**  A device that is used to extract gas or vapor from an enclosed space, leaving behind a partial vacuum in the container.

**Vacuum separator**  A separator that uses subatmospheric pressure to remove certain gases and liquids from another liquid because of their difference in vapor pressure.

**Valve lifter**  Sometimes called a *cam follower*, a component in engine designs that use a linkage system between a cam and the valve it operates. The lifter typically translates the rotational motion of the cam into a reciprocating linear motion in the linkage system.

**Vapor pressure**  Most liquids form vapors as they're heated. As temperature increases, so too does a fluid's vapor pressure. Fluids boil when their vapor pressure equals the pressure of the surrounding gas. Knowing your fluid's vapor pressure is important when specifying circulation pumps and piping. A low vapor pressure helps prevent boiling and pump cavitation. It's also an indication of how quickly a liquid evaporates; a fluid with high vapor pressure could require frequent topping up. Vapor also doesn't transfer heat as efficiently as fluid.

**Vapor pressure**  Pressure of a confined vapor in equilibrium with its liquid at specified temperature; thus a measure of a liquid's volatility.

**Variable displacement pump**  A pump in which the displacement per cycle can be varied.

**Varnish**  A soluble and insoluble contaminant made up of by-products of oil degradation. It can appear as a sticky or gel-like substance in the oil, or which plates out on the metal surfaces of lube systems.

**Viscosity**  Measure of a fluids, resistance to flow. A higher viscosity is essentially a thicker fluid. It is desirable for most high-temperature heat-transfer fluids to have viscosity in the range of 20–40 cSt at 104°F. Another aspect of a fluid's viscosity is its VI (viscosity index), or how the fluid's viscosity is affected by temperature. It's a measure of the rate of change of viscosity with temperature. In lubricants, a high VI is desirable to maintain a relatively consistent viscosity throughout the usable temperature range. In heat-transfer fluids, a lower VI is more desirable in order to allow the fluid to thin out (reduce viscosity) with temperature increases. This allows for more efficient thermal transfer properties.

**Viscosity grade**  Any of a number of systems that characterize lubricants according to viscosity for particular applications, such as industrial oils, gear oils, automotive engine oils, automotive gear oils, and aircraft piston engine oils.

**Viscosity index (VI)**  Relationship of viscosity to temperature of a fluid. High-VI fluids tend to display less change in viscosity with temperature than low-VI fluids.

**Viscosity index improvers**  Additives that increase the viscosity of a fluid throughout its useful temperature range. Such additives are polymers that possess thickening power as a result of their high molecular weight and are necessary for formulation of multigrade engine oils.

**Viscosity modifier** Lubricant additive, usually a high-molecular-weight polymer, that reduces the tendency of an oil's viscosity to change with temperature.

**Viscosity–temperature relationship** The manner in which the viscosity of a given fluid varies inversely with temperature. Because of the mathematical relationship that exists between these two variables, it is possible to predict graphically the viscosity of a petroleum fluid at any temperature within a limited range if the viscosities at two other temperatures are known. The charts used for this purpose are the ASTM Standard Viscosity–Temperature Charts for Liquid Petroleum Products, available in six ranges. If two known viscosity–temperature points of a fluid are located on the chart and a straight line is drawn through them, other viscosity–temperature values of the fluid will fall on this line; however, values near or below the cloud point of the oil may deviate from the straight-line relationship.

**Water resistance** The resistance of a lubricating grease to adverse effects due to the addition of water to the lubricant system. Water resistance is described in terms of resistance to washout due to submersion (see ASTM D1264) or spray (see ASTM D4049), absorption characteristics, and corrosion resistance (see ASTM D1743).

**Water–glycol fluid** A fluid whose major constituents are water and one or more glycols or polyglycols.

**Way** Longitudinal surface that guides the reciprocal movement of a machine part.

**Way lubricant** Lubricant for the sliding ways of machine tools such as planers, grinders, horizontal boring machines, shapers, jig borers, and milling machines. A good way lubricant is formulated with special frictional characteristics designed to overcome the stick-slip motion associated with slow-moving machine parts.

**Wear** Damage resulting from the removal of materials from surfaces in relative motion. Wear is generally described as *abrasive*, removal of materials from surfaces in relative motion by a cutting or abrasive action of a hard particle (usually a contaminant); *adhesive*, removal of materials from surfaces in relative motion as a result of surface contact (galling and scuffing are extreme cases); or *corrosive*, removal of materials by chemical action.

**Wear debris** Particles that are detached from machine surfaces as a result of wear and corrosion; also known as *wear particles*.

**Wear inhibitor** An additive that protects rubbing surfaces against wear, particularly from scuffing, if the hydrodynamic film is ruptured.

**Weld point** The lowest applied load in kilograms at which a rotating ball in the four-ball EP test either seizes and welds to the three stationary balls or at which extreme scoring of the three balls.

**White oil** Highly refined lubricant stock used for specialty applications such as cosmetics and medicines.

**Work penetration** Penetration of a sample of lubricating grease immediately after it has been brought to 77°F and then subjected to 60 stokes in a standard grease worker. This procedure and the standard grease worker are described in ASTM D217.

**Worm gear** A gear that is in the form of a screw. The screw thread engages the teeth on a worm wheel. When rotated, the worm pulls or pushes the wheel, causing rotation.

**Yield** The amount of grease (of a given consistency) that can be produced from a specific amount of thickening agent; as yield increases, percent thickener decreases.

**Yield point** (yield value or yield stress)   The minimum shear stress producing flow of a plastic material.

**ZDDP**   Zinc dialkyldithiophosphate, an antiwear additive found in many types of hydraulic and lubricating fluids.

**Zinc (ZDP)**   Commonly used name for zinc dithiophosphate, an antiwear/oxidation inhibitor.

# Appendix

This Appendix is intended to be used as a resource for answering questions as well as a depository for information that will be valuable in your job function as a machinery lubrication technician.

# Evaluation Points for Performing a Lubricant Audit

## Standards, Consolidation, and Procurement

- Are general technical standards maintained for common lubricants?
- Does the site employ a database linking lubricated components to standardized lubricants?
- Are supplier quality-assurance procedures in place and monitored routinely?
- Are lubricants for special applications properly defined and purchased with the proper documentation or purchase class?

## Storage/Handling

- Are contaminated sample wastes segregated and minimized?
- Are stored lubricants sampled and analyzed periodically?
- Are transfer containers sampled to determine cross-contamination?

## Sampling Techniques

- Do general information and guidelines for lubricant sampling exist?
- Are new lubricant deliveries sampled properly?
- Is the primary sampling port/valve located properly?
- Are sampling valves selected and installed correctly?
- Are sample containers selected and stored properly?
- Does the site employ correct machine-specific sampling procedures and use of hardware?
- Do samplers properly complete sample-bottle labels and documentation?

- Do trained and qualified technicians obtain all samples?

## Contamination Control

- Does the site change filters based on condition (particle count and/or pressure differential)?
- Is effective particulate management of hydraulic fluids included in the oil analysis program?

## Lubricant Life/Oxidation Management

- Are there proactive measures in place to monitor antioxidant additives? Is turbine oil remaining useful life evaluated routinely?
- Is effective oxidation management of hydraulic fluids included in the oil analysis program?
- Are lubricant reconditioning procedures based on oxidation and additive monitoring?

## Training, Skill Standards, and Certification

- Does the site maintain general training and development guidelines?
- Has a required sampling and oil analysis skill set been defined for
  - Lube technicians
  - PdM technicians and engineers
  - Operators
  - Supervisors and managers
  - Suppliers
- Does the program use appropriate skill competency certifications and evaluations?
- Is 5%–10% of a lube technician's time spent in lubrication training and development?
- Are the lubrication and oil analysis training and procedural documents clearly defined and accessible?

■ Does the training program include safety and environmental responsibility in handling lubricants and lubrication?

■ Is on-the-job training required (such as shadowing an experienced technician) before an individual is approved to perform a specific task?

### Analysis Activities

■ Are proper sampling frequencies defined by machine and application?

■ Are test slates defined and used correctly?

■ Are onsite oil analysis methods used properly?

■ Are proper limits and targets defined for each machine?

■ Are lubricant analysis data effectively reported and communicated throughout the organization?

■ Does the site use effective guidelines for interpreting oil analysis results for
  • Turbine oils
  • Bearing oils
  • Gear oils
  • Hydraulic oils

■ Is there an effective process for troubleshooting exception conditions?

■ Is effective testing of newly delivered and stored lubricants part of the program?

■ Are procedures in place to test and ensure lab quality?

### Program Management

■ Is there a public display of trend charts showing sampling/analysis performance?

■ Is there a public display of trend charts showing equipment performance and maintenance costs?

■ Are report findings/work requirements communicated in a format that is clear and easy to understand and act on?

### Procedures/Guidelines

■ Has the site established and does it follow lubricant sampling procedures?

■ Are lubrication and oil analysis procedures available electronically (e.g., intranet-based manual)?

■ Are lubricant analysis service providers supplied with clear directions regarding test scope, test sequence, and alarm levels?

### Program Goals/Metrics

■ Does the site employ proactive lubrication management goals and metrics?

■ Does the site employ predictive machine health goals and metrics?

■ Are the metrics compiled periodically and issued as a report to management?

■ Is cost-benefit analysis performed regularly and compiled periodically?

### Safety Practices

■ Are Material Safety Data Sheets (MSDSs) easily accessible and complete?

■ Are staff members trained to understand the health risks associated with improper lubricant handling?

■ Have precautions been taken to reduce lubricant-related injuries such as
  • Skin contact
  • Inhalation
  • Slippage
  • Fire/explosion

### Continuous Improvement

■ Are sampling points reviewed periodically and monitoring techniques adjusted as required?

■ Are alarms and targets reviewed periodically and adjusted based on operating experience?

- Are recurring problems identified and flagged for root-cause analysis?
- Are procedures and other documents evaluated and updated periodically as required?
- Are identified problems evaluated to determine whether like equipment may be affected?

- Is the program evaluated periodically and benchmarked to industry best-practice standards?
- Are program goals evaluated periodically to ensure alignment with corporate goals and expectations?

# Lubrication Application Diagnostics

*Bearings: Rolling Element[a]*

| Symptom | Possible cause | Check for |
|---|---|---|
| Excessive noise | Condition of bearing | Worn or brinelled bearing |
| Overheating | Overgreasing | Too frequent application |
| | | Bearing packed too full |
| | Starvation | Insufficient application frequency |
| | Incorrect product | Incorrect base-oil viscosity |
| | | Deficient load-carrying ability (extreme-pressure [EP] quality) |
| Excessive lubricant leakage | Seals | Mechanical damage |
| | | Excessive shrinkage or swelling |
| | | Incorrect installation |
| | Incorrect National Lubricating Grease Institute (NLGI) grade | Grease too soft for application or softening in service |
| | Incompatibility | Admixture of greases |
| Frequent bearing replacement | Excessive wear | Lack of load-carrying ability (EP rating of grease to handle shock loading) |
| | | Starvation |
| | | Contamination, dirt, rust, water |
| | | Normal bearing life exceeded |
| | | Incorrect NLGI grade |
| | Misalignment | Correct alignment |

[a]Assume that correct bearings are in service and properly installed and aligned.

*Bearings: Plain*

| Symptom | Possible cause | Check for |
|---|---|---|
| Overheating | Improper grease distribution in bearing | Incorrect NLGI grade |
| | | Incorrect bearing grooving |
| | Starvation | Lubrication frequency |
| | | Defective/plugged lubricator |
| | Improper grease application | Mechanical stability of grease in service |
| Excessive wear | Starvation | Infrequent lubrication |
| | | Defective/plugged lubricator |
| | Improper grease application | Inadequate load-carrying ability of grease |
| | | Temperature range of grease |

*Gears: Enclosed*

| Symptom | Possible cause | Check for |
|---|---|---|
| Excessive leakage | Grease too soft for application | Incorrect NLGI grade |
| | Incompatibility of greases | Contamination with incompatible greases |
| Noise | Lack of lubrication | Improper lubricant level |
| | | Incorrect NLGI grade |
| Overheating | Lack of lubrication | Improper lubricant level |
| | | Incorrect NLGI grade |
| | Churning | Overlubrication |
| | | Incorrect NLGI grade |
| Tooth breakage | Not usually lubricant related | |
| Pitting | Mostly improper design and fatigue related | While not generally lubricant related, a heavier grease or base oil may retard progression of pitting |
| Wear and scoring | Lack of lubricant film | Improper lubricant level |
| | Improper grease application | Consistency, EP quality, and base-oil viscosity |
| | Abrasive wear | Contamination |
| | Misalignment | Correct alignment |

*Gears: Open*

| Symptom | Possible cause | Check for |
|---|---|---|
| Gear wear | Lack of lubricant film | Incorrect lubricant<br>Incorrect application frequency |
| Buildup on gears or in roots | Excessive lubrication | Too frequent lubrication<br>Proper type of lubricant<br>Contamination |

*Sliding Surfaces*

| Symptom | Possible cause | Check for |
|---|---|---|
| Nonuniform motion (stick-slip) | Insufficient lubrication | Frequency of application<br>Proper lubricant type |

*Universal Joints*

| Symptom | Possible cause | Check for |
|---|---|---|
| Excessive wear | Insufficient lubrication | Proper lubricant type<br>Grease lubrication frequency<br>Slump ability |

*Electric Motors*

| Symptom | Possible cause | Check for |
|---|---|---|
| Electrical malfunction | Excessive grease leakage | Lubrication frequency |
| High temperatures | | Overlubrication |

*Couplings*

| Symptom | Possible cause | Check for |
|---|---|---|
| Dry coupling | Excessive grease leakage | Damaged seals<br>Improper NLGI grade<br>Keyway openings |
| Hardened grease | Centrifugal separation | Proper lubricant type |
| Excessive wear | Incorrect grease | Proper lubricant type |

*Centralized Lubricators*

| Symptom | Possible cause | Check for |
|---|---|---|
| No grease to points of application | Empty reservoir | Fill with proper lubricant |
| | Pump malfunction | Air/electrical supply |
| | Plugged metering blocks air-bound system | Proper lubricant type |
| High system pressure | Plugged metering devices | Contamination |
| | Malfunctioning relief valve | Check and repair |
| | Improper NLGI grade | Proper lubricant type |

*Wet Applications*

| Symptom | Possible cause | Check for |
|---|---|---|
| Noise, high wear | Insufficient lubrication | Application frequency Proper lubricant type |
| | Washout of lubricant | Proper lubricant type |
| Excessive rust | Improper grease application | Proper lubricant type |

*High-Temperature Applications*

| Symptom | Possible cause | Check for |
|---|---|---|
| Noise, high wear | Insufficient lubrication | Application frequency Proper lubricant type |
| Excessive leakage | Incompatibility of greases | Proper lubricant type |
| | Improper NLGI grade | Contamination with incompatible grease |
| | Improper base-oil viscosity | Contamination with incompatible grease |
| | Seals | Grease and seal are incompatible |
| Grease hardening | Improper grease application | Thickener type Oxidation of grease |

*Allison C-4 Test Requirements*

| Test | Requirements |
|---|---|
| Viscosity (cSt) at 40 and 100°C | Fluids must meet SAE J300 viscosity grades, except that heavy-duty diesel oils must meet MIL-L-2104E and automatic transmission fluids (ATFs) must meet DEXRON-III and MERCON. |
| Brookfield viscosity (cP) at –18°C | Report temperature at which fluid viscosity equals 3500 cP |
| Apparent viscosity | Report |
| Metals content | Within application limits |
| Nonmetals content | Within application limits |
| Acid and base number | Report |
| Infrared spectrum | Report |
| Flash point (°C), min | 160 |
| Fire point (°C), min | 175 |
| Foaming tendency | No foam at 95°C, 10 mm max foam at 135°C, 23 sec max to break |
| Rust protection | No rust or corrosion on test panels |
| Copper corrosion | No blackening with flaking |
| Rust/copper protection | No visible rust on test pins |
| Nitrile (Buna-N)[a] | |
| Volume change (%) | +1 to +5[b] |
| Hardness change (pts) | –5 to +5 |
| Dip cycle (polyacrylate)[a] | |
| Volume Change (%) | 0 to +10[b] |
| Hardness Change (pts) | 0 to +5 |
| Tip cycle (silicone)[a] | |
| Volume change (%) | 0 to +5[b] |
| Hardness change (pts) | 0 to –5 |
| Fluoroelastomer (Viton)[a] | |
| Volume change (%) | 0 to +4[b] |
| Hardness change (pts) | –4 to +4 |
| Ethylene acrylic (Vamac)[a] | |
| Volume change (%) | +12 to +28[b] |
| Hardness change (pts) | –6 to –18 |
| Vickers wear[a] | Total weight loss of cam ring and vanes not to exceed 15 mg |
| Graphite friction[a] | Defined by minimum-performance reference fluid |
| Paper friction[a] | Defined by minimum-performance reference fluid |
| Oxidation stability (300 h at 163°C, 90 mL/min air) | TAN increase 4.0 max, carbonyl absorbance difference 0.75 max, Viton seal shall not fail, copper bushings shall not cause mechanical failure, parts free of sludge and varnish |

[a]Required for MIL-L-2104F approval.
[b]Actual limits adjusted to account for batch-to-batch elastomer variation.

# American Petroleum Institute (API) Base-Oil Interchange Guidelines

Not all base oils have similar physical or chemical properties or provide equivalent engine oil performance. API Base Oil Interchangeability Guidelines were developed to ensure that the performance of engine oil products is not adversely affected when different base oils are used interchangeably. The guidelines define the minimum physical and engine testing needed to ensure satisfactory performance when substituting one base stock for another.

The API Base Oil Interchange Guidelines have been updated to establish five base oil groups and establish guidelines for interchanging very high viscosity index (VHVI) stocks with Group I and II stocks.

| Base oil category | Sulfur (%) | | Saturates (%) | Viscosity index (VI) |
|---|---|---|---|---|
| Group I | >0.03 | and/or | <90 | 80–120 |
| Group II | ≤0.03 | and | ≥90 | 80–120 |
| Group III | ≤0.03 | and | ≥90 | ≥120 |
| Group IV | All polyalphaolefins (PAOs) | | | |
| Group V | All others not included in Groups I–IV | | | |

*Passing Engine Tests Required for Interchanging Base Stocks in Original International Lubricants Standardization and Approval Committee (ILSAC) GF-1 or API-Licensed Categories SH and SJ Passenger Car Motor Oil (No Energy Conserving)*

| Original Approval Base Stock | Interchange Stocks | | | | |
|---|---|---|---|---|---|
| | Group I | Group II | Group III | Group IV | Group V |
| Group I | Seq. IIIE, VE | Seq. IIIE | ≤30%[a], none; >30%[a], all | ≤30%[a], none; >30%, all | All |
| Group II | Seq. IIIE, VE | Seq. IIIE, VE | ≤30%[a], none; >30%[a], all | ≤30%[a], none; >30%[a], all | All |
| Group III | All | All | All | ≤30%[a], Seq. VE; >30%[a], all | All |
| Group IV | All | All | ≤30%[a], Seq. VE; >30%[a], all | None[b] | All |
| Group V | All | All | All | All | All |

[a]Represents mass percentage of Passenger Car Motor Oil (PCMO) Guidelines formulation.
[b]PAOs can be interchanged one for another without engine testing as long as the original PAO has had full approval and the interchange PAO meets the original manufacturer's specifications in all physical and chemical properties.

*Passing Engine Tests Required for Interchanging Base Stocks in Original ILSAC GF-2 or API-Licensed Category SJ Passenger Car Motor Oil (Including Energy Conserving)*

| Original Approval Base Stock | Interchange Stocks | | | | |
|---|---|---|---|---|---|
| | Group I | Group II | Group III | Group IV | Group V |
| Group I | Seq. IIIE, VE | Seq. IIIE | ≤30%,[a] Seq. VI-A; >30%[a], all | ≤30%,[a] Seq. VI-A; ≥30%,[a] all | All |
| Group II | Seq. IIIE, VE, VI-A | Seq. IIIE, VE | ≤30%,[a] Seq. VI-A; >30%[a], all | ≤30%,[a] Seq. VI-A; >30%,[a] all | All |
| Group III | All | All | All | ≤30%,[a] Seq. VE, VI-A; ≥30%,[a] all | All |
| Group IV | All | All | ≤30%[a], Seq. VE, VI-A; ≥30%,[a] all | None[b] | All |
| Group V | All | All | All | All | All |

[a]Represents mass percentage of PCMO Guidelines formulation.
[b]PAOs can be interchanged one for another without engine testing as long as the original PAO has had full approval and the interchange PAO meets the original manufacturer's specifications in all physical and chemical properties.

*Passing Thermo-oxidation Engine Oil Simulation Tests (TEOST) Required for Interchanging the Base Stock in an Original API-Licensed PCMO*

| Core run on | Interchange Base Stock[a] | | | | |
|---|---|---|---|---|---|
| | Group I | Group II | Group III | Group IV | Group V |
| Group I | No | No | Yes | Yes | Yes |
| Group II | No | No | Yes | Yes | Yes |
| Group III | Yes | Yes | Yes | Yes | Yes |
| Group IV | Yes | Yes | Yes | Yes | Yes |
| Group V | Yes | Yes | Yes | Yes | Yes |

[a]Applies to SAE 5W-30 and higher viscosity grades.

*Passing Engine Tests Required for Interchanging Base Stocks in Original API-Licensed CF-4 Diesel Engine Oils*

| Original Approval Base Stock | Interchange Stocks | | | | |
|---|---|---|---|---|---|
| | Group I | Group II | Group III | Group IV | Group V |
| Group I | None[a] | None[a] | ≤30%, none; >30%, all | ≤30%, none; >30%, all | All |
| Group II | None[a] | None[a] | ≤30%, none; >30%, all | ≤30%, none; >30%, all | All |
| Group III | All | All | All | All | All |
| Group IV | All | All | All | None[b] | All |
| Group V | All | All | All | All | All |

[a]Caterpillar 1K test has been removed as a requirement for base-oil interchange in Group I and II base stocks. This change is based on substantial data gathered by the API that statistically confirm that no base-oil effects exist in the Caterpillar 1K test.
[b]PAOs can be interchanged one for another without engine testing as long as the original PAO has had full approval and the interchange PAO meets the original manufacturer's specifications in all physical and chemical properties.

*Passing Engine Tests Required for Interchanging Base Stocks in Original API-Licensed CG-4 Diesel Engine Oils*

| Original Approval Base Stock | Interchange Stocks | | | | |
|---|---|---|---|---|---|
| | Group I | Group II | Group III | Group IV | Group V |
| Group I | Seq. IIIE, Mack T-8[a] | Seq. IIIE[b] | ≤30%, 1N; >30%, all | ≤30%, 1N; >30%, all | All |
| Group II | Seq. IIIE, Mack T-8, Roller follower wear test[b,c,d] | Mack T-8,[a] Seq. IIIE | ≤30%, 1N; >30%, all | ≤30%, 1N; >30%, all | All |
| Group III | All | All | All | All[e] | All |
| Group IV | All | All | All[e] | None[f] | All |
| Group V | All | All | All | All | All |

[a]Mack T-8 waived if original oil has ≥80% saturates and interchange oil is ≥5% higher in saturates or if original oil has <80% saturates and interchange oil is >10% higher in saturates.

[b]Caterpillar 1N test has been removed as a requirement for base-oil interchange in Group I and II base stocks. This change is based on substantial data gathered by API that statistically confirm that no base-oil effects exist in the Caterpillar 1N test.

[c]Run in only one Group I base stock when interchanging from Group II to I.

[d]This test was previously called the GM 6.2L test.

[e]If <30%, CRC L-38 and Sequence IIIE are waived.

[f]PAOs can be interchanged one for another without engine testing as long as the original PAO has had full approval and the interchange PAO meets the original manufacturer's specifications in all physical and chemical properties.

*Passing Engine Tests Required for Interchanging Base Stocks in Original API-Licensed CH-4 Diesel Engine Oils*

| Original Approval Base Stock | Interchange Stocks | | | | |
|---|---|---|---|---|---|
| | Group I | Group II | Group III | Group IV | Group V |
| Group I | Seq. IIIE, M-11,[a] T-8E,[b] T-9[a] | Seq. IIIE, 1P[c] | ≤30%, 1P; >30%, all | ≤30%, 1P; >30%, all | All |
| Group II | Seq. IIIE, M-11, T-8E, T-9, Roller follower wear test[d] | Seq. IIIE, M-11,[a] T-8E,[b] T-9[a] | ≤30%, 1P; >30%, all | ≤30%, 1P; >30%, all | All |
| Group III | All | All | All | All[e] | All |
| Group IV | All | All | All[e] | None[f] | All |
| Group V | All | All | All | All | All |

[a]No testing required if interchange base stock has a saturates level greater than or equal to the original base stock and a sulfur level less than or equal to the original tested base stock within the precision of the test method.

[b]Mack T-8E waived if original oil has greater than or equal to 80% saturates and interchange oil is higher in saturates within the precision of the test method (ASTM D2007). Mack T-8E also waived if original oil has less than 80% saturates and interchange oil is greater than or equal to 5% higher in saturates at the 95% confidence level.

[c]Test pass required in one Group I base stock. Other Group II base stocks then may be interchanged without testing.

[d]Needs to be run in only one Group I base stock while interchanging from Group II to Group I. This test was previously called the GM 6.2L test.

[e]Sequence IIIE waived if less than or equal to 30% of the base oil.

[f]PAOs can be interchanged one for another without engine testing as long as the original PAO has had full approval and the interchange PAO meets the original manufacturer's specifications in all physical and chemical properties.

*Passing Engine Tests Required for Interchanging Base Stocks in Original API-Licensed CF and CF-2 Diesel Engine Oils*

| Original Approval Base Stock | Interchange Stock[a] | | | | |
|---|---|---|---|---|---|
| | Group I | Group II | Group III | Group IV | Group V |
| Group I | Cat 1M-PC[b] | Cat 1M-PC[b] | All | All[c] | All |
| Group II | None[b] | Cat 1M-PC[b] | All | All[c] | All |
| Group III | All | All | All | All | All |
| Group IV | All | All | All | None[d] | All |
| Group V | All | All | All | All | All |

[a]Caterpillar 1M-PC waived if lubricant also meets API CF-4 interchange requirements.
[b]DDC 6V92TA waived only if base oil viscosity at 100°C is greater than or equal to base oil viscosity in original approved formulation.
[c]Up to 30% mass Group IV base-oil substitution into an original Group I or II API-licensed CF oil is allowed without further engine testing, provided that the original oil also meets API SJ.
[d]PAOs can be interchanged one for another without engine testing as long as the original PAO has had full approval and the interchange PAO meets the original manufacturer's specifications in all physical and chemical properties.

*Requirements for Interchanging Base Stocks in the Navistar HEUI 7.3L Engine Oil Aeration Test*

| Original Approval Base Stock | Interchange Stock | | | | |
|---|---|---|---|---|---|
| | Group I | Group II | Group III | Group IV | Group V |
| Group I | No | No | ≤30%, no; >30%, yes | ≤30%, No; > 30%, Yes | Yes |
| Group II | No | No | ≤ 30%, 1P; > 30%, Yes | ≤ 30%, 1P; > 30%, Yes | Yes |
| Group III | Yes | Yes | Yes | Yes | Yes |
| Group IV | Yes | Yes | Yes | No[a] | Yes |
| Group V | Yes | Yes | Yes | Yes | Yes |

[a]PAOs can be interchanged one for another without engine testing as long as the original PAO has had full approval and the interchange PAO meets the original manufacturer's specifications in all physical and chemical properties.

*API Gear Oil Service Designations*

| API Classification | Type | Applications |
|---|---|---|
| GL-1 | Straight mineral oil | Truck manual transmissions |
| GL-2 | Usually contains fatty materials | Worm gear drives, industrial gear oils |
| GL-3 | Contains mild EP additives | Manual transmissions and spiral bevel final drives |
| GL-4 | Equivalent to obsolete MIL-L-2105 specification; usually satisfied with 50% GL-5 additive level | Manual transmissions, and spiral bevel and hypoid gears in moderate service |
| GL-5 | Virtually equivalent to present MIL-L-2105D; primary field service recommendation of most passenger car and truck builders worldwide | Moderate and severe service in hypoid and other types of gears. May also be used in manual transmissions |
| GL-6 | Obsolete | Severe service involving high-offset hypoid gears |
| MT-1 | Contains thermal stability and EP additives | Nonsynchronized manual transmissions in heavy-duty service |

*API GL-4, GL-5, and MT-1 Performance Criteria*

| Performance | API GL-4 MIL-L-2105 | API GL-5 MIL-L-2105D | API MT-1 |
|---|---|---|---|
| Scoring resistance under high-speed shock-load conditions | CRC L-19[a] test or FTM 6504T: equal to or better than RGO-105 | L-42 test: gear/pinion coast side scoring equal to or better than RGO-110 | No requirement |
| Resistance to gear distress under high-torque, low-speed conditions | CRC L-20 test: no tooth disturbance such as rippling, ridging, pitting, or severe wear | L-37 test: no tooth disturbance such as rippling, ridging, pitting, or severe wear | No requirement |
| Corrosion resistance in the presence of water | 1. CRC L-13 or FTM 5313.1 2. CRC L-21: no evidence of rusting | L-33 test: no evidence of rusting after 7 days of exposure on any working surface; maximum 0.5 in.² rust on cover plate (1% of surface area) | No requirement |
| Thermal and oxidation stability/component cleanliness | No requirement | L-60-1 test: 100% max viscosity increase; 3% max pentane insolubles; 2% max toluene insolubles | L-60-1 test: 100% max viscosity increase; 3% max pentane insolubles; 2% max toluene insolubles; 7.5 min carbon/varnish rating on large gear; 9.4 min sludge rating on all gears |
| Antiwear | No requirement | No requirement | FZG (A/8.3/90): minimum 10 stage pass |
| High-temperature lubricant stability | No requirement | No requirement | Mack Transmission Test T-2180: equal to or better than reference |
| Oil seal compatibility | No requirement | No requirement | Similar to ASTM D 471: Polyacrylate, –60% to 0% elongation change, –20 to +5 pts hardness change, –5% to +30% volume change; fluoroelastomer, –75% to 0% elongation change, –5 to +10 pts hardness change, –5% to +15% volume change |
| Antifoaming characteristics | CRC L-12 test: readings taken immediately after 5 min aeration; sequence I: 23.9°C, 650 mL; sequence II: 93.3°C, 650 mL | ASTM D892: readings taken immediately after 5 min aeration; sequence I: 20 mL max; sequence II: 50 mL max; sequence III: 20 mL max | ASTM D 892: readings taken immediately after 5 min aeration; sequence I: 20 mL max; sequence II: 50 mL max; sequence III: 20 mL max |

*(continued on next page)*

*API GL-4, GL-5, and MT-1 Performance Criteria (continued)*

| Performance | API GL-4 MIL-L-2105 | API GL-5 MIL-L-2105D | API MT-1 |
|---|---|---|---|
| Copper corrosion | ASTM D130: 3b max after 1 h at 121.1°C | ASTM D130: 3 max after 3 h at 121.1°C | ASTM D130: 2A max after 3 h at 121.1°C |
| Channeling characteristics | No requirement | FTM 3456.1 Modified: SAE 75, −45°C max; SAE 80W-90, −35°C max; SAE 85W-140, −40°C max | No requirement |
| Compatibility with existing gear lubricants | SS&C FED-STD-791 | SS&C FED-STD-791 | SS&C FED-STD-791 |
| Solubility: measure separated material after centrifuging oil stored for 30 days at room temperature (29.4 ±9.5°C) | FTM 3430: 0.25% wt max of original nonpetroleum material in sample | FTM 3430: 0.25% wt max of original nonpetroleum material in sample | FTM 3430: 0.25% wt max of original nonpetroleum material in sample |
| Compatibility: same as solubility except mixed 50/50 with each of six reference oils | FTM 3440: 0.50% wt max of original nonpetroleum material in sample | FTM 3440: 0.50% wt max of original nonpetroleum material in sample | FTM 3440: 0.50% wt max of original nonpetroleum material in sample |

[a]Equipment no longer available; impossible to conduct test per original procedure.

*Automatic Transmission Fluid Requirements*

| Property | DEXRON III | MERCON |
|---|---|---|
| Color | 6.0–8.0 (red) | 6.0–8.0 (red) |
| Miscibility | No separation or color change at end of test using reference fluid | No separation or color change at end of test |
| Kinematic viscosity (cSt) | Report at 40°C and at 100°C | 6.8 min at 100°C |
| Brookfield viscosity (cP) | Report at −10°C; 1500 max at −20°C; 5000 max at −30°C; 20,000 max at −40°C | 1500 max at −20°C; 20,000 max at −40°C |
| Evaporation loss Noack 1 h at 150°C | Not required | Report Report EOT Brookfield at −40°C |
| Flash point (°C), min | 170 | 177 |
| Fire point (°C), min | 185 | Not required |
| Copper strip | 1b max | 1b max |
| Rust and corrosion | No rust or corrosion on any test surface | No visible rust |

| Property | DEXRON III | | MERCON | |
|---|---|---|---|---|
| Foam test | 1. No foam at 95°C<br>2. 5 mm max at 135°C<br>3. 15 sec break time at 135°C | | ASTM D892<br>Seq. I: 100/0 mL<br>Seq. II: 100/0 mL<br>Seq. III: 100/0 mL<br>Seq. IV: 100/0 mL | |
| Elastomer compatibility | Vol. change (%) | Hardness (%) | Vol. change (%) | Hardness (%) |
| Polyacrylate (A) | +5 to +12 | −8 to +1 | +3 to +8 | −5 to +5 |
| Buna-N | +1 to +6 | −3 to +6 | +1 to +6 | −5 to +5 |
| Polyacrylate (C) | +2 to +7 | −4 to +4 | — | — |
| Silicone | +23 to +45 | −30 to −13 | +28 to +48 | −15 to −40 |
| Viton | +0.5 to +5 | −5 to +6 | Report | Report |
| Vamac | +13 to +27 | −17 to −7 | Report | Report |
| Elemental analysis | Report Al, B, Ba, Ca, Cl, Cu, Fe, Mg, N, Na, P, Pb, S, Si, Zn | | Report Al, B, Ba, Ca, Cl, Cu, Fe, Mg, N, Na, P, Pb, S, Si, Zn | |
| Infrared spectrum | ASTM E168: Report | | Not required | |
| Wear test | ASTM D2882 (80°C, 6.9 MPa) 15 mg max wt. loss | | ASTM D2882 (80°C, 6.9 MPa) 10 mg max wt. loss | |
| Band clutch test | Between 10 and 100 h of operation: no unusual wear or flaking on drum and band; midpoint dynamic torque between 185 and 230 N-m; delta torque <80 N-m; end torque >170 N-m; stop time between 0.35 and 0.55 sec; report maximum torque | | Not required | |
| Plate clutch friction test | Between 10 and 100 h of operation: no unusual plate wear or flaking; midpoint dynamic torque between 150 and 180 N-m; delta torque (max-mid) <30 N-m; maximum torque >150 N-m; stop time between 0.50 and 0.60 s; report end torque | | Satisfactory operation for 15,000 cycles: midpoint dynamic coefficient 0.130–0.160, 25–15,000 cycles (127–157 N-m); static breakaway coefficient 0.100–0.150, 100–15,000 cycles (97–147 N-m); engagement time 0.75–1.0 sec, 25–15,000 cycles; low-speed dynamic ratio 0.9–1.0, 200–15,000 cycles; report breakaway/midpoint ratio, 200–15,000 cycles; low-speed dynamic coefficient 0.120–0.160, 25–15,000 cycles (117–158 N-m); used fluid analysis: report pentane insolubles, TAN change, IR change, viscosity at 100°C and −40°C. | |

*(continued on next page)*

*Automatic Transmission Fluid Requirements (continued)*

| Property | DEXRON III | MERCON |
|---|---|---|
| Oxidation test | 4L60 Transmission: satisfactory operation for 300 h; transmission part cleanliness and physical condition equal to or better than that obtained with reference fluid; TAN increase <3.25; carbonyl absorbance increase <0.45; min 4% oxygen content of transmission effluent gas; used fluid viscosity at −20°C <2000 cP; used fluid viscosity at 100°C and GT5.5 cSt; no cooler braze alloy corrosion | ABOT — Satisfactory operation for 300 h; pentane insolubles 1.0% max at 250 h; TAN change 4.0 max at 250 h; IR change 40 max at 250 h; viscosity increase 40% max at 250 h; copper strip 3b max at 50 and 300 h; report EOT Brookfield at -40°C; report gravimetric sample loss |
| Cycling test | 4L60 Transmission (TPI): operating temperature 275°C; satisfactory operation for 20,000 cycles; transmission part cleanliness and physical condition equal to or better than that obtained with reference fluid; 1–2 shift times between 0.30 and 0.75 sec; 2–3 shift times between 0.30 and 0.75 sec; report 3–4 shift time; TAN increase <2.0; carbonyl absorbance increase <0.30; used fluid viscosity at −20°C < 2000 cP; used fluid viscosity at 100°C > 5.0 cSt | 4L60 Transmission (TPI): satisfactory operation for 20,000 cycles; transmission part cleanliness and physical condition equal to or better than that obtained with reference fluid; 1–2 shift times between 0.35 and 0.75 sec; 2–3 shift times between 0.30 and 0.75 sec; TAN increase <2.5; carbonyl absorbance increase <0.35; used fluid viscosity at −20°C <3000 cP; used fluid viscosity at 100°C >5.0 cSt; report EOT Brookfield at −40°C |
| Vehicle performance | 1991 Chevrolet Caprice with 4L60 transmission: shift performance essentially equal to that obtained with reference fluid | Ford Taurus: shift performance essentially equal to that obtained with reference fluid |

*Caterpillar TO-4 Test Requirements*

| Test | 10W | 30 | 50 | Comments |
|---|---|---|---|---|
| Viscosity at 100°C (cSt) | 4.1 min | 9.3–12.5 | 16.3–21.9 | |
| Max temp (°C) for Brookfield viscosity of 150,000 cP | –35 | –25 | –15 | |
| Max temp (°C) for low-temp pumpability of 30,000 cP max | –25 | –15 | –5 | |
| Min high-temperature/high-shear viscosity (cP) at 150°C | 2.4 | 2.9 | 4.5 | |
| Flash point (°C) | | X | | 160 min |
| Fire point (°C) | | X | | 175 min |
| Rust Control, BT-9 | | X | | Pass 2 of 3 rods at 175 h |
| Fluid compatibility | | X | | No sediment/precipitation |
| Homogeneity | | X | | No sediment/precipitation |
| Copper corrosion (2 h at 100°C) | | X | | 1A = pass; 1B–4C = fail |
| Foaming | | | | |
| Sequence I | | X | | 25/0 |
| Sequence II | | X | | 50/0 |
| Sequence III | | X | | 25/0 |
| Sequence I with 0.1% water | | X | | 25/0 |
| Sequence II with 0.1% water | | X | | 50/0 |
| Sequence III with 0.1% water | | X | | 25/0 |
| Fluoroelastomer seal test | | X | | Less than or equal to reference plus 10% |
| Allison C-4 seals | | | | Limits based on batch |
| Nitrile (Buna-N) | | X | | |
| Dip cycle (polyacrylate) | | X | | |
| Tip cycle (silicone) | | X | | |
| Fluoroelastomer (Viton) | | X | | |
| FZG gear wear | X | X | | Average of 3 separate runs <100 mg |
| Vickers pump | X | X | | 15 mg max vane weight loss, 75 max ring weight loss |
| C-4 THOT | | X | | |
| TAN increase | | | | 7.0 max |
| Carbonyl absorbance | | | | 0.9 max |
| Viton seal | | | | Pass |
| Sludge | | | | None |
| VC70 friction | X | X | X | Compared to reference |

X = must be run

*Engine Oil Classification System for Automotive Gasoline Engine Service "S"—Service Oils*

| API Automotive Gasoline Engine Service Categories | Previous API Engine Service Categories | Related Industry Definitions | Engine Test Requirements |
|---|---|---|---|
| SA | ML | Straight mineral oil | None |
| SB | MM | Inhibited oil only | CRC L-4[a] or L-38; Sequence IV[a] |
| SC | MS (1964) | 1964 models | CRC L-38; Sequence IIA[a]; Sequence IIIA[a]; Sequence IV[a]; Sequence V[a]; Caterpillar L-1[a] (1% sulfur fuel) |
| SD | MS (1968) | 1968 models | CRC L-38; Sequence IIB[a]; Sequence IIIB[a]; Sequence IV[a]; Sequence VB[a]; Falcon rust[a]; Caterpillar L-1[a] or 1H[a] |
| SE | None | 1972 models | CRC L-38; Sequence IIB[a]; Sequence IIIC[a] or IIID[a]; Sequence VC[a] or VD[a] |
| SF | None | 1980 models | CRC L-38; Sequence IID; Sequence IIID[a]; Sequence VD[a] |
| SG | None | 1989 models | CRC L-38; Sequence IID; Sequence IIIE; Sequence VE; Caterpillar 1H2[a] |
| SH | None | 1994 models | CRC L-38; Sequence IID; Sequence IIIE; Sequence VE |
| SJ | None | 1997 models | CRC L-38; Sequence IID; Sequence IIIE; Sequence VE |

SA = Formerly for utility gasoline and diesel engine service (obsolete). Category SA denotes service typical of older engines operated under such mild conditions that the protection afforded by compounded oils is not required. This category has no performance requirements, and oils in this category should not be used in any engine unless specifically recommended by the equipment manufacturer. SB = Minimum-duty gasoline engine service (obsolete). Category SB denotes service typical of older engines operated under such mild conditions that only minimum protection afforded by compounding is desired. Oils designed for this service have been used since the 1930s and provide mild antiscuff capability and resistance to oil oxidation and bearing corrosion. They should not be used in any engine unless specifically recommended by the equipment manufacturer. SC = 1964 gasoline engine service (obsolete). Category SC denotes service typical of gasoline engines in 1964 through 1967 models of passenger cars and some trucks operating under engine manufacturers' warranties in effect during those model years. Oils designed for this service provide control of high- and low-temperature deposits, wear, rust, and corrosion in gasoline engines. SD = 1968 gasoline engine service (obsolete). Category SD denotes service typical of gasoline engines in 1968 through 1970 models of passenger cars and some trucks operating under engine manufacturers' warranties in effect during those model years. This category may also apply to certain 1971 or later models as specified (or recommended) in the owners' manuals. Oils designed for this service provide more protection against high- and low-temperature deposits, wear, rust, and corrosion in gasoline engines than oils that are satisfactory for API Engine Service Category SC and may be used when API Engine Service Category SC is recommended. SE = 1972 gasoline engine service (obsolete). Category SE denotes service typical of gasoline engines in passenger cars and some trucks beginning with 1972 and certain 1971 through 1979 models operating under engine manufacturers' warranties. Oils designed for this service provide more protection against oil oxidation, high-temperature engine deposits, rust, and corrosion in gasoline engines than oils that are satisfactory for API Engine Service Categories SD or SC and may be used when either of these categories is recommended. SF = 1980 gasoline engine service (obsolete). Category SF denotes service typical of gasoline engines in passenger cars and some trucks beginning with 1980 through 1989 models operating under engine manufacturers' recommended maintenance procedures. Oils developed for this service provide increased oxidation stability and improved antiwear performance relative to oils that meet the minimum requirements of API Service Category SE. These oils also provide protection against engine deposits, rust, and corrosion. Oils meeting API Service Category SF may be used when API Engine Service Categories SE, SD, and SC are recommended. SG = 1989 gasoline engine service (obsolete). Category SG denotes service typical of gasoline engines in passenger cars, vans, and light trucks operating under manufacturers' recommended maintenance procedures. Category SG oils include the performance properties of API Service Category CC. (Certain manufacturers of gasoline engines require oils that also meet the higher diesel engine category CD.) Oils developed for this service provide improved control of engine deposits, oil oxidation, and engine wear relative to oils developed for previous categories. These oils also provide protection against rust and corrosion. Oils meeting API Service Category SG may be used when API Engine Service Categories SF, SE, SF/CC, and SE/CC are recommended. SH = 1994 gasoline engine service. Category SH was adopted in 1992 to describe engine oil first mandated in 1993. It is for use in service typical of gasoline engines in present and earlier passenger cars, vans, and light trucks operating under vehicle manufacturers' recommended maintenance procedures. Engine oils developed for this category provide performance exceeding the minimum requirements of API Service Category SG, which it is intended to replace, in the areas of deposit control, oil oxidation, wear, rust, and corrosion. Oils meeting API SH requirements have been tested according to the Chemical

Manufacturers Association (CMA) Product Approval Code of Practice and may use the API Base Oil Interchange and Viscosity Grade Engine Testing Guidelines. They may be used where API Service Category SG, and earlier categories are recommended. Effective August 1, 1997, API SH cannot be used except with API CF, CF-2, CF-4, or CG-4 when displayed in the API service symbol, and the C category must appear first. SJ: 1997 gasoline engine service. Category SJ was adopted in 1996 to describe engine oil first mandated in 1997. It is for use in service typical of gasoline engines in present and earlier passenger cars, vans, and light trucks operating under vehicle manufacturers' recommended maintenance procedures. Oils meeting API SJ requirements have been tested according to the Chemical Manufacturers Association (CMA) Product Approval Code of Practice and may use the API Base Oil Interchange and Viscosity Grade Engine Testing Guidelines. They may be used where API Service Category SH and earlier categories are recommended.

[a]This test is obsolete; engine parts, test fuel, or reference oils are no longer generally available, or the test is no longer monitored by the test developer or ASTM.

## Engine Oil Performance Comparison

| Engine Failure Source | Engine Oil Property | Engine oil test methods, descriptions, test result values and what they mean | Conventional 10W30 and 15W40 Engine Oil | Advanced 10W30 and 15W40 Engine Oil |
|---|---|---|---|---|
| Oil thins out, operating temperature increases allowing bearings and pistons to seize | Engine oil operating viscosity | ASTM D3945: Shear Stability Index<br><br>Measures the percent viscosity loss at 100°C of polymer-containing fluids when evaluated by using the fuel injector shear stability test (FISST). The less viscosity loss, the better protection and lubrication. | 20% loss | 7% loss |
| | | ASTM D 2270: Viscosity Index<br><br>Measures variation in viscosity due to changes in temperature. The higher the number, the more stable the oil will be at varying temperatures and conditions. | 110–130 | 132 and 145 |
| | | ASTM D2602: Viscosity by Cold Cranking Simulator<br><br>Measures the apparent viscosity of the oil at cold temperatures. The results are related to the cranking characteristics of the oil. The lower the viscosity reading, the easier the oil will flow at low temperatures. | 3500–4900 cP | 2850 and 3270 cP |
| | | ASTM D4683: High Shear/High Temperature<br><br>Viscosity at the shear rate and temperature of this test method is the condition encountered in the bearings of automotive engines in severe service. Should have a minimum viscosity of 3.5 cP according to SAE. | >3.7 cP | >3.7 cP |

*(continued on next page)*

*Engine Oil Performance Comparison (continued)*

| Engine Failure Source | Engine Oil Property | Engine oil test methods, descriptions, test result values and what they mean | Conventional 10W30 and 15W40 Engine Oil | Advanced 10W30 and 15W40 Engine Oil |
|---|---|---|---|---|
| Main bearings, rings, rod eyes, and cam shafts develop deposits and/or wear out | Engine oil antiwear and contamination control performance | ASTM D2896: Total Base Number (TBN)<br><br>Measures the engine oil's ability to neutralize acid formation, which commonly occurs. The highest TBN for engine oil is 14. The higher the TBN, the more acid will be neutralized and the longer the oil will last. | 8 | 14 |
| | | ASTM D5158: Phosphorous/Zinc Content<br><br>Measures the amount of additive elements, wear metals, and contaminants in lubricating oils. Phosphorous and zinc are part of the antiwear package. The greater the concentration in parts per million of each in the oil, the better the oil will protect various engine components. | 130 ppm phosphorus<br><br>150 ppm zinc | 1200 ppm phosphorus<br><br>1100 ppm zinc |
| | | ASTM D874: Sulfated Ash Content<br><br>Measures the amount of sulfated ash from unused lubricating oils containing additives. The less the amount of sulfated ash, the better. | 1.5%–2.8% | <1.3% |
| Difficulty starting in cold temperature | Engine oil low-temperature fluidity and pumpability | ASTM D3829: Borderline Pumping Temperature of Engine Oil<br><br>Measures the lowest temperature at which engine oil can be continuously and adequately supplied to the oil pump inlet of an automotive engine. In this case, measures viscosity at 20°C. | −15°F | −31°F |
| | | FTM 203C: Stable Pour Point<br><br>Measures the lowest temperature at which movement of the fluid is observed. The lower the pour point, the better utility the fluid has for certain applications at low temperatures. | −10°F | −22°F and −9°F |

*Ford MERCON V Specification*

| Test | Method | Requirement |
|---|---|---|
| Kinematic viscosity | ASTM D445 | 6.8 cSt min at 100°C |
| Viscosity stability | FISST (ASTM D5273) | Kinematic viscosity 6.8 cSt at 100°C after 40 passes |
| | Savant TBS VLT (ASTM D4683) (Shear rate 200/sec, 150°C) | Undegraded: rate and report Degraded: rate and report |
| | Savant TBS VLT (ASTM D4683) (Shear rate 100,000/sec, 150°C) | Undegraded: rate and report Degraded: 2.6 cP min after 40 passes |
| | Savant TBS VLT (ASTM D4683) (Shear rate 100,000/sec, 100°C) | Degraded: 5.4 cP min after 40 passes |
| Flash point | ASTM D92 | 180°C min |
| Brookfield viscosity | ASTM D2983 | 1500 cP at –20°C; 5000–13,000 cP at –40°C |
| Evaporative loss | NOACK | 10% weight loss max (2 h at 150°C) Brookfield at –40°C EOT: 2000 cP max change |
| Copper strip | ASTM D130 (3 h at 150°C) | 1b max |
| Foam test | ASTM D892 | Sequence I: 50/0 mL Sequence II: 50/0 mL Sequence III: 50/0 mL Sequence IV: 50/0 mL |
| Elemental analysis | | Report Al, B, Ba, Ca, Cl, Cu, Fe, Mg, N, Na, P, Pb, S, Si, Zn |
| Infrared spectrum | | Report |
| Miscibility | Ford | No separation or color change at end of test |
| Rust protection | ASTM D665 Procedure A | No visible rust |
| Color | ASTM D1500 | 6.5–7.5 |
| Elastomer tests (ASTM D412/D2240) | Nitrile (NBM) ATRR-100 (168 h, 150°C) | Volume change +1% to +6% Hardness +7 to –7 |
| | Polyacrylic (ACM) ATRR-200 (70 h, 163°C) | Volume change +3% to +8% Hardness +5 to –5 |
| | Silicone (VMQ) ATRR-300 (240 h, 163°C) | Volume change +33% to +43% Hardness +5 to –5 |
| | Fluoroelastomer (FKM) ATRR-400 (240 h, 163°C) | Volume change +1% to +6% Hardness –8 to +8 |
| | Ethylene-acrylic (AEM) ATRR-500 (240 h, 163°C) | Volume change +6% to +16% Hardness -5 to +5 |
| | Epichlorohydrin hypalon (CSM) ATRR-600 (240 h, 163°C) | Volume change +17% to +37% Hardness –34 to –10 |
| | Hydrogenated nitrile (HNBR) ATRR-700 (240 h, 163°C) | Volume change 0% to +10% Hardness –5 to +5 |

*(continued on next page)*

*Ford MERCON V Specification (continued)*

| Test | Method | Requirement |
|---|---|---|
| Wear tests | Vickers pump wear (ASTM D2882) | Max weight loss 10 mg at 80 and 150°C |
| | FZG (ASTM D5182) | 11 load stage pass at 150°C, 1450 rpm |
| | Four ball | 0.61 mm max wear scar 2 run avg at 100 and 150°C, 600 rpm |
| | Timken | At 150°C, 0.6 mm 2 run avg burnish width; no scoring |
| | Falex EP | No seizure load at 100 and 150°C, 750 lb (2 run avg) |
| Oxidation test | Aluminum beaker (ABOT) | Pentane insolubles 0.5% max; TAN change 4.0 max; differential IR 40 max; viscosity increase 40%; copper strip 3b max at 50 and 300 h; report EOT Brookfield at −40°C; report gravimetric sample loss |
| Vehicle performance | Ford Taurus | Shift performance essentially equal to that obtained with reference fluid |
| Cycling test | 4L60 transmission tuned port injected (TPI) engine | Transmission must shift properly; no excess wear; no clutch plate deterioration; "clean" test parts; minimum viscosity at 100°C, 6.0 cSt; max −20°C viscosity 1500 cP |
| Friction durability | SAE 2 SD-1777 plates | Satisfactory operation for 20,000 cycles; midpoint dynamic coefficient 0.140–0.160, 25–20,000 cycles; static breakaway coefficient 0.120–0.150, 100–20,000 cycles; low-speed dynamic coefficient 0.135–0.160, 25–20,000 cycles; stop time 0.70–0.90 sec, 25–20,000 cycles; low-speed dynamic ratio 0.9–1.0, 25–20,000 cycles; report breakaway/midpoint ratio, 200–15,000 cycles; individual plate wear 0.1524 mm max; average plate wear 0.0762 mm max; used fluid analysis, report pentane insolubles, TAN change, IR change, viscosity at 100 and −40°C |

| Test | Method | Requirement |
|------|--------|-------------|
| Friction durability | SAE 2 BW-4400 plates | Satisfactory operation for 20,000 cycles; dynamic coefficient 0.105–0.140, 100–20,000 cycles; static breakaway coefficient 0.105–0.135, 100–20,000 cycles; low-speed dynamic coefficient 0.120–0.155, 25–20,000 cycles; stop time 0.5-1.05 sec, 25–20,000 cycles; low-speed dynamic ratio 1.10–1.35, 25–20,000 cycles; report breakaway/midpoint ratio 200–20,000 cycles; individual plate wear 0.1524 mm max; average plate wear 0.0762 mm max; used fluid analysis, report pentane insolubles, TAN change, IR change, viscosity at 100 and –40°C |
| Antishudder friction test | | Torque ratio R1 (2 rpm/20 rpm) >1.0; torque ratio R2 (40 rpm/120 rpm) >1.0; negative slope 135 hours' duration; individual plate wear 0.1524 mm max; average plate wear 0.0762 mm max |

*Hydraulic Fluid Specifications*

| | Denison | | | | Vickers | | Cincinnati Milacron | | | U.S. Steel | | DIN 51524 Part 2 | GM (LS-2) LH, 03, 04, 06 |
|---|---|---|---|---|---|---|---|---|---|---|---|---|---|
| | HF-0/T6C | HF-0/T5D | HF-1 | HF-2 | M-2950-S | I-286-S | P-68 | P-69 | P-70 | 127 | 136 | | |
| **Denison T-6C Vane Pump Test (Phase 1: 300 h, 1700 rpm, 176°F, 150–3625 psi; phase 2: phase 1 + 1% water, 300 h)** | | | | | | | | | | | | | |
| Denison assessment | Pass | — | — | — | — | — | — | — | — | — | — | — | — |
| **Denison T-5D Vane Pump Test (2500 psi, 2400 rpm, 60 h at 160°F, 40 h at 210°F)** | | | | | | | | | | | | | |
| Denison assessment | — | Pass | Pass | Pass | — | — | — | — | — | — | — | — | Pass |
| **Denison P-46 Piston Pump Test (5000 psi, 2400 rpm, 60 h at 160°F, 40 h at 210°F)** | | | | | | | | | | | | | |
| Denison assessment | Pass | Pass | Pass | — | — | — | — | — | — | — | — | — | Pass |
| **Vickers 35VQ-25 Vane Pump Test (3000 psi, 2400 rpm, 200°F)** | | | | | | | | | | | | | |
| Ring weight loss (mg) | — | — | — | — | 75 | — | — | — | — | — | — | — | 50 |
| Vane weight loss (mg) | — | — | — | — | 15 | — | — | — | — | — | — | — | 10 |
| Total ring and vane weight loss (mg) | — | — | — | — | 90 | — | — | — | — | — | — | — | 60 |
| **Vickers V-104C Vane Pump Test (2000 psi, 1200 rpm, 175°F)** | | | | | | | | | | | | | |
| Ring and vane weight loss (mg) | — | — | — | — | — | 50 max | 50 max | 50 max | 50 max | 90 max, 0.01% | — | 150[1] max | — |
| **Turbine Oil Oxidation Test (ASTM D943)** | | | | | | | | | | | | | |
| Time to 2.0 NNA (h), min | 1000 | 1000 | b | 1000 | — | — | — | — | — | — | — | — | 1500 |
| **1000-h Sludge (ASTM D4310)** | | | | | | | | | | | | | |
| NNA (mg KOH), max | 2 | 2 | 0.2 increase | 2 | — | — | — | — | — | — | — | 2 | — |
| Insoluble sludge (mg), max | 200 | 200 | 100 | 400 | — | — | — | — | — | — | — | — | — |
| Total copper (mg), max | 50 | 50 | — | 200 | — | — | — | — | — | — | — | — | — |

| | Denison | | | | Vickers | | Cincinnati Milacron | | | U.S. Steel | | DIN 51524 Part 2 | GM (LS-2) LH, 03, 04, 06 |
|---|---|---|---|---|---|---|---|---|---|---|---|---|---|
| | HF-0/T6C | HF-0/T5D | HF-1 | HF-2 | M-2950-S | I-286-S | P-68 | P-69 | P-70 | 127 | 136 | | |
| Total iron (mg), max | 50 | 50 | — | 100 | — | — | — | — | — | — | — | — | — |
| **Rotary Bomb Oxidation (ASTM D2272)** | | | | | | | | | | | | | |
| Time to 25-psi loss (min) | — | — | — | — | — | — | — | — | — | >120 | >120 | — | — |
| **Turbine Oil Demulsibility (ASTM D1401, 130°F, ISO VG 32/46)** | | | | | | | | | | | | | |
| Oil-water cuff (mL) | 40-37-3 | — | — | — | b | b | — | — | — | 40-37-3 | Use D2711 | 40ᶜ max | 40-40-0 |
| Separation time (min) | 30 | — | — | — | — | — | — | — | — | 30 | Use D2711 | — | 30 |
| **Turbine Oil Rust Test (ASTM D665)** | | | | | | | | | | | | | |
| A. Distilled water | Pass | Pass | Pass | Pass | b | b | Pass | Pass | Pass | Pass | Pass | — | — |
| B. Synthetic seawater | Pass | Pass | Pass | Pass | b | b | — | — | — | — | — | Pass | Pass |
| **Hydrolytic Stability (ASTM D2619)** | | | | | | | | | | | | | |
| NNA (mg KOH), max | 4 | 4 | — | 6 | — | — | — | — | — | — | — | — | 4 |
| Copper weight loss (mg/cm²), max | 0.2 | 0.2 | — | 0.5 | — | — | — | — | — | — | — | — | 0.2 |
| **Thermal Stability: Cincinnati Milacron (168 h, 275°F)** | | | | | | | | | | | | | |
| Viscosity change (%), max | — | — | — | — | — | — | 5 | 5 | 5 | — | — | — | 5 |
| Neutralization number (%), max | — | — | — | — | — | — | ±50 | ±50 | ±50 | — | — | — | ±50 |
| Sludge (mg/100 mL), max | 100 | 100 | — | — | — | — | 25 | 25 | 25 | — | — | — | 25 |
| Copper weight loss (mg), max | 10 | 10 | — | — | — | — | 5 | 5 | 5 | — | — | — | 10 |

*(continued on next page)*

Hydraulic Fluid Specifications (continued)

| | Denison | | | | Vickers | | Cincinnati Milacron | | | U.S. Steel | | DIN 51524 Part 2 | GM (LS-2) LH, 03, 04, 06 |
|---|---|---|---|---|---|---|---|---|---|---|---|---|---|
| | HF-0/T6C | HF-0/T5D | HF-1 | HF-2 | M-2950-S | I-286-S | P-68 | P-69 | P-70 | 127 | 136 | | |
| Copper rod appearance | Report | Report | — | — | — | — | — | — | — | — | — | — | 5 |
| Iron rod appearance | — | — | — | — | — | — | No discoloration | | | — | — | — | No discol-oration |
| **Filterability** | | | | | | | | | | | | | |
| TP 02100 A. No water (sec), max | 600 | 600 | — | — | — | — | — | — | — | — | — | — | 600 |
| TP 02100 B, 2% water (sec), max | 2xA | 2xA | — | — | — | — | — | — | — | — | — | — | 2xA |
| Pall bench | 80 min | — | — | — | — | — | — | — | — | — | — | — | — |
| AFNOR 1F1 | Report | — | — | — | — | — | — | — | — | — | — | — | — |
| AFNOR 1F2 | Report | — | — | — | — | — | — | — | — | — | — | — | — |
| **Foam (ASTM D892)** | | | | | | | | | | | | | |
| after 10 min | None | None | None | None | — | — | — | — | — | None | None[d] | None | 50/0 |
| **FZG (DIN 51354, Part 2)** | | | | | | | | | | | | | |
| Damage load stage, min | Report | Report | — | — | — | — | — | — | — | — | 10 | 10 | 10 |
| **Four-Ball Wear (ASTM D2266) (20 kg, 1800 rpm, 1 h, 54°C)** | | | | | | | | | | | | | |
| Scar diameter (mm), max | — | — | — | — | — | — | — | — | — | 0.5 at 40 kg | 0.5 at 20 kg | — | — |
| **Four-Ball EP (ASTM D 2783)** | | | | | | | | | | | | | |
| Weld load (kg), min | — | — | — | — | — | — | — | — | — | — | 150 | — | — |
| LWI (kg), min | — | — | — | — | — | — | — | — | — | — | 30 | — | — |
| **Copper Strip (ASTM D130, 3 h, 100°C)** | | | | | | | | | | | | | |
| Rating | — | — | — | — | — | — | — | — | — | — | — | 2 max | 1b max |
| **Air Separation (DIN 51381): Separation time (min)** | | | | | | | | | | | | | |
| ISO VG 46/68 | 7 max (ISO46) | — | — | — | — | — | — | — | — | — | — | 10 max | 10 max |

|  | Denison | | | | Vickers | | Cincinnati Milacron | | | U.S. Steel | | DIN 51524 Part 2 | GM (LS-2) LH, 03, 04, 06 |
|---|---|---|---|---|---|---|---|---|---|---|---|---|---|
|  | HF-0/T6C | HF-0/T5D | HF-1 | HF-2 | M-2950-S | I-286-S | P-68 | P-69 | P-70 | 127 | 136 |  |  |
| ISO VG 32 | — | — | — | — | — | — | — | — | — | — | — | 5 max | 5 max |
| **Seal (DIN 53538, Part 1): Volume change (%)** | | | | | | | | | | | | | |
| ISO 32/46 | — | — | — | — | — | — | — | — | — | — | — | 0 to 12 | 0 to 12 |
| ISO VG 68 | — | — | — | — | — | — | — | — | — | — | — | 0 to 10 | 0 to 10 |
| **Seal (DIN 53538, Part 1): Shore A hardness change (%)** | | | | | | | | | | | | | |
| ISO 32/46 | — | — | — | — | — | — | — | — | — | — | — | 0 to –7 | 0 to –7 |
| ISO VG 68 | — | — | — | — | — | — | — | — | — | — | — | 0 to –6 | 0 to –6 |
| *Viscosity (ASTMD 88)* | | | | | | | | | | | | | |
| Viscosity (cSt) at 40°C | Viscosity sufficient for application | | | | | | 32 | 68 | 46 | Viscosity sufficient for application | | | |
| **Minimum Viscosity Index (ASTM D567)** | | | | | | | | | | | | | |
| VI | 90 | 90 | 90 | 90 | — | — | 90 | 90 | 90 | 90 | 90 | — | 95 |
| **Aniline Point (ASTM D611)** | | | | | | | | | | | | | |
| Temperature (°C) | 100 | — | — | — | — | — | — | — | — | — | — | — | — |

aV-104C or V-105C10: 2030 psi, 1440 rpm, 250 h.
bEvidence of satisfactory performance required; no tests specified.
cISO VG 68, 60 max.
dZero break at 10 min, 100°F.

*Marine Diesel Engine Oil Requirements*

| Low-Speed Crosshead | | Medium-Speed Trunk Piston Crankcase |
|---|---|---|
| Cylinder | System | |
| Sulfur acid neutralization | Lubrication of bearings, crankshaft, chains, and running gear | Piston deposit control and ring stick prevention |
| Detergency and thermal stability | Detergency and thermal stability for piston undercrown cooling | Sludge and lacquer control |
| Good film strength and scuffing protection | Good film strength | Stability in the presence of fuel contamination |
| Good spreadability | Control of drip oil and combustion trash | Ability to control asphaltene-related deposits in engines |
| Piston, ring, and port cleanliness | Release of insolubles and water to purifiers | Keep oil scraper rings clear of deposits and sludge |
| Antiwear properties | Low emulsibility and good water tolerance | Thermal stability and oxidation control |
| System oil compatibility | Rust and oxidation prevention | Rust control and alkalinity retention |
| SAE 50 grade | SAE 30 grade | Combustion acid neutralization |
| — | — | Bearing corrosion protection |
| — | — | EP properties |
| — | — | Good water tolerance and low emulsibility |
| — | — | SAE 30/40 grade |

## Marine Diesel Engine Oils

The performance of lubricants for marine engines is not defined by a standard classification system. Standard dynamometer test methods or testing protocols leading to industry-wide approvals are not available.

Oil performance levels and the approval process are driven by the major engine builders, most of which publish lists of lubricants approved for use in their engines. They encourage their customers to use lubricants appearing on the approval list. In general, original equipment manufacturers (OEMs) require a full-scale ship trial lasting 5000 running hours (about one year) to approve a lubricant. Engine builders approve only formulation-specific lubricants; blanket approvals for additives are not available.

Marine engines generally use residual fuels with high sulfur content (2%–5% by mass) and high asphaltene content (5%–10% mass), although fuel quality can vary widely around the world. Because the cost of fuel is a significant portion of a ship's total operating cost, engine designers are optimizing their engines to cut fuel consumption, and shipowners purchase the lowest-cost fuels available. These two factors are placing greater demands on the quality of the oils used for marine engine lubrication.

Two types of engines dominate propulsion on large oceangoing vessels. Parameters for each type are listed next.

| Type | Speed (rpm) | Bore size (mm) | Output (bhp/cylinder) |
|------|-------------|----------------|-----------------------|
| Slow speed | 50–250 | 260–980 | 500–7500 |
| Medium speed | 400–2000 | 180–580 | 150–1900 |

Two-stroke slow-speed crosshead engines require two lubricants: one for the upper cylinder (cylinder oil) and the other for the crankcase (system oil). MAN B&W and Sulzer dominate the market for these engines. (Together they account for 90% of the market.) Their requirements for cylinder oils are generally SAE 50 viscosity grade and 70 TBN. System oils are usually SAE 30 viscosity grade and 5–10 TBN.

Four-stroke medium-speed trunk piston engines require only one lubricant because they have a common sump for the crankcase and cylinder. The market for these engines is more fragmented than that for two-stroke engines, and several builders hold a significant position. Wartsila, NSD, Pielstick, MAN B&W, and Mak are among the largest OEMs in this market. Lubricants for medium-speed engines are generally SAE 30 or 40 viscosity grade; TBN requirements vary from 12 to 55 depending on fuel sulfur content and engine oil consumption. Newer oils have been formulated to deal with problems caused by the use of fuel with high levels of asphaltene.

*Metalworking Fluid Tests*

| Test and Procedure | Purpose |
|--------------------|---------|
| **Corrosion** | |
| Copper (ASTM D130) 3 h at 100°C | Measures fluid's nonferrous compatibility |
| Turbine Oil Rust (ASTM D665) A. Distilled water B. Synthetic seawater | Measures the ability of inhibited mineral oils to aid in preventing the rusting of ferrous metals in the presence of water |
| Aqueous Cutting Fluid (IP 125) | Measures corrosion protection of aqueous cutting fluids |
| Filter Paper Chip Breakpoint (IP 287) | Evaluates rust-inhibition properties of aqueous cutting fluids compared with a reference fluid |
| Aluminum Cup Stain: 20 min at 350°C | Cup appearance indicates nonferrous corrosion properties |
| Humidity Cabinet Rust (ASTM D1748) | Measures ability of preservative oils to protect metal parts from rusting under conditions of high humidity |
| Salt Spray (MIL-B-117-64) | Steel part corrosion protection measured after exposure to 5% salt spray for 24 hours |
| Cleveland Condensing Humidity Cabinet (ASTM D2247) | Measures antirust properties of metal preservative fluids on steel panels; considered more severe than ASTM D1748 humidity test |
| **Extreme pressure** | |
| Four-Ball Wear (ASTM D2266) 40 kg, 1200 rpm, 75°C, 1 h; average coefficient of friction; maximum scar diameter (mm) | Evaluates antiwear and antiweld properties of lubricants |

*(continued on next page)*

*Metalworking Fluid Tests (continued)*

| Test and Procedure | Purpose |
|---|---|
| Timken (ASTM D2782) OK load (lb) | Measures abrasion resistance and load-carrying capacity of lubricants |
| Four-Ball EP (ASTM D2783) Seizure (kg), weld (kg), LWI (kg) | Evaluates extreme pressure and antiweld properties of lubricants |
| Falex EP (ASTM D3233) | Measures load-carrying capacity and wear properties of lubricants |
| **Stability** | |
| Foam (ASTM D892, IP 312) Tendency/stability (ml) | Determines foaming characteristics of lubricating oils at specific temperatures |
| Panel Coker 4 h at 260°C, continuous splash | Determines relative stability of lubricants in contact with hot metal surfaces |
| Demulsibility (mL oil/mL water/mL emulsion) | Measures separation of oil and water emulsion over time |
| Emulsion Stability (IP 263) | Measures emulsion stability in water |
| Aquarium biostability | Measures foam, bacteria, fungus, and odor over time in controlled aqueous environment |
| **Miscellaneous** | |
| Color (ASTM D1500) | Visual determination of fluid color based on colorimetric readings |
| GM Quenchometer (ASTM D3520) R. C. index | Determines heat removal speed of a quench oil |
| Thread Tapping (Lubrizol test) | Evaluates fluid efficiency by measuring torque required during tapping operation in steel |
| Pipe Threading (Lubrizol test) | Evaluates fluid efficiency by measuring torque required during threading operation on cast iron or stainless steel |
| Stick-Slip (Cincinnati Milacron test) | Measures static and dynamic coefficients of friction in slideway lubricants |
| Bijur Filtration | Determines compatibility of lubricants with Bijur setup (specific to Bijur filter design) |
| Falex 8 | Evaluates fluid efficiency by measuring torque required during tapping operation in steel |
| SLT (draw bead simulator) | Evaluates friction generated in a drawing process |
| Reichert | Measures load-carrying and wear-resistance properties of lubricants |

*Physical Characteristics of Farm Tractor Hydraulic/Transmission Fluids*

| Specification | Viscosity at 100°C (cSt) | Viscosity index (VI) | Brookfield Viscosity (cP), max | | | | Pour Point (°C) max | Flash Point (°C) min |
|---|---|---|---|---|---|---|---|---|
| | | | At –18°C | At –9°C | At –35°C | At –40°C | | |
| **AGCO: Deutz-Allis** | | | | | | | | |
| PF821 XL | 9.0 min | 135 (ref) | 4000 | 19,000 (ref) | — | — | –32 | 193 |
| **AGCO: Massey–Ferguson** | | | | | | | | |
| M1110[a] | 5.75 min | — | 2400 | — | — | — | –30 | 160 |
| M1127A[a] | 11.1 max | 120 min | 6000 | — | — | — | –30 | — |
| M1127b[a] | 9.6 max | 120 min | 4000 | — | — | — | –30 | — |
| M1129a[a] | 9.65 max | 120 min | 4000 | — | — | — | –30 | — |
| M1135 | 10.0–11.5 | 95 min | 9000 | — | — | — | –25 | — |
| M1139 | 10.2–11.9 | — | 8000 | — | — | — | –30 | — |
| M1141 | 9.6 max | 130 min | 4000 | — | — | 70,000 | 37 | 200 |
| **AGCO: White Farm** | | | | | | | | |
| Q1826 | 8.5–9.6 | — | 4000 | — | — | — | –37 | 193 |
| **Allison** | | | | | | | | |
| C-4 | Conform to SAE J 300 Grade | | | | | | | |
| **J.I. Case** | | | | | | | | |
| MS 1204[a] | 8.75 min | 130 min | 2700 (CCS) | — | — | — | –34 | 193 |
| MS 1205[a] | 11.1 min | 140 min | 5600 | — | — | — | –32 | 193 |
| MS 1206[a] | 8.8 min | 140 min | 4000 (–20°C) | — | — | — | –34 | 190 |
| MS 1207 | 6.2 min | 95-115 | 3500 (–20°C) | 15,000 (–30°C) | — | — | –37 | 195 |
| MS 1210 | 6.65 min | 120 min | 1800 (CCS) | — | — | — | –46 | 182 |
| Hy-Tran[b] | 6.2 min | 95–115 | 2600 (CCS) | 14,000 (CCS) | — | — | –37 | 195 |
| **John Deere** | | | | | | | | |
| J14C[a] | 5.4 min | — | 3500 | 20,000 | — | — | –35 | 180 |
| J20C | 9.1 min | — | 5500 (–20°C) | — | 70,000 | — | –36[c] | 200 |
| J20D | 7.0 min | — | 1500 (–20°C) | — | — | 20,000 | –45[c] | 150 |
| J21A[a] | 5.0 min | — | 1400 | — | — | — | –40 | 180 |
| J27 | Conform to SAE J 300 Grade | | | | | | | |
| **New Holland** | | | | | | | | |
| M2C41B | 7.0 min | — | 1400 | — | — | — | –35 | 177 |
| M2C48B | SAE 10W-30 | 135 | — | — | — | — | –30 | 190 |
| M2C48C | ISO 32 | — | — | — | — | — | — | — |
| M2C86B | 10.5–11.6 | 105 min | 9230 | — | — | — | –27 | 219 |
| M2C134D | 9.0 min | — | 4000 | — | — | — | –37 | 190 |
| M2C159-B1/C1 | SAE 10W-30 | — | — | — | — | — | — | 190 |
| M2C159-B2/C2 | SAE 15W-30 | — | — | — | — | — | — | 190 |
| M2C159-B3/C3 | SAE 20W-40 | — | — | — | — | — | — | 190 |
| **Kubota** | | | | | | | | |
| UDT | 8.8 min | 140 | 4800 (–20°C) | — | 70,000 (–34°C) | — | –37 | 200 |

[a]Obsolete specifications, CCS = cold cranking simulator.
[b]THP fluid specification for transmission, differential, wet brake, power steering and hydraulic systems.
[c]Stable pour point.

## Physical Requirements for Engine Oils

SAE has established that 12 viscosity grades are suitable for engine lubricating oils. The physical requirements for these viscosity grades are described in SAE J300, which is intended for use by engine manufacturers in determining engine oil viscosity grades suitable for use in their engines. Current low- and high-temperature viscosity requirements for these viscosity grades are shown next. The U.S. military imposes additional requirements on the oils it purchases.

*SAE Viscosity Grades for Engine Oils[a]: SAE J300 January 2015*

| SAE viscosity grade | Low-temperature cranking (CCS) viscosity (cP) max @temp °C[a] | Low-temperature pumping viscosity (cP) max @temp °C[b] | Kinematic viscosity (cSt) @100°C[c] | Kinematic viscosit (cSt) @100°C[c] | High shear rate (HTHS) viscosity (cP) @150°C[v] |
|---|---|---|---|---|---|
| | ASTM D5293 | ASTM D4684 | Minimum | Maximum | Minimum |
| 0W | 6,200 @ –35°C | 60,000 @ –40°C | 3.8 | | |
| 5W | 6,600 @ –30°C | 60,000 @ –35°C | 3.8 | | |
| 10W | 7,000 @ –25°C | 60,000 @ –30°C | 4.1 | | |
| 15W | 7,000 @ –20°C | 60,000 @ –25°C | 5.6 | | |
| 20W | 9,500 @ –15°C | 60,000 @ –20°C | 5.6 | | |
| 25W | 13,000 @ –10°C | 60,000 @ –15°C | 9.3 | | |
| 8 | | | 4.0 | <6.1 | 1.70 |
| 12 | | | 5.0 | <7.1 | 2.0 |
| 16 | | | 6.1 | <8.2 | 2.3 |
| 20 | | | 5.6 | <9.3 | 2.6 |
| 30 | | | 9.3 | <12.5 | 2.9 |
| 40 | | | 12.5 | <16.3 | 2.9[e] |
| 40 | | | 12.5 | <16.3 | 3.7[f] |
| 50 | | | 16.3 | <21.9 | 3.7 |
| 60 | | | 21.9 | <26.1 | 3.7 |

[a]ASTM D5293. CCS or cold cranking simulator, a test method for apparent viscosity of engine oils and base stocks between –10°C and –35°C.
[b]ASTM D4684 (no yield stress), a test method for determination of yield stress and apparent viscosity of engine oils at low temperature.
[c]ASTM D445. Test method for kinematic viscosity of transparent and opaque liquids.
[d]ASTM D4683, CEC-L-36-A-90 (ASTM D4741) or ASTM D5481. A test method for measuring viscosity of new and used engine oils at high shear rate and high temperature.
[e]0W-40, 5W-40 and 10W-40. (For 0W, 5W, and 10W, HTHS limit is 2.9 cSt).
[f]15W-40, 20W-40, 25W-40 and 40. (For 15W, 20W, and SAE 40 HTHS limit is 3.7 instead of 2.9).

*Military Grades*

| Specification: MIL-PRF-CID | 2104G | 2104G | 2104G | A-A-52039B | A-A-52039B | 2104G A-A-52306A |
|---|---|---|---|---|---|---|
| Viscosity grade | 10W | 30 | 40 | 5W-30 | 10W-30 | 15W-40 |
| **Cranking Viscosity[a] (cP) at temperature °C** | | | | | | |
| min | 3500 at −25 | — | — | 3500 at −30 | 3500 at −25 | 3500 at −20 |
| max | 3500 at −20 | — | — | 3500 at −25 | 3500 at −20 | 3500 at −15 |
| Pumping viscosity[b] (cP) at temp °C, max | 30,000 at −25 | — | — | 30,000 at −30 | 30,000 at −25 | 30,000 at −20 |
| **Viscosity[c] (cSt) at 100°C** | | | | | | |
| min | 5.6 | 9.3 | 12.5 | 9.3 | 9.3 | 12.5 |
| max | <7.4 | <12.5 | <16.3 | <12.5 | <12.5 | <16.3 |
| Viscosity index, min | — | 80 | 80 | — | — | — |
| HTHS viscosity (cP), min | 2.9 | — | — | 2.9 | 2.9 | 3.7 |
| Pour point (°C), max | −30 | −18 | −15 | −35 | −30 | −23 |
| Stable pour point (°C), max | −30 | — | — | −35 | -30 | −23 |
| Flash point (°C), min | 205 | 220 | 225 | 200 | 205 | 215 |
| Evaporative loss[d] (%), max | 18 | — | — | 20 | 17 | 15 |

[a]ASTM D2602 modified.
[b]ASTM D4684, allows no detectable yield stress.
[c]ASTM D445.
[d]Not required for all military specifications.

*Physical Requirements for Gear Lubricants Intended for Axle and Manual Transmission Applications*

| SAE Viscosity Brade | Max Temperature for Viscosity of 150,000 cP (°C)[a,b] | Kinematic Viscosity at 100°C (cSt)[c] | |
|---|---|---|---|
| | | Min[d] | Max |
| 70W | −55[e] | 4.1 | — |
| 75W | −40 | 4.1 | — |
| 80W | −26 | 7.0 | — |
| 85W | −12 | 11.0 | — |
| 80 | — | 7.0 | <11.0 |
| 85 | — | 11.0 | <13.5 |
| 90 | — | 13.5 | <24.0 |
| 140 | — | 24.0 | <41.0 |
| 250 | — | 41.0 | — |

[a]Using ASTM D2983.
[b]Additional low-temperature viscosity requirements may be appropriate for fluids intended for use in light-duty synchronized manual transmissions.
[c]Using ASTM D445.
[d]Limit also must be met after testing in CEC L-45-T-93, Method C (20 hours).
[e]The precision of ASTM D2983 has not been established for determinations made at temperatures below −40°C. This fact should be taken into consideration in any producer–consumer relationship.

*MIL-PRF-2105E*

| SAE Viscosity Grade | Max Temperature for Viscosity of 150,000 cP (°C) | Kinematic Viscosity at 100°C (cSt) | | Channel Point (°C), max | Flash Point (°C) min, |
|---|---|---|---|---|---|
| | | Min | Max | | |
| 75W | −40 | 4.1 | — | −45 | 150 |
| 80W-90 | −26 | 13.5 | 24.0 | −35 | 165 |
| 85W-140 | −12 | 24.0 | 41.0 | −20 | 180 |

*Railroad Engine Builder Lubricant Requirements*

| Engine Builder | SAE Grade | TBN (ASTM D2896) | Sulfated ash (% mass), max | Zinc Content | | API Classification | Road Test Requirement |
|---|---|---|---|---|---|---|---|
| | | | | Max | Min | | |
| General Motors, EMD | 40 or 20W-40 | 10 to 20 | — | 10 ppm | — | — | 3 locomotives, 1 year |
| General Electric, United States | 40 or 20W-40 | 13 to 20 | — | — | — | — | 3 locomotives, 100,000 miles |
| MTU | 30 | — | 1.5 | — | 0.05% | SE/CC, SE/CD | Required |
| | 40 | — | 1.5 | — | 0.05% | SE/CD | Required |
| | 15W-40 | — | 1.8 | — | 0.05% | SE/CD | Required |
| GE Locomotives, Canada (Alco 251 engines) | 40 | 7 to 13 | — | — | — | CD | — |
| Sulzer | 40 | — | — | — | — | CD | Required |
| SEMT Pielstick | 40 | 10 min | — | — | — | CD | Required |
| SACM | 40 | 10 min | — | — | — | CD | Required |

## STOU Performance Requirements

Viscosity          SAE 10W-30
                   9.3–12.5 cSt at 100°C
                   3500 cP max at –20°C (cold cranking simulator)

| Application | Specification | Test |
|---|---|---|
| Engine | API CD[a]/CE, API SF[a] | Caterpillar 1G2,[a] Sequence IID, Sequence IIID,[a] Sequence VD,[a] L-38, Mack T6 and T7, NTC 400[a] |
|  | CCMC D-4 | M-B OM 364A, CCMC quality must be demonstrated with API or MIL data as background information |
| Transmission/wet brakes | API GL-4, Massey-Ferguson M1139, New Holland M2C159B/C | L-20 high-torque, low-speed, independent power takeoff, 4-square axle, pump |
|  | John Deere J27 | Brake chatter/capacity, powershift transmission, high-torque axle |
|  | Caterpillar TO-2,[a] Allison C-4 | Powershift transmission friction, oxidation, wear, friction |
| Hydraulic equipment | Vickers vane pump | Wear (ASTM D2882) |
|  | Denison HF2 | Vane and piston pump |
|  | Vickers sheet I-286-S | Vickers 35VQ25 |

[a]This test or category is obsolete; parts, fuel, or reference oils are no longer generally available, or the test is no longer monitored by the test developer or ASTM.

## UTTO Performance Requirements

Viscosity    Summer    SAE 10W-30: 9.0–11.6 cSt at 100°C
                       50,000–70,000 cP at –35°C Brookfield

             Winter    SAE 5W-20:  6.2–7.0 cSt at 100°C
                       20,000 cP at –40°C Brookfield

| Application | Specification | Test |
|---|---|---|
| Transmission/wet brakes | API GL-4 | L-20 high-torque, low-speed |
|  | Massey-Ferguson M1141/ M1135 | Independent power takeoff, 4-square axle, pump |
|  | New Holland FNHA-2-C-201.00 (M2C134) | Jenkins tractor transmission stall and high energy |
|  | John Deere J20C/D | Brake chatter/capacity, powershift transmission, high-torque axle |
|  | Caterpillar to 2,[a] Allison C-4 | Powershift transmission friction, wear, oxidation, friction |
| Hydraulic equipment | Vickers vane pump | Wear (ASTM D2882) |
|  | Denison HF0 | Vane and piston pump |
|  | Vickers Sheet I-286-S | Vickers 35VQ25 |

[a]This test or category is obsolete; parts, fuel, or reference oils are no longer generally available, or the test is no longer monitored by the test developer or ASTM.

*STOU Performance Requirements*

Viscosity          SAE 10W-30, 9.3–12.5 cSt at 100°C, 500 cP max at –20°C (cold cranking simulator)

| Application | Specification | Test |
|---|---|---|
| Engine | API CD[a]/CE, API SF[a] | Caterpillar 1G2,[a] Sequence IID, Sequence IIID,[a] Sequence VD,[a] L-38, Mack T6 and T7, NTC 400[a] |
| | CCMC D-4 | M-B OM 364A, CCMC quality must be demonstrated with API or MIL data as background information |
| Transmission/wet brakes | API GL-4, Massey-Ferguson M1139, New Holland M2C159B/C | L-20 high-torque, low-speed, independent power takeoff, 4-square axle, pump |
| | John Deere J27 | Brake chatter/capacity, powershift transmission, high-torque axle |
| | Caterpillar TO-2,[a] Allison C-4 | Powershift transmission friction, oxidation, wear, friction |
| Hydraulic equipment | Vickers vane pump | Wear (ASTM D2882) |
| | Denison HF2 | Vane and piston pump |
| | Vickers Sheet I-286-S | Vickers 35VQ25 |

[a]This test or category is obsolete; parts, fuel, or reference oils are no longer generally available, or the test is no longer monitored by the test developer or ASTM.

*Tests for Characterizing Gear Oils*

| Property | Test Method | Description | Test Result Values, What They Mean |
|---|---|---|---|
| Viscosity | ASTM D2983 | Brookfield viscosity test | Measures low-temperature viscosity of automotive fluid lubricants using the Brookfield viscometer. |
| | ASTM D2270 | Viscosity index | Measures variation in viscosity due to changes in temperature; the higher the number the better. |
| Extreme pressure and antiwear | ASTM D3233 | Falex load test | Measures extreme pressure properties (in pounds) of a fluid lubricant using the falex pin method. Greater amount equals greater extreme pressure characteristics. |
| | ASTM D2783 | Film strength | Uses 4-ball method to determine load-carrying properties of lubricating fluid. The higher number the better. |
| | ASTM D2783 | Weld point | Measures the lowest applied load in kilograms at which the rotating ball welds to the three stationary balls. The higher the number the better extreme pressure characteristics. |
| | ASTM D2783 | Load wear index | An index of the ability of a lubricant to minimize wear at applied loads. The higher the better. |
| | ASTM D2782 | Timken method | Line contact, similar to roller bearing, the higher the number the greater load-carrying capabilities |
| | ASTM D4172 | Four-ball (wear scar) | Measures the relative wear preventive properties of lubricating fluids in sliding contact, using the 4-ball method. The lower the number the more protection the oil provides. |

| Property | Test Method | Description | Test Result Values, What They Mean |
|---|---|---|---|
| Corrosion | ASTM D130 | Copper corrosion | 1A is the best rating, most are 1B, measures ability to protect yellow metals. |
| Oxidation Resistance | ASTM D2070 | Thermal stability test | Measures thermal stability of oil. Pass is good. Also measures viscosity increase and insolubles recorded. The lower percent the better. |
| Foaming Tendency | ASTM D892 | Foam test | Measures the amount of foam produced at specified temperatures and different air blowing intervals; no foam is best. |
| Residue | ASTM D189 | Conradson carbon residue | Measures the residue formed by evaporation and thermal degradation of the oil; the less percentage of carbon residue the better. |
| Fluidity | FTM 3456 | Channel point | When a gear passes through the gear oil, the oil flows in behind it, filling in the channel left by the gear; this method records the temperature at which the gear oil no longer flows back into the channel. |
| | ASTM D97 | Pour point | The lowest temperature at which movement of the fluid is observed is recorded as the pour point; the lower the pour point, the better utility the fluid has for certain applications at low temperature. |
| Safety and precautions | ASTM D92 | Flash point | Measures the lowest temperature at which application of a test flame causes the vapors above the surface of the liquid to ignite; the higher the temperature the better (safer). |
| | ASTM D92 | Fire point | Measures the lowest temperature at which the flame causes the oil to ignite and burn for at least 5 seconds; the higher the temperature the better (safer). |

*Tests for Characterizing Greases*

| Property | Test Method | Description | Test Result Values, What They Mean |
|---|---|---|---|
| Shear stability | ASTM D217 | Multistroke penetration | The lower the percent change in the number, the more mechanically stable is the grease |
| | ASTM D1831 | Roll stability | The lower the percent change in the number, the more mechanically stable is the grease |
| | ASTM D1263 | Wheel bearing leakage | Measures percent loss in a wheel bearing application; the lower the number the better; above 5% will cause brake problems |
| Oxidation resistance | ASTM D942 | Bomb oxidation | Measures the oxidative life of the grease; this is used to help determine shelf life |
| | ASTM D3527 | Wheel bearing life | The higher the hours, the longer the grease will last in high-temperature applications |
| | ASTM D3336 | High-temperature performance | The higher the temperature, the better the grease will perform at high temperatures |

*(continued on next page)*

*Tests for Characterizing Greases (continued)*

| Property | Test Method | Description | Test Result Values, What They Mean |
|---|---|---|---|
| Water resistance | ASTM D1264 | Water washout | The lower the percent, the less likely it will wash out |
| | ASTM D4049 | Water spray-off | The lower the percent, the less likely it will wash out |
| Bleed resistance | FTM 321.3 | Oil separation (static) | Measures the percent oil that may separate during storage and idle time |
| | ASTM D1742 | Pressure oil separation | Measures the percent oil that will separate when grease is under load |
| Extreme pressure/ antiwear | ASTM D2596 | Four-ball | Point contact, similar to ball bearings; the higher the number the greater load carrying |
| | ASTM D2509 | Timken method | Line contact, similar to roller bearings, the higher the number the greater load carrying |
| | ASTM D2266 | Four-ball (wear scar) | The lower the number the more protection |
| Corrosion | ASTM D1743 | Rust test | Determines how well the grease keeps water and corrosives away from the metal surface; static test |
| | — | Emcor | Determines how well the grease keeps water and corrosives away from the metal surface; dynamic test |
| | ASTM D130 | Copper corrosion | 1A is the best rating; most are 1B; measures the ability to protect yellow metals |
| Pumpability | ASTM D4693 | Low-temperature torque | Measures the effort required to move the grease in a bearing at low temperatures; the lower the number the better |
| | US Steel LT37 | Mobility | Measures the grease flow at a given temperature at 150 psig; the higher the number the better; critical is 2 g/min |
| Identification and quality control | ASTM D2265 | Dropping point | Measures the temperature at which the soap melts; used to help determine the upper usable temperature range |

*Tests for Characterizing Hydraulic Oils*

| Property | Test Method | Description | Test Result Values, What They Mean |
|----------|-------------|-------------|-------------------------------------|
| Viscosity | ASTM D2602 | Viscosity by cold cranking simulator | Measures the apparent viscosity of the oil at 0°C; the results are related to the cranking characteristics of the oil; the lower the viscosity reading the better |
| | ASTM D2270 | Viscosity index | Measures the variation in kinematic viscosity due to changes in temperature; the higher the number the better |
| Antiwear | ASTM D2882 | Vickers vane pump wear test | Measures wear characteristics of hydraulic fluids; the lower the number the more protection |
| Fluidity | ASTM D97 | Pour point | The lowest temperature at which movement of the fluid is observed is recorded as the pour point; the lower the pour point, the better utility the fluid has for certain applications at low temperatures |
| Emulsification | ASTM D1401 | Water separation | Measures the time it takes for water to separate from oils; the less amount of time it takes the better |
| Safety precautions | ASTM D92 | Flash point | Measures the lowest temperature at which application of the test flame causes the vapors above the surface of the liquid to ignite; the higher the temperature the better (safer) |
| | ASTM D92 | Fire point | Measures the lowest temperature at which the flame causes the oil to ignite and burn for at least 5 seconds; the higher the temperature the better (safer) |
| Residue | ASTM D189 | Carbon residue | Measures the residue formed by evaporation and thermal degradation of the oil; the less percentage of carbon residue the better |
| Dielectric | ASTM D877-76 | Dielectric breakdown voltage | Measures the dielectric, in kilovolts, for a substance; the higher the dielectric number the better |

| Test | Typical Results | JI Case MS1207 | John Deere J20C/D | John Deere J27 | New Holland FNHA-2-C-201.00[a] | New Holland M2C 159B/C | AGCO Massey-Ferguson M1135 | AGCO Massey-Ferguson M1139 | AGCO Massey-Ferguson M1141 |
|---|---|---|---|---|---|---|---|---|---|
| **Four-Ball Wear** | | | | | | | | | |
| Scar diam (mm) | 0.35 | — | — | 0.45 max | 0.40 max | 0.40 max | 0.40 max | 0.40 max | 0.40 max |
| **Load-Carrying Capacity (ASTM D2733)** | | | | | | | | | |
| Wear index (kg) | 57 | — | — | — | — | — | 55 min | 55 min | 35 min |
| Weld load (kg) | 251 | — | — | — | — | — | — | — | 200 min |
| **Wet Brake Chatter/Squawk** | | | | | | | | | |
| Various tests | Pass | Pass | Pass | Pass | Pass | Pass | Pass | Pass | Pass |
| **PTO Clutch** | | | | | | | | | |
| Various tests | Pass | Pass | Pass | Pass | Pass | Pass | Pass | Pass | — |
| **High-Torque Axle** | | | | | | | | | |
| Various tests | Pass | Pass | Pass | Pass | Pass | Pass | Pass | Pass | — |
| **Allison C-4** | | | | | | | | | |
| Oxidation | Pass | — | Pass | Pass | — | Pass | — | — | — |
| Vickers pump | Pass | — | Pass | Pass | — | Pass | — | — | — |
| Graphite friction | Pass | — | Pass | Pass | Pass | Pass | — | — | — |
| Paper friction | Pass | — | Pass | Pass | — | Pass | — | — | — |
| **Rust Protection** | | | | | | | | | |
| Result | Pass | Pass | Pass | Pass | Pass | — | Pass | Pass | Pass |
| **Falex Pin Corrosion** | | | | | | | | | |
| Result | Pass | — | — | — | — | Pass | — | — | — |
| **Copper Strip Corrosion** | | | | | | | | | |
| 3 h at 150°C | 1a | — | 1b max | 1b | 2b | 1b | 1a | 1a | 1b |
| **Foaming (Tendency/Stability)** | | | | | | | | | |
| Sequence I (mL) | 0/0 | 50/10 max | 25/0 | 25/0 max | 20/0 max | 20/0 | 100/0 max[b] | 100/0 max[b] | 50/0 max[b] |
| Sequence II (mL) | 30/0 | 50/10 max | 50/0 | 25/0 max | 50/0 max | 50/10 | 100/0 max | 100/0 max | 50/0 max |
| Sequence III (mL) | 0/0 | 50/10 max | 25/0 | 25/0 max | 20/0 max | 20/10 | 100/0 max | 100/0 max | 50/0 max |

| Test | Typical Results | JI Case MS1207 | John Deere J20C/D | John Deere J27 | New Holland FNHA-2-C-201.00[a] | New Holland M2C 159B/C | AGCO Massey-Ferguson M1135 | M1139 | M1141 |
|---|---|---|---|---|---|---|---|---|---|
| **Oxidation (100 h at 149°C)** | | | | | | | | | |
| Evaporation loss (%) | 0.6 | Pass | 5 max (J20C) 10 max (J20D) | 5 max | — | — | — | — | — |
| Viscosity increase at 99°C (%) | 2.5 | Pass | 10 max (J20C) 20 max (J20D) | 10 max | 10 max | 10 max | 10 max | 10 max | 10 max |
| Separation/sludge | None | Pass | None | Trace/none | — | — | None | None | None |
| **Seal (70 h at 100°C)** | | | | | | | | | |
| Vol. change (%) | +0.9 | 0 to +5 | Allison C-4 test | 0 to +5 | — | — | — | — | — |
| Hardness change (%) | −2 | — | | 0 to −5 | — | — | — | — | — |
| Precipitation | None | — | | None | — | — | — | — | — |
| **Water Tolerance** | | | | | | | | | |
| Water volume (%) | — | 10 | 0.4 | 0.4 | 0.5 | 0.5 | — | — | — |
| Sediment volume (% max) | — | — | 0.1 | 0.1 | 0.1 | 0.1 | — | — | — |
| Additive loss (% m) | — | — | 15 max | — | — | — | — | — | — |
| Water separation | — | — | — | — | Trace | Trace | — | — | — |

[a]Same as Ford M2C134D.
[b]1.0% volume water added to Massey-Ferguson foam tests.

311

*Typical U.S. and European R&O Specifications*

| Performance Test | U.S. Steel 126 | Cincinnati Milacron P-54 | Denison HF-1 | General Electric GEK 32568A | AFNOR NFE 48-600 HL | DIN 51524 Part 1 |
|---|---|---|---|---|---|---|
| **Vane Pump** (ASTM D2882) | | | | | | |
| Ring and vane wt. loss (mg), max | 500 | | | | | |
| **Four-Ball Wear** (ASTM D2266) | | | | | | |
| Scar diameter (mm), max | 0.8 | | | | | |
| **Denison P-46 Piston Pump** | | | | | | |
| All tests | | | Pass | | | |
| **Rotary Bomb Oxidation** (ASTM D2272) | | | | | | |
| Time to 25-psi loss (min) | 120 min | | 450 min | | | |
| **Turbine Oil Oxidation** (ASTM D943) | | | | | | |
| Time to 2.0 NNA (h), min | | | | 2000[a] | | 1000 |
| **1000-Hour Sludge** (ASTM D4310) | | | | | | |
| Sludge (mg), max | | | 100 | | | |
| NNA Increase, max | | | 0.2 | | | 2.0 |
| **Rust** (ASTM D665) | | | | | | |
| A. Distilled water | Pass | Pass | Pass | Pass | | Pass |
| B. Synthetic seawater | Pass | | Pass | | | |
| **Turbine Oil Demulsibility** (ASTM D1401) | | | | | | |
| Oil–water cuff (mL) | 40-37-3 | | | | 40-37-3 | |
| Separation time (min) | 30 | | | | 30 | 30–60[b] |
| **Cincinnati Milacron Heat** | | | | | | |
| Copper, max | | 5 | | | | |
| Steel, max | | 2 | | | | |
| Sludge (mg/100 mL), max | | 25 | | | | |
| Viscosity change (%), max | | ±5 | | | | |
| Acid number change, max | | ±0.15 | | | | |
| **Neutralization Number** | | | | | | |
| Neutralization Number, max | | 0.2 | | | Report | Report |
| **Copper Strip** (ASTM D 30, 3 h, 100°C) | | | | | | |
| Rating | | | | 1 | 1 max | 2 max |

| Performance Test | U.S. Steel 126 | Cincinnati Milacron P-54 | Denison HF-1 | General Electric GEK 32568A | AFNOR NFE 48-600 HL | DIN 51524 Part 1 |
|---|---|---|---|---|---|---|
| **Foam tendency/stability (ASTM D892)** | | | | | | |
| Sequence I (mL) | | | | 10/0 | 200/10 | |
| Sequence II (mL) | | | | 20/0 | 200/10 | |
| Sequence III (mL) | | | | 10/0 | 200/10 | |
| **Air Release (DIN 51381, ASTM D3427)** | | | | | | |
| Time (min) | | | | | <5–10[b] | 5–10[b] |

[a]Additional oxidation test FTMS 5308.6: –5 to +20% viscosity change, 3.0 max TAN increase.
[b]Depends on viscosity grade.

### Typical U.S. and European Turbine Oil Requirements

| Performance Test | Non-EP oils | | | EP oils |
|---|---|---|---|---|
| | British Standard BS 489 | DIN 51515 | General Electric GEK 32568A | Brown Boveri HTGD 90117 |
| **Viscosity Index** | | | | |
| VI, min | 80 | | | 90 |
| **Flash Point** | | | | |
| Temperature (°C), min | 168 | 160–215[a] | 215 | 185 |
| **Pour Point** | | | | |
| Temperature (°C), min | –6 | –6 | –12 | –6 |
| **Neutralization Number** | | | | |
| Neutralization number | 0.2 | Report | | Report |
| **Air Release (DIN 51381, ASTM D3427)** | | | | |
| Time (min) | 5–10[a] | 5–6[a] | | 5 |
| **Foam tendency/stability (ASTM D892)** | | | | |
| Sequence I (ml) | 450/0[b] | | 10/0 | 450/10 |
| Sequence II (ml) | 50/0 | | 20/0 | 50/10 |
| Sequence III (ml) | 450/0 | | 10/0 | 450/10 |
| **Demulsibility** | | | | |
| Steam (s) max | 300 | 300 | | 5 |
| **Turbine Oil Demulsibility (ASTM D 1401)** | | | | |
| Separation time (min) | | | | 30 max |
| **Oxidation Stability** | | | | |
| ASTM D943 Time to 2.0 TAN (h) | | 2000 | 2000[c] | 2000 |
| ASTM D4310 Sludge (mg), max | | | | 100 |
| IP 280 TOP, max | 1.0 | | | 1.8 |
| IP 280 Sludge (%), max | 0.4 | | | 0.4 |
| ASTM D2272 Time to 25-psi loss (min) | | | 450 | |

(continued on next page)

*Typical U.S. and European Turbine Oil Requirements (continued)*

| Performance Test | Non-EP oils | | | EP oils |
|---|---|---|---|---|
| | British Standard BS 489 | DIN 51515 | General Electric GEK 32568A | Brown Boveri HTGD 90117 |
| **Copper Strip** (ASTMD 130, 3 h, 100°C) | | | | |
| Rating | 2 max | 2 max | 1 max | 2 max |
| **Rust** (ASTM D665) | | | | |
| A. Distilled water | | Pass | Pass | |
| B. Synthetic seawater | Pass[d] | | | Pass |
| **FZG** (A/8.3/90) | | | | |
| Pass load stage, min | | | | 6–7[a] |

[a]Depends on viscosity grade.
[b]ISO VG 68/100 requirements: Sequence I 450/40, Sequence II 100/10, Sequence III 450/40.
[c]Additional oxidation test FTMS 5308.6: –5 to +20% viscosity change, 3.0 max TAN increase.
[d]Modified procedure.

*UTTO Performance Requirements*

| Viscosity | Summer | SAE 10W-30: 9.0–11.6 cSt at 100°C |
|---|---|---|
| | | 50,000–70,000 cP at –35°C Brookfield |
| | Winter | SAE 5W-20: 6.2–7.0 cSt at 100°C |
| | | 20,000 cP at –40°C Brookfield |

| Application | Specification | Test |
|---|---|---|
| Transmission/wet brakes | API GL-4 | L-20 high-torque, low-speed |
| | Massey-Ferguson M1141/M1135 | Independent power take-off, four-square axle, pump |
| | New Holland FNHA-2-C-201.00 (M2C134) | Jenkins tractor transmission stall and high energy |
| | John Deere J20C/D | Brake chatter/capacity, powershift transmission, high-torque axle |
| | Caterpillar TO-2,[a] Allison C-4 | Powershift transmission friction, wear, oxidation, friction |
| Hydraulic equipment | Vickers vane pump | Wear (ASTM D2882) |
| | Denison HF0 | Vane and piston pump |
| | Vickers Sheet I-286-S | Vickers 35VQ25 |

[a]This test or category is obsolete; parts, fuel, or reference oils are no longer generally available, or the test is no longer monitored by the test developer or ASTM.

# Practice Exam
with Correct
Answers and
Statements
of Truth

1. The American Petroleum Institute (API) has defined the categories of lubricant base stocks as:

   A. I, II, II+, III, III+, IV, V
   B. Groups I, II, III, and synthetics
   C. Groups I, II, III, IV, V
   D. Mineral, synthetic, semisynthetic

   Correct answer is C.
   The American Petroleum Institute (API) has defined five categories of lubricant base stocks: Groups I, II, III, IV, V.

2. The _____ chemical reaction of the oil with hydrogen changes the polar compounds slightly but retains them in the oil.

   A. hydrofinishing
   B. hydrolysis
   C. hydrogenation
   D. saturation

   Correct answer is A.
   Hydrofinishing by chemical reaction of the oil with hydrogen changes the polar compounds slightly but retains them in the oil. In this process, the most obvious result is oil of a lighter color.

3. _____ offers methods to further aid in the removal of aromatics, sulfur, and nitrogen from the lube base stocks.

   A. Hydrofinishing
   B. Hydrolysis
   C. Hydrogenation
   D. Hydroprocessing

   Correct answer is D.
   Hydroprocessing offers methods to further aid in the removal of aromatics, sulfur, and nitrogen from the lube base stocks.

4. Two of the _____ methods use the feedstock from the solvent refining process are hydrofinishing and hydrotreating.

   A. synthesis
   B. hydrolysis
   C. hydrogenation
   D. hydroprocessing

   Correct answer is D.
   Two of the hydroprocessing methods use the feedstock from the solvent refining process (hydrofinishing and hydrotreating), and a third, hydrocracking, uses the vacuum gas oil (VGO) from the crude distillation unit.

5. _____ are chemical compounds added to lubricating oils to impart specific properties to the finished oils.

   A. Mistifiers
   B. Chelating agents
   C. Additives
   D. Organics

   Correct answer is C.
   Additives are chemical compounds added to lubricating oils to impart specific properties to the finished oils.

6. Pour-point depressants function by inhibiting the formation of _____ that would prevent oil flow at low temperatures.

   A. ice
   B. wax
   C. high-molecular-weight polymers
   D. sludge

   Correct answer is B.
   Pour-point depressants are high-molecular-weight polymers whose function is

inhibiting the formation of a wax crystal structure that would prevent oil flow at low temperatures.

7. The concentration of organic acids in the oil increases when _____

    A. the oil is mixed with another oil.
    B. the oil is mixed with water.
    C. the oil is heated.
    D. the oil is cooled.

Correct answer is C.
When oil is heated in the presence of air, oxidation occurs. As a result of this oxidation, both the oil viscosity and the concentration of organic acids in the oil increase, and varnish and lacquer deposits may form on hot-metal surfaces exposed to the oil. In extreme cases, these deposits may be further oxidized to form hard, carbonaceous materials.

8. _____ additives are used in many lubricating oils to reduce friction, wear, scuffing, and scoring under boundary-lubrication conditions, that is, when full lubricating films cannot be maintained.

    A. Antiwear
    B. Extreme pressure
    C. Antioxidant
    D. Anticorrosion

Correct answer is A.

Antiwear additives are used in many lubricating oils to reduce friction, wear, scuffing, and scoring under boundary-lubrication conditions, that is, when full lubricating films cannot be maintained.

9. At high temperatures or under heavy loads where more severe sliding conditions exist, _____ additives are required to reduce friction, control wear, and prevent severe surface damage.

    A. antiwear
    B. extreme pressure
    C. antioxidant
    D. anticorrosion

Correct answer is B.
At high temperatures or under heavy loads where more severe sliding conditions exist, compounds called extreme pressure (EP) additives are required to reduce friction, control wear, and prevent severe surface damage.

10. _____ agents are usually compounds containing sulfur, chlorine, or phosphorus, either alone or in combination.

    A. Antiwear
    B. Extreme pressure
    C. Antioxidant
    D. Anticorrosion

Correct answer is B.
Extreme pressure agents are usually compounds containing sulfur, chlorine, or phosphorus, either alone or in combination.

11. The principal thickeners used in greases are _____

    A. clay based.
    B. aluminum.
    C. metallic soaps.
    D. animal fat.

Correct answer is C.
The principal thickeners used in greases are metallic soaps.

12. _____ is a combination of a conventional metallic soap-forming material with a complexing agent.

   A. Complex grease
   B. Organic-based grease
   C. Organoclay-based grease
   D. Synthetic grease

   Correct answer is A.
   Modifications of metallic soap greases, called complex greases, are continuing to gain popularity. These complex greases are made by using a combination of a conventional metallic soap-forming material with a complexing agent.

13. Additives and modifiers commonly used in lubricating greases are _____ lubricating oils.

   A. much the same as similar materials added to
   B. specific and very different from those in
   C. only the same for extreme-duty
   D. only the same for food-grade (H1)

   Correct answer is A.
   Additives and modifiers commonly used in lubricating greases are oxidation or rust inhibitors, pour-point depressants, extreme-pressure additives, antiwear agents, lubricity- or friction-reducing agents, and dyes or pigments. Most of these materials have much the same function as similar materials added to lubricating oils.

14. Molybdenum disulfide is used in many greases for applications in which _____

   A. the grease has to be black.
   B. high vibrations exist.
   C. food-grade grease is needed.
   D. there are low speeds.

   Correct answer is D.
   Molybdenum disulfide is used in many greases for applications in which loads are heavy, surface speeds are low, and restricted or oscillating motion is involved.

15. _____ impart specific properties to grease.

   A. Base oils
   B. Thickeners
   C. Bearings
   D. Conditions

   Correct answer is B.
   Certain types of thickeners usually impart specific properties to a finished grease.

16. The penetration point is a measure of the relative hardness or softness of a grease and may indicate _____

   A. flow and dispensing properties.
   B. low-temperature flow.
   C. dropping point.
   D. contamination.

   Correct answer is A.
   The penetration point is a measure of the relative hardness or softness of a grease and may indicate something of flow and dispensing properties.

17. The _____ of a grease is the temperature at which material falls from the orifice of a test cup under prescribed test conditions.

    A. flow and dispensing properties
    B. low-temperature flow
    C. dropping point
    D. water washout contamination

    Correct answer is C.
    The dropping point of a grease is the temperature at which a drop of material falls from the orifice of a test cup under prescribed test conditions.

18. A _____ is produced by combining or building individual units into a unified entity.

    A. synthetic oil
    B. biobased base oil
    C. grease
    D. conspiracy

    Correct answer is A.
    A synthesized oil is one that is produced by combining or building individual units into a unified entity. The production of synthetic lubricants starts with synthetic base stocks that are often manufactured from petroleum.

19. _____ have outstanding flow characteristics at low temperatures, and their stability at high temperatures marks the preferred use of these lubricants.

    A. Group III base oils
    B. Soy oil base oils
    C. Polyglutarimides
    D. Synthetic base oils

    Correct answer is D.
    The primary performance advantage of synthetic lubricants is the extended service life capability and handling a wider range of application temperatures. Their outstanding flow characteristics at low temperatures and their stability at high temperatures mark the preferred use of these lubricants.

20. The single most important physical characteristic of a hydraulic fluid is _____

    A. Viscosity.
    B. antifoaming.
    C. pump wear protection.
    D. API classification.

    Correct answer is A.
    The single most important physical characteristic of a hydraulic fluid is its viscosity.

21. _____ means that entrained air is released from an oil.

    A. Antifoam
    B. Air separation
    C. Air release
    D. Antientrapment

    Correct answer is B.
    Air causes spongy or erratic motion, which results in poor system performance, particularly during the production of close-tolerance parts. Antifoaming and air-separation characteristics are two different concepts, although somewhat connected. *Air separation* means that the entrained air is released from the oil, whereas *antifoam* means that the air bubbles getting to the surface of the oil are readily dissipated.

**22.** The $nD_m$ is commonly referred to as the

_____

A. bearing speed factor.
B. pitch-line velocity.
C. Avogedro's value.
D. REO Speed Wagon.

Correct answer is A.
The speed at which the surfaces of a rolling-element bearing roll together is a fairly complex calculation, so an approximation called the *bearing speed factor $nD_m$* is usually used.

**23.** The _____ is determined by multiplying the rotational speed in revolutions per minute $n$ by the pitch diameter in millimeters $D_m$.

A. bearing speed factor
B. pitch-line velocity
C. Avogedro's value
D. REO Speed Wagon

Correct answer is A.
The bearing speed factor is determined by multiplying the rotational speed in revolutions per minute $n$ by the pitch diameter in millimeters $D_m$.

**24.** Under _____ conditions, the effect of load on film thickness is not as great as the effect of speed or oil viscosity.

A. boundary
B. mixed-film
C. hydrodymanic
D. EHL

Correct answer is D.
Under EHL conditions, the effect of load on film thickness is not as great as the effect of speed or oil viscosity.

**25.** Most open gears and wire ropes, many drive chains and rolling-element bearings, and some cylinders, bearings, and enclosed gears are lubricated by _____ methods.

A. spray
B. all-loss
C. boundary
D. viscoelastic

Correct answer is B.
Most open gears and wire ropes, many drive chains and rolling-element bearings, and some cylinders, bearings, and enclosed gears are lubricated by all-loss methods.

**26.** What is the advantage of an engine oil with a high viscosity index?

A. Less viscous drag during starting
B. Provides thinner oil films
C. Stabilizes oil consumption
D. Reduced oxidation rates

Correct answer is A.
Viscosity index is important in engines that must be started and operated over a wide temperature range. In these cases, all other factors being equal, oils with higher viscosity indexes give less viscous drag during starting and provide thicker oil films for better sealing and wear prevention, and oil consumption at operating temperatures is lower.

27. Many additives can have a few functions. The _____ can have the ability to neutralize the acidic end products of fuel combustion and oil oxidation.

    A. antifoam agents
    B. chelating agents
    C. detergents
    D. polyalphaolefins

    Correct answer is C.
    Most detergents and, to a lesser extent, many dispersants have the ability to neutralize the acidic end products of fuel combustion and oil oxidation. When a considerable ability to neutralize acids is required, as in oils for diesel engines burning high-sulfur fuels, however, highly alkaline (overbased) detergent-type materials are used.

28. Filter ratings are important to any system. Which factor is used to calculate filter efficiency?

    A. Media type
    B. Micron rating
    C. Beta ratio
    D. Flow rate

    Correct answer is C.
    Filter efficiency is calculated as $(\beta - 1)/\beta \times 100\%$, the beta ratio of the quantity of particles larger than the micron-rating size upstream of the filter to the quantity of particles larger than the micron-rating size downstream of the filter.

29. When determining the grease quantity for bearings, the following formula is used: $G = 0.005DB$. What does the $D$ represent in this formula?

    A. Bearing outside diameter (mm)
    B. Bearing inside diameter (mm)
    C. Bearing width (mm)
    D. Bearing roller diameter (mm)

    Correct answer is A.
    Grease quantity $G = 0.005DB$, where $G$ is grease quantity in grams (g), $D$ is bearing outside diameter in millimeters (mm), and $B$ is total bearing width in millimeters (mm) or $H$ for thrust bearings.

30. A centralized system for rotating equipment needs to be lubricated within the steel industry over a large area. The system must be designed to have good filtration (contamination is a challenge), reduce oil usage (by as much as 30%), and lower operating costs. Which would be the best type of design for this system?

    A. Wick oilers
    B. Oil mist lubrication
    C. Constant-level oilers
    D. Splash lubrication

    Correct answer is B.
    Oil mist lubrication can drastically reduce the volume of oil used and control contamination. It will lower the overall operating costs as well.

31. When establishing a lube route, it is important to identify the _____

    A. lubrication points.
    B. lubricants in storage.
    C. lubrication costs.
    D. lubricants not in use.

    Correct answer is A.
    Lubrication points must be noted when planning a lube route as well as the lubrication and inspection tasks.

32. Storage and handling of lubricants are critical. Which of the following is *not* a best practice when receiving lubricants at your site?

    A. Keep lubricants in clean, cool, dry storage at all times.
    B. Place identification on lubricants when in storage.
    C. Use any available containers for lubricant transfers.
    D. Control transfer of lubricant from the delivery vehicle.

    Correct answer is C.
    Dedicated storage containers should be used when transferring lubricants to avoid cross-contamination.

33. Automatic grease systems need to be maintained. Which of the following is the *most appropriate* maintenance technique?

    A. Inspections can be done once per year.
    B. No calibration is required for these grease guns.
    C. Inspection is not required for the applicators.
    D. Visual inspections can detect anomalies.

    Correct answer is D.
    Calibration must be done for automatic grease applicators. These should be com-pleted as part of a formal maintenance schedule and can be incorporated into the predictive maintenance strategy.

34. As per the Stribeck curve, which of the following is in the correct order?

    A. Boundary, mixed film, elastohydrodynamic, hydrodynamic
    B. Boundary, mixed film, hydrodynamic, elastohydrodynamic
    C. Mixed film, boundary, elastohydrodynamic, hydrodynamic
    D. Mixed film, boundary, hydrodynamic, elastohydrodynamic

    Correct answer is A. See the diagram on the next page.

35. Which of the following is a characteristic required of biodegradable lubricants?

    A. Viscosity of the biodegradable products
    B. Speed at which the products biodegrade
    C. Operating temperature of the products
    D. Energy produced to create the products

    Correct answer is B.
    Biodegradable lubricants must meet two important environmental characteristics: speed at which the products biodegrade and the toxicity that can affect bacteria or aquatic life.

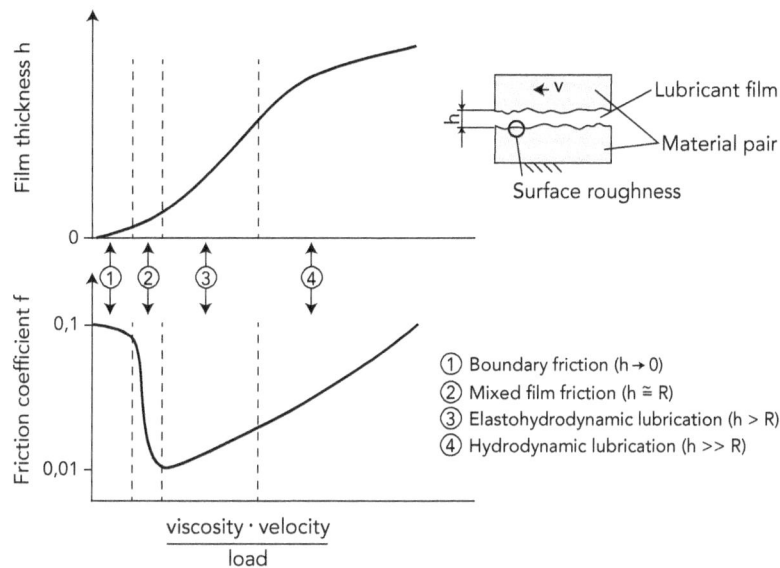

Source: Bharat Bhushan, Stribeck Graph according to H. Czichos and K. H. Habig, *Introduction to Tribology*, Second Edition, John Wiley & Sons, Ltd., 2013, 403.

36. When selecting a lubricating oil, which is *not* a factor to be considered?

   A. Cost
   B. Operating environment
   C. Load
   D. Temperature

   Correct answer is A.
   Cost is not a factor to consider when determining the type of lubricant. One should consider type of motion, speed, operating environment, load, and temperature.

37. Fire-resistant fluids are required for certain applications. Which of the following is *not* a consideration when selecting a fire-resistant fluid?

   A. Cost
   B. Function
   C. Load
   D. Risk

   Correct answer is C.
   Considerations for fire-resistant fluids are cost, function, and risk. A balance should be reached between the func-tional sacrifices sometimes made by using a fire-resistant oil instead of a normal oil, the properties of the fire-resistant oil, the need for the oil being fire resistant, and cost and risk.

38. Industrial gears can experience different types of action between the gear teeth depending on the type of gear. Which of the following gears would experience sliding at right angles to the lines of contact?

   A. Helical gears
   B. Spur gears
   C. Herringbone gears
   D. Worm gears

   Correct answer is B.
   For bevel and conventional spur gears, sliding occurs at right angles to the lines of contact (across the tooth faces). In contrast, helical, herringbone, and spiral bevel gears do not have sliding at right angles to the lines of contact. For worm gears, sliding and rolling are slow because of the low rotational speed of the worm wheel.

39. When selecting the right type of grease for a chassis, it is important that the product contains _____

    A. antifoaming additives.
    B. antiwear additives.
    C. viscosity improvers.
    D. detergents.

    Correct answer is B.
    The product should contain anti-oxidants, corrosion inhibitors, and extreme-pressure and antiwear addi-tives. Because of its special formulation and semifluid consistency, grease should exhibit excellent flow even at very low temperatures and in long pipelines.

40. The flash point of a lubricant is the lowest temperature at which the lubricant must be heated before its vapor when mixed with air will

    _____

    A. ignite and continue to burn.
    B. ignite but not continue to burn.
    C. spark and fully combust.
    D. spark spontaneously.

    Correct answer is B.
    The flash point is the lowest tempera-ture to which a lubricant must be heated before its vapor, when mixed with air, will ignite but not continue to burn. The fire point is the temperature at which lubricant combustion will be sustained.

41. When taking an oil sample, which of the following should be used?

    A. Sample from a nonturbulent area.
    B. Sample when the machine is hot.
    C. Sample tubing can be reused.
    D. Sample from the bottom of the sump.

    Correct answer is B.
    Proper sampling procedures include:
    - Sample when the machine is hot.
    - Sample while operating or after shut-down within a couple of hours max.
    - Sample from pressurized stream or the midpoint of the reservoir, not from the bottom.
    - Sample in a region of turbulence.
    - Sample point must be free of dirt.
    - Sample directly to clean dry sample bottle.
    - Use new plastic tube for each sample.
    - Flush sampling tube.
    - Flush oil bottle.

42. Oxidation degrades the quality of an oil. Which of the following is *not* a contributing factor to oxidation?

    A. High temperature
    B. Load speed
    C. Wear metal
    D. Oil aeration

    Correct answer is B.
    The load speed does not directly con-tribute to oxidation. Water contamina-tion is another contributing factor to oxidation.

43. The NLGI grade of a grease relates to which of its properties?

    A. Thickener
    B. Base oil
    C. Consistency
    D. Additive

    Correct answer is C.
    The NLGI grade denotes the consistency of a grease and ranges from 000 (fluid) to 6 (block).

44. Coupling greases must be able to withstand _____

    A. centripetal forces.
    B. centrifugal forces.
    C. low-torque loads.
    D. low rotational speeds.

    Correct answer is B.
    Coupling greases must be able to withstand the centrifugal forces that are applied to them. Additionally, they must be able to perform in very high rotational speeds and carry high-torque loads.

45. Additive depletion can occur in several ways. Which of the following is *not* a mode of additive depletion?

    A. Decomposition
    B. Contamination
    C. Adsorption
    D. Separation

    Correct answer is B.
    Contamination can promote additive depletion, but it is not a mode of depletion.

46. Food-grade lubricants must provide the same functions as oils but be approved by the USDA. Which category best describes H1 food-grade oils?

    A. Possibility of incidental food contact
    B. No possibility of food contact
    C. Edible oils
    D. Nonedible oils

    Correct answer is A.
    H1 lubricants are food-grade lubricants used in food-processing environments where there is the possibility of incidental food contact. H2 lubricants are food-grade lubricants used on equipment and machine parts in locations where there is no possibility of contact with food. H3 lubricants are food-grade lubricants, typically edible oils, used to prevent rust on hooks, trolleys, and similar equipment.

47. Polyalkylene glycols are also known as PAGs. Which base oil group do they fall under?

    A. Group II
    B. Group III
    C. Group IV
    D. Group V

    Correct answer is D.
    Group IV represents PAOs, while Group V represents all other synthetics.

**48.** Hydrodynamic lubrication is usually achieved during _____

    **A.** startup.
    **B.** shutdown.
    **C.** operation.
    **D.** the highest load.

Correct answer is C.
Hydrodynamic lubrication is also called full-film or full fluid-film lubrication. Moving surfaces are fully separated by a hydrodynamically formed wedge. No contact is expected between the asperities of the two surfaces in relative motion. The formation of the oil wedge depends on surface geometry, speed, load, and oil viscosity. Hydrodynamic lubrication is not achieved during startup and shutdown conditions.

**49.** Pitting is a form of wear characterized by the presence of _____

    **A.** fractures.
    **B.** surface cavities.
    **C.** extensive grooves.
    **D.** scratches.

Correct answer is B.
Pitting is a form of wear characterized by the presence of surface cavities, the formation of which is attributed to processes such as fatigue, local adhesion, or cavitation.

**50.** What should be avoided when handling full drums of lubricant at your site?

    **A.** Use of appropriate lifting gear
    **B.** Working under suspended loads
    **C.** Bending knees when lifting
    **D.** Use of two people to move drums

Correct answer is B.
Bigger storage containers such as 200-liter drums are heavy. Normal care should be taken when handling these drums to ensure safety of personnel:

- Use two people to move one drum.
- Bend knees when lifting.
- Use appropriate lifting gear.
- Do not work under suspended loads.
- For smaller cans, use a safety locker with ventilation.
- Have Material Safety Data Sheets available for stored product and understand the content.
- Fire extinguishers (inspected and maintained) should be available in storage areas.
- If solvent containers are stored with oil containers, these should be grounded to prevent sparks from static electricity.
- Avoid contact of oil with the skin.
- In case of contact with skin, clean with warm water and soap, not with solvents.
- Preferably use barrier cream and gloves.

# Index